METHODS IN CELL BIOLOGY

VOLUME XV

Contributors to This Volume

Barbro Andersson

Stratis Avrameas

Karyn E. Bird

L. M. Black

Michel Bornens

Joseph E. Cummins

Richard L. Davidson

Alan W. Day

T. Ege

A. J. Faber

S. Fakan

Park S. Gerald

Peter J. Goldblatt

Nicholas K. Gonatas

H. Hamberg

R. Hancock

Reinhold G. Herrmann

Richard H. Hilderman

Masakatsu Horikawa

I. R. Johnston

Joseph R. Kates

R. R. Kay

Claudia Kent

Alois Knüsel

LeRoy Kuehl

Clive C. Kuenzle

M. Lanneau

Teresa H. Liao

M. Loir

Joseph J. Lucas

F. Machicao

Gerardo Martínez-López

Marvin L. Meistrich

Volker Mell

Masami Muramatsu

Toshio Onishi

Ram Parshad

N. R. Ringertz

William S. Runyan

Marlene Sabbath

Takashi Sakamoto

R. J. Sanderson

Steven D. Schimmel

Rupert Schmidt-Ullrich

Jürgen M. Schmitt

Daniel Schümperli

Horst Senger

E. Sidebottom

Johann Sonnenbichler

William G. Taylor

J. E. Thompson

P. Roy Vagelos

Donald F. H. Wallach

Wayne Wray

I. Zetl

Methods in

Cell Biology

Edited by

DAVID M. PRESCOTT

DEPARTMENT OF MOLECULAR, CELLULAR AND
DEVELOPMENTAL BIOLOGY
UNIVERSITY OF COLORADO
BOULDER, COLORADO

VOLUME XV

1977

ACADEMIC PRESS • New York   San Francisco   London
A Subsidiary of Harcourt Brace Jovanovich, Publishers

ACADEMIC PRESS, INC.
111 Fifth Avenue, New York, New York 10003

United Kingdom Edition published by
ACADEMIC PRESS, INC. (LONDON) LTD.
24/28 Oval Road, London NW1

LIBRARY OF CONGRESS CATALOG CARD NUMBER: 64–14220

ISBN 0–12–564115–X

PRINTED IN THE UNITED STATES OF AMERICA

CONTENTS

11.   *Fractionation of Cell Organelles in Silica Sol Gradients*
Jürgen M. Schmitt and Reinhold G. Herrmann

12.   *Preparation of Photosynthetically Active Particles from Synchronized Cultures of Unicellular Algae*
Horst Senger and Volker Mell

13.   *Rapid Isolation of Nucleoli from Detergent-Purified Nuclei of Tumor and Tissue Culture Cells*
Masami Muramatsu and Toshio Onishi

14. *Isolation of Plasma Membrane Vesicles from Animal Cells*
Donald F. H. Wallach and Rupert Schmidt-Ullrich

15. *Rapid Isolation of Nuclear Envelopes from Rat Liver*
R. R. Kay and I. R. Johnston

16. *Isolation of Plasma Membranes from Cultured Muscle Cells*
Steven D. Schimmel, Claudia Kent, and P. Roy Vagelos

17. *Preparation of Plasma Membranes from Amoebae*
J. E. Thompson

18.   *Induction of Mammalian Somatic Cell Hybridization*
      by Polyethylene Glycol
      Richard L. Davidson and Park S. Gerald

19.   *Preparation of Microcells*
      T. Ege, N. R. Ringertz, H. Hamberg, and E. Sidebottom

20.   *Nuclear Transplantation with Mammalian Cells*
      Joseph J. Lucas and Joseph R. Kates

25.   Peptones as Serum Substitutes for Mammalian Cells in Culture
William G. Taylor and Ram Parshad

26.   In Situ Fixation and Embedding for Electron Microscopy
Marlene Sabbath and Barbro Andersson

27.   Genetic and Cell Cycle Analysis of a Smut Fungus (Ustilago violacea)
Joseph E. Cummins and Alan W. Day

LIST OF CONTRIBUTORS

Numbers in parentheses indicate the pages on which the authors' contributions begin.

BARBRO ANDERSSON, Department of Pathology, School of Medicine, Leahi Hospital, Honolulu, Hawaii (435)

STRATIS AVRAMEAS, Immunocytochemistry Unit, Pasteur Institute, Paris, France (387)

KARYN E. BIRD, Webb-Waring Lung Institute, University of Colorado Medical Center, Denver, Colorado (1)

L. M. BLACK, Department of Genetics and Development, University of Illinois, Urbana, Illinois (407)

MICHEL BORNENS, Department of Molecular Biology, Pasteur Institute, Paris, France (163)

JOSEPH E. CUMMINS, Department of Plant Sciences, The University of Western Ontario, London, Ontario, Canada (445)

RICHARD L. DAVIDSON, Division of Human Genetics, Children's Hospital Medical Center, and Department of Microbiology and Molecular Genetics, Harvard Medical School, Boston, Massachusetts (325)

ALAN W. DAY, Department of Plant Sciences, The University of Western Ontario, London, Ontario, Canada (445)

T. EGE, Institute for Medical Cell Research and Genetics, Medical Nobel Institute, Karolinska Institute, Stockholm, Sweden (339)

A. J. FABER, Swiss Institute for Experimental Cancer Research, Lausanne, Switzerland (127)

S. FAKAN, Swiss Institute for Experimental Cancer Research, Lausanne, Switzerland (127)

PARK S. GERALD, Division of Human Genetics, Children's Hospital Medical Center, and Department of Pediatrics, Harvard Medical School, Boston, Massachusetts (325)

PETER J. GOLDBLATT, Department of Pathology, University of Connecticut Health Center, Farmington, Connecticut (371)

NICHOLAS K. GONATAS, Department of Pathology (Division of Neuropathology), University of Pennsylvania, School of Medicine, Philadelphia, Pennsylvania (387)

H. HAMBERG, Department of Pathology, Sabbatsberg Hospital, Stockholm, Sweden (339)

R. HANCOCK, Swiss Institute for Experimental Cancer Research, Lausanne, Switzerland (127)

REINHOLD G. HERRMANN, Institut für Botanik der Universität Düsseldorf, Düsseldorf, West Germany (177)

RICHARD H. HILDERMAN, Department of Biochemistry, Clemson University, Clemson, South Carolina (371)

MASAKATSU HORIKAWA, Division of Radiation Biology, Faculty of Pharmaceutical Sciences, Kanazawa University, Kanazawa, Japan (97)

I. R. JOHNSTON, Biochemistry Department, University College, London, England (277)

JOSEPH R. KATES, Department of Microbiology, State University of New York at Stony Brook, Stony Brook, New York (359)

R. R. KAY, Imperial Cancer Research Fund, London, England (277)

CLAUDIA KENT, Department of Biochemistry, Purdue University, West Lafayette, Indiana (289)

ALOIS KNÜSEL, Department of Pharmacology and Biochemistry, School of Veterinary Medicine, University of Zürich, Zürich, Switzerland (89)

xv

LeRoy Kuehl, Department of Biochemistry, University of Utah, Salt Lake City, Utah (79)

Clive C. Kuenzle, Department of Pharmacology and Biochemistry, School of Veterinary Medicine, University of Zürich, Zürich, Zwitzerland (89)

M. Lanneau, I.N.R.A. – Laboratoire de Physiologie de la Reproduction, Nouzilly, France (55)

Teresa H. Liao, Department of Chemistry, Nutritional Science Program, University of Maryland, College Park, Maryland (381)

M. Loir, I.N.R.A. – Laboratoire de Physiologie de la Reproduction, Nouzilly, France (55)

Joseph J. Lucas, Department of Microbiology, State University of New York at Stony Brook, Stony Brook, New York (359)

F. Machicao, Max-Planck-Institut für Biochemie, Martinsried, West Germany (149)

Gerardo Martínez-López, Virus Laboratory, Instituto Colombiano Agropecuario, Bogota, Colombia (407)

Marvin L. Meistrich, Section of Experimental Radiotherapy, The University of Texas System Cancer Center, M.D. Anderson Hospital and Tumor Institute, Houston, Texas (15)

Volker Mell, Botanisches Institut der Universität Marburg, Marburg, West Germany (201)

Masami Muramatsu, Department of Biochemistry, Tokushima University School of Medicine, Tokushima, Japan (221)

Toshio Onishi, Department of Biochemistry, Tokushima University School of Medicine, Tokushima, Japan (221)

Ram Parshad, Department of Pathology, Howard University College of Medicine, Washington, D.C. (421)

N. R. Ringertz, Institute for Medical Cell Research and Genetics, Medical Nobel Institute, Karolinska Institute, Stockholm, Sweden (339)

William S. Runyan, Department of Food and Nutrition, Iowa State University, Ames, Iowa (381)

Marlene Sabbath, Department of Pathology, Lenox Hill Hospital, New York, New York (435)

Takashi Sakamoto,[1] Division of Radiation Biology, Faculty of Pharmaceutical Sciences, Kanazawa University, Kanazawa, Japan (97)

R. J. Sanderson, Webb-Waring Lung Institute, University of Colorado Medical Center, Denver, Colorado (1)

Steven D. Schimmel, Department of Biochemistry, University of South Florida, College of Medicine, Tampa, Florida (289)

Rupert Schmidt-Ullrich, Tufts-New England Medical Center, Department of Therapeutic Radiology, Division of Radiobiology, Boston, Massachusetts (235)

Jürgen M. Schmitt, Institut für Botanik der Universität Düsseldorf, Düsseldorf, West Germany (177)

Daniel Schümperli, Department of Pharmacology and Biochemistry, School of Veterinary Medicine, University of Zürich, Zürich, Switzerland (89)

Horst Senger, Botanisches Institut der Universität Marburg, Marburg, West Germany (201)

E. Sidebottom, Sir William Dunn School of Pathology, Oxford University, Oxford, England (339)

[1] Present address: Research Laboratories, NISSUI Pharmaceutical Co., Ltd., 1805 Inari-chō, Souka-shi, Saitama 340, Japan.

JOHANN SONNENBICHLER, Max-Planck-Institut für Biochemie, Martinsried, West Germany (149)

WILLIAM G. TAYLOR, Cell Physiology and Oncogenesis Section, Laboratory of Biochemistry, National Cancer Institute, Bethesda, Maryland (421)

J. E. THOMPSON, Department of Biology, University of Waterloo, Waterloo, Ontario, Canada (303)

P. ROY VAGELOS, Merck, Sharp & Dohme Research Laboratories, Rahway, New Jersey (289)

DONALD F. H. WALLACH, Tufts-New England Medical Center, Department of Therapeutic Radiology, Division of Radiobiology, Boston, Massachusetts (235)

WAYNE WRAY, Department of Cell Biology, Baylor College of Medicine, Houston, Texas (111)

I. ZETL, Max-Planck-Institut für Biochemie, Martinsried, West Germany (149)

PREFACE

Volume XV of this series continues to present techniques and methods in cell research that have not been published or have been published in sources that are not readily available. Much of the information on experimental techniques in modern cell biology is scattered in a fragmentary fashion throughout the research literature. In addition, the general practice of condensing to the most abbreviated form materials and methods sections of journal articles has led to descriptions that are frequently inadequate guides to techniques. The aim of this volume is to bring together into one compilation complete and detailed treatment of a number of widely useful techniques which have not been published in full detail elsewhere in the literature.

In the absence of firsthand personal instruction, researchers are often reluctant to adopt new techniques. This hesitancy probably stems chiefly from the fact that descriptions in the literature do not contain sufficient detail concerning methodology; in addition, the information given may not be sufficient to estimate the difficulties or practicality of the technique or to judge whether the method can actually provide a suitable solution to the problem under consideration. The presentations in this volume are designed to overcome these drawbacks. They are comprehensive to the extent that they may serve not only as a practical introduction to experimental procedures but also to provide, to some extent, an evaluation of the limitations, potentialities, and current applications of the methods. Only those theoretical considerations needed for proper use of the method are included.

Finally, special emphasis has been placed on inclusion of much reference material in order to guide readers to early and current pertinent literature.

DAVID M. PRESCOTT

Chapter 1

Cell Separations by Counterflow Centrifugation

R. J. SANDERSON AND KARYN E. BIRD

Webb-Waring Lung Institute,
University of Colorado Medical Center,
Denver, Colorado

I. Introduction

Counterflow centrifugation has been available as a technique for cell separation since 1948, when Lindahl first published an account of the method in *Nature* (Lindahl, 1948). In that same year, a patent was also filed in California by MacLeod (1948) for a counterflow centrifugation apparatus designed for a nonbiological application. More recently, Beckman Instruments developed and marketed a rotor which has been used by a number of investigators (Glick *et al.*, 1971; Flangas, 1974; Grabske *et al.*, 1974).

The original apparatus of Lindahl was exceedingly complex, and that of MacLeod was specialized for the concentration of particles of high specific gravity. The Beckman apparatus, however, is quite simple. We have modified it to increase its sensitivity, so that it can be used to separate cells whose physical properties differ from one another only subtly (e.g. erythrocytes of varying age) as well as for grossly varying cell systems.

Counterflow centrifugation depends on the ability to balance the outwardly directed centrifugal force acting on a cell with inwardly directed fluid dynamic and buoyancy forces. The fluid dynamic forces are virtually independent of the composition of the medium used, at least for media that are compatible with cell integrity. This means that the medium used can be selected solely for reasons of compatibility with the cells. Gravitational forces involved in counterflow centrifugation are mild, being at most around 600 g; however, cells are subjected to viscous shear forces that can be damaging under certain circumstances. The damage can be eliminated or minimized through the addition of protein to the medium. These features are in contrast to other methods of separation. For example, in differential flotation (Danon and Marikovsky, 1964) specific gravity of the separation medium must be rigidly controlled at the expense of osmolality, while gradient centrifugation (see, for example, Piomelli et al., 1967) requires large centrifugal forces.

In this chapter the general theory of the dynamics of counterflow centrifugation is discussed, and a rational approach to the design of a separation chamber is established. Finally, the result of a number of separations using a chamber built according to this design are presented. These include the fractionation of human erythrocytes according to age, the separation of the various types of leukocytes in peripheral blood and the spleen, and the isolation of narrow-range uniform cells from tissue culture.

II.   Principles of Counterflow Centrifugation

In a counterflow centrifuge, fluid is pumped into the center of the centrifuge rotor through a rotating seal. This fluid moves through internal tubing to the outside of the rotor and is then turned toward the center through a divergent separation chamber. At the end of this chamber, the flow reconverges into a tube of small diameter through which it moves back to the center of the rotor. It then leaves the system through a second rotating seal. The basic elements of the system are shown in Fig. 1.

The cell suspension to be fractionated is introduced into the inlet tube at some convenient point. If the pump rate is correctly chosen, cells move through the system until they reach the separation chamber, where they will come to rest relative to the rotor at a point where the outwardly directed inertial forces are balanced by the inwardly directed fluid dynamic and buoyancy forces. Forces in the tangential plane, namely, the Coriolis force

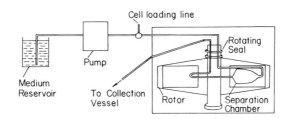

FIG. 1.   The basic elements of the counterflow centrifugation system.

and the opposing pressure gradients, are not important under the operating conditions of this system. The small circulation resulting from these forces is, in any case, in the tangential plane. The cell population distributes itself along the separation chamber in accordance with certain physical properties and the velocity gradient within the chamber. To elute a fraction of cells from the centrifuge, the pump rate is increased until the innermost group of cells reaches the end of the divergent chamber. At this point the fluid velocity is increased by the converging walls and the cells are swept downstream to escape from the rotor.

III.   Theory of the Mechanics of Counterflow Centrifugation

A.   Notation and Coordinate System

The following notation will be used in the analysis below:

A, B	Constants
d	Cell diameter
F	External forces acting on cell
k, k_1, k_2	Cell shape constants
m	Cell mass
$\hat{n}$	Unit vector normal to cell surface, positive inward
p	Pressure
P	Pump rate
$r, \dot{r}, \ddot{r}$	Distance from center of rotation; velocity and acceleration relative to center of rotation of cell
S	Cell surface area
t	Time
$\mathscr{V}$	Cell volume
V	Velocity of fluid relative to cell
V'	Velocity of fluid relative to inertial axes

Ox, y, z Set of right-handed rotating rectangle Cartesian axes with origin at center of rotation and Ox lying along the axis of symmetry of the centrifugation chamber.

η Coefficient of viscosity of eluting medium

$\rho - \rho_F$ Cell density, fluid density

ρ' $\rho - \rho_F$

$\omega, \dot{\omega}$ Rotational velocity, acceleration

x, y, z Used as subscripts indicate components in x, y, z directions, taken as positive along Ox, Oy, Oz

∇ The vector operator ($i\, \partial/\partial x + j\, \partial/\partial y + k\, \partial/\partial z$), i, j, k being unit vectors in x, y, z direction

B.   A Derivation of the Equation of Motion of a Particle under Combined Centrifugal and Hydrodynamic Fields

The equation of motion of a particle relative to a set of moving axes may be written, using vector notation, as follows:

$$\mathbf{F}_p = m\,\{\ddot{\mathbf{r}} + [\omega \times (\omega \times \mathbf{r}) + \dot{\omega} \times \mathbf{r} + 2\omega \times \dot{\mathbf{r}}]\} \tag{1}$$

where $\mathbf{F}_p$ is the sum of the external forces acting on the particle. In the system that concerns us, the only forces of significance are the hydrodynamic forces, $\mathbf{F}_D$, and the buoyancy forces, $\mathbf{F}_B$. We will assume that these are independent of one another and that the buoyancy forces can be calculated from a knowledge of the pressure distribution in the absence of the particle. Then, Stokes' law gives

$$\mathbf{F}_D = k_1 \eta\, d\mathbf{V} \tag{2}$$

for the hydrodynamic forces, while the buoyancy forces are given by

$$\mathbf{F}_B = \int_S p\hat{n}\, dS \tag{3}$$

Expression of the buoyancy term as a surface integral is not convenient, as we do not have an explicit equation for the pressure, p. However, we may use Gauss' theorem (see, for example, Rutherford, 1949) to replace the surface integral by a volume integral that used the gradient of p instead of p, that is

$$\int_S p\hat{n}\, dS = -\int_{\mathscr{V}} \nabla p\, d\mathscr{V} \tag{4}$$

∇p follows from the equation of motion of the fluid, which in the absence of large shears may be written

$$DV'/Dt = -1/\rho_F \text{ grad } p \tag{5}$$

Here, V' has to be the velocity relative to inertial axes and D/Dt is the substantial derivative $D/Dt \equiv d/dt + V \cdot \nabla$. Thus,

$$V' = V + \omega \times r \tag{6}$$

Expansion of (5) and (6) gives

$$\text{grad } p = \nabla p = \rho_F \{F - [dV/dt + (V \cdot \nabla)V + 2\omega \times V + \omega \times (\omega \times r)]\} \tag{7}$$

Equations (1), (2), (3), (4), and (7) may then be combined to yield the equation of motion of the particle:

$$\ddot{r} = (k_1 \eta dV)/m + (\rho_F/m) \int_{\mathscr{V}} \{[dV/dt + (V \cdot \nabla)V + 2\omega \times V + \omega$$
$$\times (\omega \times r)] - F\} \, d\mathscr{V} - \omega \times (\omega \times r) - 2\omega \times \dot{r} \tag{8}$$

As the particle under consideration is small, the lengths, velocities, velocity gradients and angular velocities within the integral sign may be taken to be constants. Equation (8) then simplifies to

$$\ddot{r} = (k_1 \eta dV)/m + (\rho_F \mathscr{V}/m) [\partial V/\partial t + (V \cdot \nabla)V + 2\omega \times V + \omega \times (\omega \times r)]$$
$$- \omega \times (\omega \times r) - 2\omega \times \ddot{r} \tag{9}$$

If we consider rotation only about the z axis, Eq. (9) may be expanded and resolved into components as follows:

$$\ddot{r} = [(k_1 \eta d)/m \ V_x + (\rho_F \mathscr{V}/m) \{V_x(\partial V_x/\partial x) + V_y(\partial V_x/\partial y) + V_z(\partial V_x/\partial z)$$
$$- 2\omega V_y - \omega^2 x\} + \omega^2 x + 2\omega \dot{y}]\hat{x} + [(k_1 \eta d/m)V_y + (\rho_F \mathscr{V}/m)$$
$$\times \{V_x \, \partial V_y/\partial x + V_y \partial V_y/\partial y + V_z \partial V_y/\partial z + 2\omega V_x - \omega^2 y\} + \omega^2 y - 2\omega \dot{x}]\hat{y}$$
$$+ [(k_1 \eta d/m) \ V_z + (\rho_F \mathscr{V}/m) \{V_x \ (\partial V_z/\partial x) + V_y \ (\partial V_z/\partial y) + V_z(\partial V_z/\partial z)\}]\hat{z} \tag{10}$$

where x, y, z are the components of r; V_x, V_y, and V_z are the components of V. In particular, the component of acceleration in the x direction is given by

$$\ddot{r}_x = (k_1 \eta d/m) \ V_x + (\rho_F \mathscr{V}/m) \{V_x \ (\partial V_x/\partial x) + V_y(\partial V_x/\partial y) + V_z \ (\partial V_x/\partial z)$$
$$- 2\omega V_y - \omega^2 x\} + \omega^2 x + 2\omega \dot{y} \tag{11}$$

It is useful at this point to examine Eq. (11) from the physical point of view. The physical significance of each term is as follows:

$(k_1 \eta d/m) V_x$ = acceleration due to viscous forces, acting in the direction of V_x
$\omega^2 x$ = the centrifugal acceleration on the particle

$2\omega\dot{y}$ = the Coriolis acceleration on the particle, due to the component of particle velocity in the y direction

$(\rho_F \mathscr{V}/m)\omega^2 x$ = acceleration due to buoyancy forces associated with centrifugal acceleration of fluid

$(\rho_F \mathscr{V}/m)2\omega V_y$ = acceleration due to buoyancy forces associated with Coriolis acceleration of fluid

$(\rho_F \mathscr{V}/m)\ [V_x(\partial V_x/\partial x) + V_y(\partial V_y/\partial y) + V_z(\partial V_z/\partial z)]$ = acceleration due to buoyancy forces associated with local velocity (or pressure) gradients.

If we assign to x the radial direction and assume (a) that the channel is of slowly varying cross section such that fluid and particle velocity components other than V_x are small, (b) that all velocity gradients are small, (c) that the dimensions of the chamber are small compared to the radius, and (d) that we may drop the Coriolis terms, Eq. (11) becomes

$$\ddot{r}_r = (k_1\eta \, dV_r/m) - (\rho_F \mathscr{V}/m)\,\omega^2 r + \omega^2 r \tag{12}$$

when we replace x by r and V_x by V_r. The physical justification for dropping the Coriolis terms is apparent in Fig. 2A, where we see a stable planar cell front. Radial equilibrium would not be possible if the Coriolis accelerations on elemental fluid volumes were significant. Motion of cells in the tangential plane is not precluded, nor is it of practical importance if it does occur. Note that Eq. (12) with $\ddot{r}_r = 0$, i.e., with the cell at equilibrium, may be written intuitively on the assumption that the only forces acting are the viscous, buoyant, and centrifugal forces. Approximations corresponding to those above cause the tangential and vertical components of acceleration to vanish. Under these circumstances we are left with the equation

$$k_1\eta \, dV_r - \rho_F \mathscr{V}\omega^2 r + m\omega^2 r = 0$$

or, writing $\rho \mathscr{V}$ for m

$$k_1\eta \, dV_r + (\rho - \rho_F)\mathscr{V}\omega^2 r = 0 \tag{13}$$

In the analysis below, we will concern ourselves only with inward velocities, and it is convenient to write $V = -V_r$, as we will take the velocity of the fluid relative to the particle to be positive. Making this substitution, and rearranging Eq. (13), we obtain an expression for the radius at which a particle will equilibrate. Thus

$$r = k_1\eta \, dV/\rho \mathscr{V}\omega^2 \tag{14}$$

or, because we may write $\mathscr{V} = k_2 d^3$,

$$r = k\eta V/(\omega^2 \rho' d^2) \tag{15}$$

where $k = k_1/k_2$.

IV. Derivation of a Suitable Chamber Shape

Because the velocity in the chamber, for a given flow rate, is proportional to the cross sectional area of the chamber, it follows from Eq. (15) that it is possible to compute a distribution of area to give an arbitrary separation of cells having differing values of the parameter $\rho'd^2/k$. For example, we could make a chamber give a distribution of cell equilibrium position which would be linearly related to the magnitude of $\rho'd^2/k$. This concept leads from Eq. (15) to the differential equation

$$d/dr\,(\rho'd^2/k) = \eta/\omega^2 \cdot d/dr \cdot V/r = A, \quad \text{a constant} \tag{16}$$

Integration of Eq. (16) gives

$$V = (Ar^2 + Br)\,\omega^2/\eta \tag{17}$$

The desired variation in cross-sectional areas S of the chamber then follows. For any pump rate, P (measured in volume per unit time), we have

$$S = P/V = P\eta/\omega^2\,(Ar^2 + Br) \tag{18}$$

A convenient value of A may be obtained from the range of $\rho'd^2/k$ of the cell population to be fractionated. B may be obtained from an arbitrary inlet cross-sectional area and pump rate P.

A chamber designed according to these principles will have a volume that is small if the inlet dimensions are small. Large inlet dimensions lead to pump rates that are unacceptably high. However, we may look for a chamber shape in which cells having high values of $\rho'd^2/k$ and those having low values are not separated significantly in the outer portion of the chamber. This condition is fulfilled by making $(d/dr)\,(\rho'd^2/k)$ large, which necessarily, from Eq. (16), leads to the condition that rdV/dr must be large compared with V, because $(d/dr)\,V/r = (1/r^2)\,[(rdV/dr) - V]$. Physically, this is achieved by making the outer portion of the chamber cross-sectional area diverge rapidly.

In the chamber we have designed, the zone of rapid expansion occupies the outer third of its length. This is run smoothly into an inverse square contour, as given by Eq. (18) with $B = 0$, which occupies the remaining two-thirds of the chamber length. The exact contours are not critical, so long as the outer portion of the chamber diverges rapidly and the inner portion slowly. The two photographs shown in Fig. 2, which were taken using the centrifuge's stroboscopic lamp, demonstrate the functional characteristics of the two regions during a red cell separation. In Fig. 2A, cells occupy only the outer region where the divergence is large. The cells farthest downstream form a clearly defined front. When the pump rate is increased, as in Fig. 2B,

FIG. 2.   Photographs taken, using a stroboscopic flash, of the described separation chamber during the fractionation of erythrocytes (A) with the cell front in the rapidly divergent portion of the chamber and (B) with the leading cells in the slowly divergent portion.

the cells move into the slowly divergent portion of the chamber and the front disappears because the cells disperse radially according to their values of the separation parameter.

V.   Application of Counterflow Centrifugation to Cell Separations

In using the system described above to fractionate cell suspensions, there are three major variables to be fixed: the medium, the rotor speed, (which determines the pump speed for any fraction), and the temperature.

For separating the cell types in peripheral blood and spleen, we have found Hanks' balanced salt solution (Ca^{2+}, Mg^{2+} free) to which 0.1–0.2% bovine serum albumin (BSA) has been added, to be satisfactory. Addition of a small amount of EDTA may be beneficial in preventing adhesion of some types of cells. Cultured cells are best fractionated in a medium similar to that in which they are grown. For the separations reported here, Spinner's medium with 5% fetal calf serum (FCS) was used. Recently, however, we have obtained satisfactory results using buffered saline with only 0.2% FCS.

In selecting the rotor speed, a compromise between sensitivity of control and minimization of the forces acting on a cell, which potentially cause irreversible damage, is required. For red cell fractionations in which it is desired to separate the population into five fractions, rotor speeds below 1500 rpm (225 g at a radius of 9 cm) lead to inconveniently small pump speed increments. Speeds above 2200 rpm, or less if the protein is omitted from the medium, combined with the flow rates required to elute later fractions, lead to damage. This may be associated with shear forces in excess of some critical range, which appears to be about 3000 dynes cm^{-2}, as reported by Sutera and Mehrjardi (1975). We have obtained satisfactory results using 2000 rpm.

There are a number of factors influenced by the temperature at which a separation is performed. For both erythrocytes and leukocytes, much lower pump rates are required at 20°C than at 30°C. This is not because of any temperature effect on medium viscosity, but apparently because of changes induced in the physical properties of the cell membrane. We have not, however, examined the cells for reduced damage at lower temperatures. At 4°C, we have not been able to obtain satisfactory fractionation. For mouse spleen fractionation, the large amount of cellular debris in the preparation loaded into the centrifuge made separation at 4°C impossible because of clumping in the apex of the chamber, which led to blockage. This effect was much less serious at 37°C.

For all cell separations, we have used a Coulter Model ZH particle counting system to give data on cell volume and numbers, and as a monitor to determine when a fraction was completely eluted. Depending on the types of cells, the total number that can be loaded into the centrifuge is between 10^7 and 10^8.

A. Age Distribution of Red Cells

The chamber described has been used with the Beckman Elutriator Rotor system to fractionate human red cells into different age groups. As red cells age, they become more dense (Danon and Marikovsky, 1964), smaller (Canham and Burton, 1968), and lose activity of certain enzymes (Sass *et al.*, 1964; Chapman and Schaumberg, 1967). Shape and membrane flexibility also change significantly (Canham and Burton, 1968). The fluid medium used was Hanks' balanced salt solution containing 0.2 % BSA. Centrifuge speed was 2000 rpm. The assay systems used to establish the effectiveness of the separation were the decrease in activity of GOT (asparate aminotransferase, EC 2.6.1.1), expressed as a ratio of activity to hemoglobin, and the measurement of cell volume distribution for each fraction using the Coulter ZH sizing apparatus. Results of these assays for a number of separations are combined in Fig. 3.

Old cells constitute early fractions, while younger cells make up later fractions. Although $\rho'd^2$ is supposedly lower for young than old cells, the

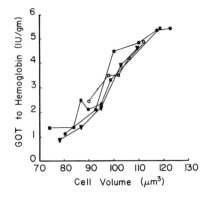

FIG. 3.   Results of human erythrocyte fractionation. The activity of glutamate oxaloacetate transaminase (GOT), expressed as a ratio to intracellular hemoglobin concentration, is plotted against mean cell volume for each separated fraction for a number of donors. Both the enzyme activity and the volume decrease with cell age.

young cell has a lower drag, either because of its basic shape or because its flexibility allows it to conform to a low drag configuration under the system of prevailing forces. This results in a value of $\rho'd^2/k$ which is higher in young cells. Consequently, higher fluid velocities are required to elute them. Alternatively, it is possible that the values of the buoyant densities of young and old cells are not so different from one another as those given by Danon and Marikovsky (1964).

B.   Peripheral Blood Leukocyte Separation

When counterflow centrifugation is used to isolate the various leukocyte species from peripheral blood, it is generally advantageous to use a preliminary separation to eliminate cell types not of interest. We are currently concerned with the mononuclear cells, for which separation on a Ficoll–Hypaque column having a specific gravity between 1.077 and 1.083 is suitable. The technique is essentially that described by Boyum (1968). Whole blood is diluted to four times its initial volume in Hanks' balanced salt solution (HBSS), layered gently onto the Ficoll–Hypaque column and spun at 400 g for 20 minutes at room temperature in an IEC Model PR2 centrifuge. Erythrocytes and polymorphonuclear leukocytes sink to the bottom of the tube, leaving the monocytes between the Ficoll–Hypaque and the mixture of HBSS and plasma, where they are easily collected in a Pasteur pipette with a 90° bent tip. The collected cells should be washed twice in HBSS. Boyum was originally able to obtain yields in excess of 90 % by this method both for lymphocytes and monocytes. Our yields have been lower, but are improved by use of a Plasmagel column prior to the Ficoll–Hypaque. This eliminates the bulk of the red cells and reduces entrapment of monocytes in the erythrocyte mass.

If it is desired to isolate the polymorphs, the preparation obtained from the Plasmagel can be loaded directly into the counterflow centrifuge. Although comparatively large numbers of erythrocytes remain, these are eluted at much lower pump speeds than all cells except some of the smaller lymphocytes. We have not, up to the present time, established the exact degree of overlap between monocytes and polymorphs, but it is clear that polymorph preparations are contaminated by only small numbers of large monocytes and possibly a few plasma cells.

Other preliminary separation methods suitable for leukocyte preparations include methyl cellulose–Hypaque columns and red cell lysis in hypertonic ammonium chloride. These techniques leave both mononuclear and polymorphonuclear cell populations intact. All these techniques leave platelets in the suspension, but these are eluted at very low pump speeds. Table I shows the ranges of flow rates of HBSS containing 0.1 % bovine serum albumin required for eluting the various leukocyte species against a gravitational field of 400 g at 30°C.

Figure 4 shows micrographs of a separated fraction containing polymorphonuclear leukocytes.

C. Spleen Cell Separation

Suspensions of mouse spleen cells were prepared by passing the organ as gently as possible through a fine wire screen. The cells were suspended in

TABLE I

FLOW RATES FOR ELUTING VARIOUS
LEUKOCYTE SPECIES

Type	Flow velocity at downstream end of chamber[a] (cm sec^{-1})
Lymphocytes	0.07–0.1
Monocytes	0.09–0.12
Polymorphs	0.11–0.14

[a] 400 g corresponds to 2000 rpm at the downstream end of the separation chamber, which is 9 cm from the center of rotation. A velocity of 0.1 cm sec^{-1} corresponds to 14.5 ml of medium per minute, since the cross section of the chamber measures 2.54×0.95 cm.

ation, the enhancement was still greater. Because the original population was heterogeneous, it would be expected that large diploid cells would remain with the tetraploids. The second fractionation of a cultured suspension involves a population of hybrid cells whose parents were an adenine-requiring mutant of fibroblastlike morphology, and a small spherical proline-requiring cell. The initial fusion was caused by incubation with inactivated Sendai virus, and a population of tetraploid cells which required neither adenine nor proline was obtained by culturing the hybrids in a medium lacking those ingredients. These cells were then separated according to their volume in the counterflow centrifuge. It is anticipated that karyotyping cells from the lowest volume fraction will lead to identification of the chromosome carrying the gene responsible for the mutation to the spherical morphology.

Counterflow centrifugation may also be a convenient method for isolating hybrid cells, provided that the parents are of comparable size and produce larger offspring. It should be noted that, when it is desired to put separated cells back into culture, it is necessary to keep the system sterile. We have achieved satisfactory results by initially soaking the rotor and tubing in 70% ethyl alcohol, followed by flushing with sterile distilled water. The elution medium used under these circumstances contained per milliliter 25 μg of the fungicide Amphotericin B and 50 μg of Gentamycin to arrest bacterial growth.

Acknowledgments

The authors thank Dr. D. W. Talmage of the Webb-Waring Lung Institute and Dr. B. E. Favara of Children's Hospital, Denver, for support given throughout this project. Research was funded by NIH Grants HL 54458 and CA 1224703, NIH Contract NCI–CB–53900, the University of Colorado Program Project in Basic Oncology (CA 13419–02), and the Children's Hospital Pathology Research Fund.

References

Boyum, A. (1968). *Scand. J. Clin. Lab. Invest.* **21**, Suppl. 97, 1–109.
Canham, P. B., and Burton, A. C. (1968). *Cir. Res.* **22**, 405–422.
Chapman, R. G., and Schaumberg, L. (1967). *Br. J. Haematol.* **13**, 665–678.
Danon, D., and Marikovsky, Y. (1964). *J. Lab. Clin. Med.* **64**, 668–674.
Flangas, A. L. (1974). *Prep. Biochem.* **4**, 165–177.
Glick, D., von Redlich, D., Juhos, E. T., and McEwen, C. R. (1971). *Exp. Cell Res.* **65**, 23–26.
Grabske, R. J., Lake, S., Gledhill, B. L., and Meistreich, M. L. (1974). *J. Cell Biol.* **63**, 119a.
Lehman, J. M., and Defendi, V. (1970). *J. Virol.* **6**, 738–749.
Lindahl, P. E. (1948). *Nature (London)* **161**, 648–649.
MacLeod, N. A. (1948). U.S. Patent 2,616,619.
Piomelli, S., Lurinsky, G., and Waserman, L. R. (1967). *J. Lab. Clin. Med.* **69**, 659–674.
Rutherford, D. E. (1949). "Vector Methods." Oliver & Boyd, Edinburgh.
Sass, M. D., Vorsanger, E., and Spear, P. W. (1964). *Clin. Chim. Acta.* **10**, 21–26.
Sutera, S. P., and Mehrjardi, M. H. (1975). *Biophys. J.* **15**, 1–10.

Chapter 2

Separation of Spermatogenic Cells and Nuclei from Rodent Testes

MARVIN L. MEISTRICH

Section of Experimental Radiotherapy, The University of Texas System Cancer Center,
M. D. Anderson Hospital and Tumor Institute,
Houston, Texas

I. Introduction

Studies of the molecular events of differentiation of spermatogenic cells from spermatogonia to spermatozoa have been complicated by the great number of different cell types present in the mammalian testis (Fig. 1). Traditionally, studies of these events have involved either microscopic techniques using autoradiography or histochemistry, or comparative biochemical analysis of tissue homogenates from animals having normal and incomplete spermatogenesis. Although these methods have produced much information regarding macromolecular events of spermatogenesis, they have certain limitations. Microscopic methods are not easily quantitated and cannot resolve different molecular species within a class of macromolecules unless immunological or hybridization reactions are employed. Cellular localization of biochemical events using animals with incomplete spermatogenesis is confused by the variety of cell types still present in these animals and in such abnormal situations the nongerminal and possibly the remaining germinal cells are perturbed.

In order to overcome the above limitations, methods for obtaining large quantities of homogeneous populations of spermatogenic cells at specific stages of differentiation are required. These cells can then be directly analyzed

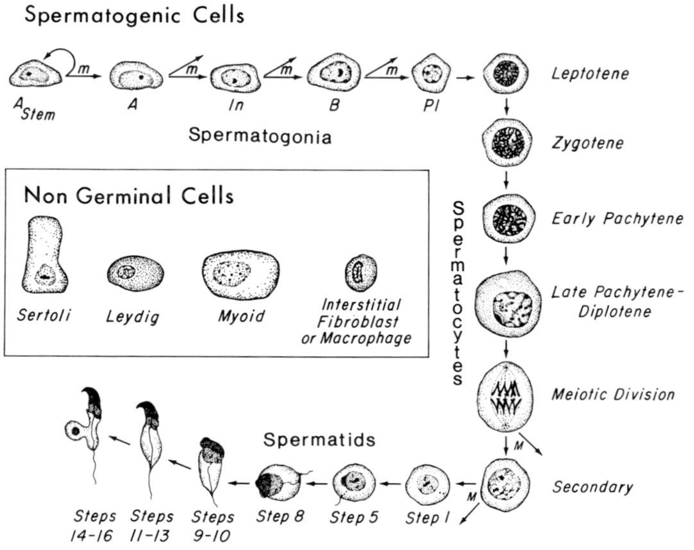

FIG. 1. Cells of mouse testis. Mitotic and meiotic divisions are indicated by m and M, respectively.

for their biochemical parameters. The separation of testicular nuclei is also of great interest since many of the important biochemical events occur within the nucleus.

In this chapter separation methods employing two physical parameters, density and sedimentation rate, as well as the use of ultrasound or enzymes to destroy cells and nuclei selectively, will be described and evaluated. The techniques, as currently being used in this laboratory, will be outlined. Some further details may be found in the following original papers: Meistrich (1972), Meistrich and Eng (1972), Meistrich et al. (1973a, 1975a, 1976), Bruce et al. (1973), Letts et al. (1974a), Meistrich and Trostle (1975), Grabske et al. (1975), Platz et al. (1975), and Grimes et al. (1975). The important factors and principles involved in the development of the separation methods will be emphasized. Methods developed in other laboratories will be referenced. The results obtained with the different procedures will be compared.

These studies have been performed primarily with the mouse and rat. Some separations have also been done on the testis cells from Syrian and Chinese hamsters. Because of the similarities in spermatogenesis in different rodents (and even in most mammals), the methods used and the results obtained are essentially the same with the mouse, rat, and hamsters.

II.   Principles of Separation

The principles of separation to be discussed here are based on Stokes' law of slow sedimentation of a sphere through a viscous noncompressible fluid medium. A more generalized form of Stokes' law (Wyrobek et al., 1976) in which the coefficient in this relationship varies with the shape of the object is:

$$S = (\rho_c - \rho_0) \, V^{2/3} g / \eta f \tag{1}$$

S is the sedimentation rate of the particle with respect to the fluid; ρ_c and ρ_0 are the densities of the particle and fluid, respectively; V is the volume of the particle; g is the gravitational acceleration due to the earth or centrifugation; η is the viscosity of the fluid; f is the frictional coefficient, which is dependent only on the shape of the particle and equal to 11.7 for a sphere.

When the density of the fluid is less than that of the particle, the particle will continuously sediment in the direction of the gravitational field. If the fluid is at rest, as in the Staput method, the particle will eventually sediment to the bottom of the chamber. However, when the flow of the fluid is equal but opposite to the sedimentation of the particle, the particle will remain stationary. This principle is the basis of centrifugal elutriation. A gradient of flow rates is established (Fig. 2), and the particle moves to a position where the fluid flow rate exactly balances the sedimentation rate. If the sedimentation rate of the particle is less than the minimum fluid flow rate in the chamber, it will be carried out of the chamber in a centripetal direction and collected. The maximum sedimentation velocity, S (mm/hr per g), of cells elutriated from the chamber can be calculated from the dimensions of the chamber, the fluid flow rate, f (ml/min), and the rotor speed, R (10^3 rpm) (Grabske *et al.*, 1975):

$$S = 1.93 \, f/R^2 \qquad (2)$$

The differences in size, rather than density or shape, are the major factors in the separation of different classes of testicular cells by velocity sedimentation. The volume difference between spermatozoa and pachytene spermatocytes is about 100-fold producing a 22-fold variation in the $V^{2/3}$ term. Variations in density of spermatogenic cells yield less than a 4-fold variation in $\rho_c - \rho_0$. The effect of cell shape is minor, amounting only to about a 2-fold change in frictional coefficients between a sphere and an object of such an extreme shape as a spermatozoon.

The principle of equilibrium density gradient centrifugation is also based on Eq. (1). The motion of the particle will be in the direction of the gravitational force when $\rho_c > \rho_s$, opposite to the gravitational force when $\rho_c < \rho_s$ and zero when $\rho_c = \rho_s$. Therefore, if a density gradient is created in a centrifuge tube

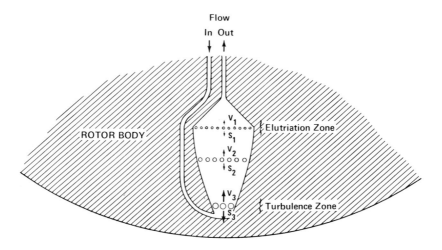

FIG. 2.   Sedimentation of particles and flow within the elutriator chamber. S_1, S_2, and S_3 are sedimentation rate vectors of small, intermediate, and large particles, respectively. V_1, V_2, and V_3 are the flow vectors in different regions of the chamber.

such that somewhere in the tube $\rho_c = \rho_s$, the particle will move to that position and remain there.

The density of a cell or nucleus, however, is dependent on the separation medium employed. The density of a cell is the total mass divided by the volume enclosed within the plasma membrane. However, the high concentrations of solutes required to obtain $\rho_s = \rho_c$ often cause these solutions to be hypertonic. Water is withdrawn from the cell, thereby increasing the cell density. Nuclei differ from cells in that they do not respond to the osmotic pressure (Anderson and Wilbur, 1952). The pores in the nuclear membrane, varying from 100 Å to 600 Å in diameter depending on the cell type, allow passage of relatively large macromolecules. If the gradient material does not penetrate into the nucleus, as for example with colloidal silica particles 160 Å in diameter (Loir and Wyrobek, 1972), the density of the nucleus will be the mass enclosed by the nuclear membrane, including all the water and buffer solution, divided by the volume contained within it. If the solute is, however, freely permeable across the nuclear membrane, the apparent density is that of the complex of chromatin, membrane, and other nuclear matrix elements. A further dependence on the nature of the gradient material occurs with variation of hydration of the macromolecules within the nucleus (Rickwood and Birnie, 1975).

III.   Preparation of Cell Suspensions

In any cell separation method the preparation of the single-cell suspension from a solid tissue is of critical importance. In general, it is important to maximize the number of intact or viable cells, disperse clumps of cells, and minimize reaggregation. In the case of testicular cells, it is also important to minimize the formation of multinucleate cells and control the loss of cytoplasm and the flagellum from the late spermatids. The effect of the method of preparation on the separation will be discussed in Section VII.

A.   Mechanical Plus Trypsin

Single-cell suspensions obtained by purely mechanical dissociation of testicular tissue (Lam *et al.*, 1970) yield at most 80 % intact cells as determined by the trypan blue dye exclusion (Meistrich, 1972). Furthermore, mechanical dissociation of testicular tissue selectively damages Sertoli cells, Leydig cells, spermatogonia, and young primary spermatocytes and removes the cytoplasm and flagella from many of the elongated spermatids. In addition, cells prepared mechanically aggregate more readily than do trypsinized cells, thereby reducing the quality of separation. However, treatment of a mechanical cell suspension with trypsin and DNase completely destroys the damaged cells with the exception of late spermatids. This method has the added advantage of reducing the numbers of certain cell classes in the preparation.

The mechanical cell preparation described by Lam *et al.* (1970) is usually employed as follows. After removal of the tunica albuginea, two mouse testes, or equivalent amounts of rat or hamster testis, are placed on a battery of chilled razor blades. The assembly consists of 25 double-edged blades, each seperated by a blade with one edge cut off as a spacer, bolted together between 2 steel plates. The tissue is pressed into the blade assembly by rolling a wooden dowel across the blades. This procedure cuts the seminiferous tubules into 200-μm segments. The tissue is washed from the blades and the cells are dispersed by pipetting with cold phosphate-buffered saline (PBS) which consists of 8 gm of NaCl, 0.2 gm KCl, 1.15 gm of $Na_2HPO_4 \cdot 2H_2O$, 0.2 gm of KH_2PC_4, 0.1 gm of $MgCl_2 \cdot 6H_2O$, and 0.1 gm of $CaCl_2$ per liter of H_2O. Tubule fragments are removed by passage through an 80-μm stainless steel screen.

A more gentle mechanical preparation method, used with rat testes, yields a high percentage of elongated spermatids, many with intact flagella (Sanborn *et al.*, 1975). Removal of interstitial tissue is first accomplished by incubation of gently teased tissue with 0.25% trypsin at 33°C for 20 minutes and then placing the tissue on a 125-μm stainless steel grid at room temperature and washing with a forceful stream of buffer. The tubules are then finely minced on the screen and washed thoroughly with PBSG (PBS containing 0.1% glucose). The freed germ cells pass through the screen.

The cell suspension, prepared by either of the above methods is treated with trypsin and 18 μg/ml of crude pancreatic DNase (Sigma, DN-25) for 10 minutes at 31°C. Either a crude preparation of trypsin could be used at 0.25% w/v (GIBCO, 2.5% trypsin in saline) or, when cellular proteins are to be analyzed, a purified preparation at 0.1% w/v is used (Worthington, TRL). Fetal calf serum is then added to a concentration of 8% to inhibit the trypsin. This treatment results in lysis of damaged cells, digestion of subcellular debris, and about a 50% reduction in the number of sperm with intact flagella.

B. Trypsin

Testes freed of the tunica are placed in a plastic petri plate at room temperature. A drop of PBS is added to keep them moist. The testes are chopped into 1–2 mm cubes using a single razor blade, first making a series of cuts in one direction and then in a perpendicular direction. Testis tissue is gently transferred to PBSG containing trypsin and crude DNase as described above.

A minimum of 12 ml of PBSG per gram of testis is used. The suspension is incubated at 31°C for 30 minutes with stirring. Fetal calf serum is added to 8%, and the sample is filtered through an 80-μm pore screen. Although in in earlier experiments we found that centrifugation of the sample in the cold caused damage to spermatozoa, we now routinely centrifuge the sample (500 g, 15 minutes) at room temperature with little or no loss of sperm. The sample is resuspended in PBS containing 20 μg/ml of crude DNase. If analysis of proteins is desired, 2 μg/ml of purified DNase (Worthington D) is used instead and 0.02% purified soybean trypsin inhibitor (Worthington SI) is also added. When more than 3×10^8 cells are used, 5 mM NDA (2-naphthol-6,8-disulfonic acid, dipotassium salt, Eastman Chemicals) is included in this buffer. The sample is incubated at room temperature for 5 minutes to disperse clumps of DNA and then quickly cooled on ice.

The yield of cells measured by DNA recovery is 35%, and cell integrity assayed by trypan blue dye exclusion is 98%. The ultrastructural features of the cells are identical to cells found *in situ* with the exception of acrosomal alterations in some round spermatids and loss of cytoplasm, flagella, and membranes from some elongated spermatids (Barcellona and Meistrich, 1976). Activities of enzymes, such as lactate dehydrogenase and glycosyltransferases (Letts *et al.*, 1974a) are unaffected by the preparation of the suspension. No detectable degradation of nucleoproteins is produced by either this or the EDTA–trypsin method (Platz *et al.*, 1975).

C. EDTA–Trypsin

This method is similar to that described above except that the chopped testes are placed in calcium-magnesium-free PBSG containing 5 mM EDTA (Meistrich and Trostle, 1975). Then the following sequence of steps is performed: (1) incubation at 31°C with stirring for 10 minutes; (2) addition of trypsin; (3) incubation without stirring for 10 minutes; (4) addition of MgCl$_2$ to 10 mM and crude DNase to 17 μg/ml; (5) incubation with stirring for 10 minutes; and (6) continuation with the trypsin protocol at step of fetal calf serum addition.

The EDTA-trypsin method has the advantage of producing the highest yield of intact spermatogonia and young primary spermatocytes. The cell yield is slightly lower than with the trypsin method and the percentage of intact sperm is reduced.

D. Other Methods

Several different methods and modifications of our methods have been described. Changes in the medium used for preparation of the cell suspensions include increasing the osmolarity of the PBSG solution (Loir and

Lanneau, 1974), the substitution of more complete tissue culture media for PBSG (Pretlow *et al.*, 1974; Bellvé and Romrell, 1974), and the removal of calcium and magnesium from the PBS with addition of EDTA (Davis and Schuetz, 1975). The use of EDTA alone for dissociation results in lower cell integrity, but when used in conjunction with trypsin as described above, cell integrity is restored to about 98 %. The replacement of trypsin with Pronase has been reported to yield a greater number of viable cells (Pretlow *et al.*, 1974), but we find no significant differences in the suspensions prepared with these two enzymes. We have elected to use trypsin since Pronase binds to cell surfaces (Poste, 1971) but, in contrast to trypsin, is not readily inhibited.

Temperature may also be an important factor. The viability of ram testicular cells is increased if the tissue is maintained at 30°C throughout the trypsin preparation procedure (Loir and Lanneau, 1974).

The handling of the tissue has several effects. A reduction in the number of multinucleate cells has been achieved by more gentle handling of the testis prior to incubation with enzymes (Bellvé and Romrell, 1974). The extent of detachment of the cytoplasm from the elongated spermatids is dependent on the intensity of stirring the tissue during trypsinization (Loir and Lanneau, this volume, Chapter 3).

Selective removal of contaminating cells during preparation is often advantageous. The removal of Leydig cells may be accomplished by treatment of intact seminiferous tubules with distilled water (Galena and Terner, 1974), collagenase (Bellvé and Romrell, 1974), or trypsin (Steinberger, 1975). The removal of spermatids with flagella can be accomplished by passing the suspension through Sephadex and glass wool (Galena and Terner, 1974). Selective isolation of the cytoplasmic fragments and residual bodies from the later spermatids is achieved by homogenization followed by centrifugation in a sucrose step gradient (Nyquist *et al.*, 1973).

IV.   Preparation of Nuclear Suspensions

A.   Mechanical–Dense Sucrose

Mouse testes are homogenized with a Potter–Elvehjem type homogenizer in isotonic sucrose (0.25 M sucrose, 3 mM $CaCl_2$), filtered through an 80-μm screen and centrifuged through dense sucrose (2.2 M sucrose, 1 mM $CaCl_2$) for 1 hour at 68,000 g_{max} (Meistrich and Eng, 1972). This method produces clean nuclei without the use of detergents, but has the following disadvantages: the homogenization is insufficient to break the membranes of isolated cells in suspensions, 2 hours are required for this step, the yield is only 10 %, and the centrifugation step also results in a few nuclei sticking tightly together in small clumps. Similar methods have been described elsewhere (Utakoji *et al.,* 1968; Utakoji, 1970; Kaye and McMaster-Kaye, 1974; Kumaroo *et al.,* 1975).

B.   Cetrimide

Resuspension of testis cell suspensions in solutions containing 5 mg/ml of the cationic detergent, Cetrimide (hexadecyltrimethylammonium bromide, Eastman Kodak), and 1 mM CaCl$_2$ (pH 5.2) rapidly produces a high yield of isolated nuclei (Bruce *et al.,* 1973). These nuclei are free of cytoplasm and flagella and are suitable for separation by velocity sedimentation and equilibrium density centrifugation (Loir and Wyrobek, 1972). However, it has not been possible to extract a normal complement of nucleoproteins from these nuclei (R. B. Goldberg and C.P. Stanners, personal communication).

C.   Hypotonic Lysis–Triton

Testis cells suspended in PBS are centrifuged and resuspended at a concentration of not more than 2×10^7/ml in a hypotonic buffer (MP buffer consisting of 5 mM MgCl$_2$ and 5 mM sodium phosphate, pH 6.5) also containing 0.25% Triton X-100 and 5 mM NDA (Platz *et al.,* 1975). The addition of NDA produces more rapid lysis, permits greater concentrations of cells, and yields cleaner nuclei. If these nuclei are to be used for studies of nuclear proteins, 0.025% soybean trypsin inhibitor (STI) and 100 μM phenylmethylsulfonyl fluoride (PMSF) are included in the cell lysis mixture. Nuclei are passed twice through a 25-gauge needle. The suspension may be used as is or diluted in MP buffer. Centrifugation of the sample is avoided in order not to produce clumping. About 20% of the spermatids with elongated nuclei retain their flagella. The advantages of this method are that clean nuclei are rapidly obtained in high yield and with no clumping and no detectable loss or degradation of nuclear proteins.

D.   Sonication and Trypsin Resistance of Elongated Spermatids

When the spermatid nucleus elongates, changes occur in chromatin structure. The nucleus becomes resistant to sonication at step 12 in the mouse and rat and resistant to trypsin digestion as well at later stages (Meistrich *et al.,* 1976; Grimes *et al.,* 1976). Several different preparation procedures, based on this differential resistance, have been used depending on the purpose of the experiment.

Elongated spermatid nuclei from mouse testes have been prepared for counting these nuclei and measurement of the radioactivity contained within their DNA. Two to six mouse testes are homogenized in 8 ml of cold distilled water with a Polytron homogenizer (PT-10, Brinkman Instruments) for 5 seconds at 90% maximum power. Two milliliters of 5% Triton X-100 are added. The sample is then sonicated for 1 minute with a Branson W-185 sonifier fitted with a microprobe and set at 6 (25-W output). The sample is passed through a 10-μm nylon screen to remove debris.

For analysis of radioactivity in sonication-resistant sperm heads, the sample may be filtered onto Whatman GF/C glass fiber filters, which are then washed with water, trichloroacetic acid, and ethanol to remove contaminating radioactivity (Meistrich et al., 1975a). Contamination by DNA from other cell types, measured radiochemically, is less than 2%.

Trypsin-resistant nuclei are prepared by centrifuging (450 g, 10 minutes) the sonication resistant nuclei and resuspending in 5 ml of 0.9 mM MgCl$_2$ containing 0.23% crude trypsin and 18 μg of crude DNase per milliliter. The suspension is incubated at room temperature for 25 minutes and then placed on ice to stop the enzymic reaction. The suspension is again filtered through a 10-μm screen, and the sperm heads are collected by filtration onto GF/C filters.

This procedure results in nearly complete recovery of the step 15–16 spermatids with only 0.005% contamination by radioactive DNA from other cells. This level of contamination is equivalent to that obtained by the more intricate procedure described previously (Meistrich et al., 1976).

Sonication-resistant spermatid nuclei from rat testes have been purified for studying the biochemistry of the nuclear proteins and their changes during spermiogenesis. The method is adaptable to large-scale biochemical applications. Precautions are taken to prevent degradation of nuclear proteins.

Rat testes are homogenized with a Potter–Elvehjem type homogenizer in 5 volumes of a solution containing 0.31 M sucrose, 3 mM MgCl$_2$, 10 mM potassium phosphate, pH 6.0, 0.05% Triton X-100, and 100 μM PMSF. The sample is maintained between 0° and 6°C throughout. The homogenate is filtered through cheesecloth and centrifuged (800 g, 20 minutes). The pellet is rehomogenized in 2 volumes (with respect to the amount of tissue at start) of distilled water and is sonicated with twelve 15-second bursts of ultrasound (Bronwill Biosonik BP III, fitted with a large probe). To purify the sonication-resistant nuclei, 20 ml of the sonicated suspension is layered over 10 ml of 1.5 M sucrose in a 40-ml round-bottom centrifuge tube and centrifuged (1000 g, 30 minutes) in a swinging-bucket rotor. This step is followed by two additional cycles of homogenization in distilled water and centrifugation through sucrose.

Testicular spermatid nuclei prepared by this method are about 99% pure. Contamination by ribonucleoprotein aggregates from residual bodies amounts to 1% and contamination by other nuclei to less than 0.1%. The absence of histones in these preparations confirms their purity (Grimes *et al.*, 1975). Other workers (Marushige and Marushige, 1975) have developed similar methods involving the preparation of rat testis chromatin, which is then subjected to extensive shearing and sucrose centrifugation.

V.   Velocity Sedimentation

A.   Unit Gravity (Staput)

1.   SEPARATION OF CELLS

The Staput apparatus (Miller and Phillips, 1969; Lam *et al.*, 1970) has been used for velocity sedimentation at unit gravity. The Staput chamber is a cylindrical chamber with a conical base forming a 30° angle with the horizontal and a small baffle at the bottom to divert the fluid flow in a horizontal direction (Fig. 3). We have chosen this device, in which the cells are loaded and unloaded through the bottom of the chamber rather than unloading through the top of the chamber (Peterson and Evans, 1967) because of greater simplicity in construction and operation of the chamber and to minimize contamination of rapidly sedimenting cells by the greater number of slowly sedimenting cells. A Staput apparatus with autoclavable glass chambers is commercially available (Johns Scientific Company, Toronto, Canada). Since sterility was not required, we constructed Lucite chambers with an almost identical design except for the replacement of the stainless steel baffle at the base of the chamber by a Lucite baffle cemented into the body of the chamber (Lee and Dixon, 1972).

Chambers of three different diameters (7 cm, 12.5 cm, and 28 cm) are routinely used. The choice of chamber size is based on the number of cells one desires to separate and the desired resolution. Since the depth or thickness of the starting band of cells limits the resolution, one must first determine the desired starting zone. We routinely use a 1.6-mm starting zone; however, thicker zones could be used if only rapidly sedimenting cells are to be collected, wide fractions are taken, or the separation is run longer. The thickness of starting zone times the cross-sectional area of the chamber yields the volume of cell suspension that can be loaded. The maximum number of cells which which can be separated is the product of this volume with the maximum cell concentration at which no cell clumping or streaming occurs (i.e., 1.5×10^7 testis cells/ml). Therefore, using a 1.6 mm starting zone in a 28-cm diameter chamber, we can separate 1.5×10^9 cells.

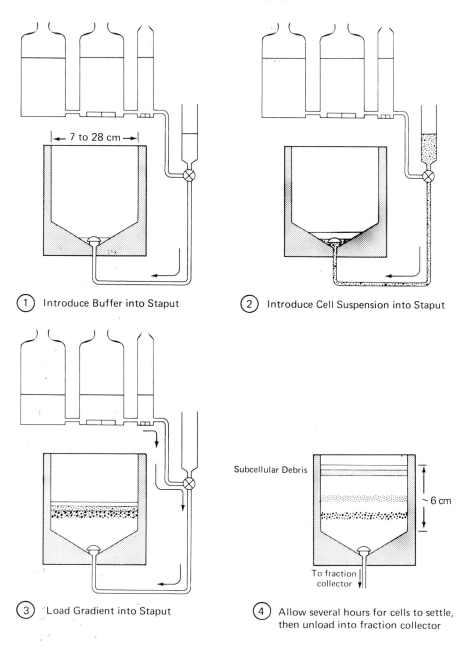

FIG. 3.   Steps involved in separation of cells or nuclei by the Staput method.

The maximum concentration of cells that can be loaded is determined by the shape and composition of the gradient and the nature of the cell surface. When the appropriate nonlinear gradient is used, the streaming phenomenon appears to be completely suppressed (Miller and Phillips, 1969). Under these circumstances, the concentration of cells (or nuclei) appears to be limited by clumping or loose aggregation of cells. Although the cells are not clumped when introduced into the chamber, when the cell concentration is too high, small clumps of cells form within 1 hour after loading the chamber and sediment more rapidly than the bands of individual cells. The formation of these clumps could be a result of collisions between cells sedimenting at different rates. Since the liquid in the chamber is extremely still, even the smallest attractive forces could maintain stable clumps. Increased cell concentrations without clumping have been achieved by three factors: (1) trypsin treatment of the cells, (2) the use of BSA instead of Ficoll for the gradient, and (3) the inclusion of 5 mM NDA in the gradient buffer. Under these conditions we can load up to 1.5×10^7 cells/ml without any appreciable clumping.

The procedure for loading and running the chamber is indicated in Fig. 3. All procedures are carried out at 4°C. When cells are loaded at concentratins of $\geq 5 \times 10^6$/ml, three gradient bottles are employed to generate a nonlinear gradient (Miller and Phillips, 1969). The larger bottles are constructed with diameters roughly two-thirds of the chamber diameter. The smaller bottle has a diameter one-third that of the larger ones. The outlet ports of the bottles must be sufficiently large (6 mm i.d.) to provide low enough flow resistance to minimize the differences in the liquid levels between bottles. We use gradients of 1% to 4% of BSA (Cohn fraction V, Reheis Chemical Company) dissolved in PBS, although any basic salt solution or culture medium could be substituted. The 1% solution is placed in the first (narrow diameter) bottle, 2% in the center bottle and 4% in the third bottle. When fewer than 5×10^6 cells/ml are used, streaming is not a problem, and only the two larger gradient bottles are used to generate a linear 1% to 3% gradient.

The gradient bottles are connected to a three-way valve (Becton-Dickinson, MS-11) and syringe barrel. The three-way valve is connected to the Staput chamber with tubing (Travenol Laboratories, 2C0050 for the 7 and 12.5-cm chambers or 2C0051 for the 28-cm chamber) which fits directly into the tapered bottom of the chamber. The flow is generated by gravity since the gradient bottles are positioned 40 cm higher than the Staput chamber.

The chamber is first loaded with buffer such that a buffer layer 1.6 mm deep will be formed when the liquid rises into the cylindrical part of the chamber. Then cells are introduced in a buffer solution containing 0.2% BSA. Next the gradient is allowed to flow in under the cells, raising them up in the chamber. Precautions must be taken to exclude air bubbles from the tubing during these steps. We routinely load the chamber 6 cm above the top of the cone. We allow 3 or 4 hours for separation of rat and mouse testis cells, respectively. During this time the pachytene spermatocytes have settled about 4 cm down from the starting zone. These times maximize separation but avoid having cells settle onto the conical part of the chamber. The major bands are clearly visible in the transparent chamber (Fig. 4), and their positions are measured to calculate the peak fractions that will contain the purest populations of specific cell types. The chamber is unloaded into a fraction collector based on volume collection (Gilson Escargot) over a period of from 15 (small chamber) to 50 minutes (medium and large chambers). Fractions are counted with a Coulter counter to determine peaks around which several fractions are pooled. Cells are concentrated by centrif-

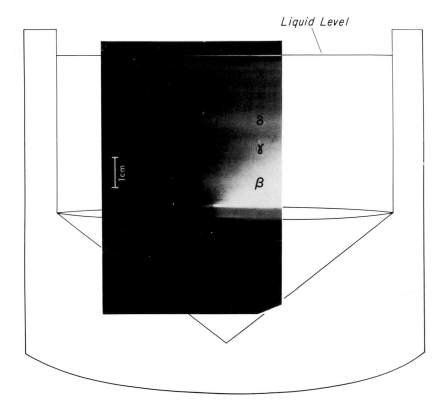

FIG. 4.   Photograph of bands of cells in Staput chamber taken with scattered light. Bands are designated with Greek letters corresponding to the notation of Meistrich (1972). Origin indicates the starting position of the cells; subcellular debris remains at origin.

ugation in plastic tubes or bottles (500 g for 10 minutes). Air-dried smears are made from pooled fractions for subsequent cytological analysis (Meistrich et al., 1973a). The smears are fixed in Bouin's solution and stained with PAS–hematoxylin.

The composition of some of the pooled fractions is indicated in Table I; data are condensed from Meistrich et al., (1973a). Very similar results have been obtained with rat (Go et al. 1971a, Platz et al., 1975) and Chinese hamster testicular cells with the exception that sedimentation rates of cells are slightly different (Table II).

TABLE I

PERCENTAGE OF CELL TYPES IN FRACTIONS B, D, F, I, AND K OF MOUSE
TESTICULAR CELLS SEPARATED BY THE STAPUT METHOD

Cell type	B	D	F	I	K
Spermatogonia and preleptotene spermatocytes	2[a]	4	2	0	0
Early primary spermatocytes[b]	2[a]	4	3	0	0
Pachytene spermatocytes[c]	75	3	0	0	0
Secondary spermatocytes[c]	0	18	0	0	0
Spermatids steps 1–8	10[a]	59[a]	81	1	0
Spermatids steps 9–13	1	2	6	21	9
Spermatids steps 14–16	0	0	0	2	37
Cytoplasmic fragments and residual bodies	1	5	5	74	54
Nongerminal cells and unknowns	8	5	3	2	0

[a] Most of the cells are multinucleates.
[b] Includes leptotene, zygotene, and early pachytene spermatocytes.
[c] Including cells in meiotic division.

TABLE II

SEDIMENTATION RATES OF TESTICULAR CELLS AND NUCLEI AT UNIT GRAVITY
AT 4°C (MM/HR)[a]

	Mouse cells	Rat		Chinese hamster[c] cells
		Cells	Nuclei	
Spermatogonia and preleptotene spermatocytes	4.4	3.6	3.2	—
Pachytene spermatocytes	9.8	12.5	5.7	14.1
Round spermatids	4.4	5.7	2.0	6.7
Sperm[b]	0.75	0.8	1.1	0.8

[a] Cells in 1–3% BSA gradients in PBS; nuclei in 2–10% sucrose gradients in MP buffer.
[b] Elongated spermatids with flagellum and little remaining cytoplasm.
[c] M. L. Mace, Jr., and M. L. Meistrich (unpublished observations).

2. Separation of Nuclei

The Staput method can also be used for the separation of testicular nuclei (Meistrich and Eng, 1972; Kaye and McMaster-Kaye, 1974). The procedures employed are, with a few exceptions, the same as those used for cells.

For separation of nuclei, the buffer (MP buffer) and gradient material (sucrose) need not be isotonic. Either a 1.3% to 10% nonlinear gradient (4% in second bottle) or a 2% to 10% linear gradient are employed. Although shallower gradients would suffice, the steeper gradient provides greater stability. The use of BSA is avoided since it binds to nuclei and contaminates nucleoprotein extracts.

Despite their greater density, nuclei sediment more slowly than cells because of their smaller size (Table II). Longer times must therefore be allowed for separation. We allow 4 hours for separation of pachytene spermatocyte nuclei, 8 hours for round spermatid nuclei and up to 16 hours for elongated spermatid nuclei. After collection, fractions are counted in a hemacytometer to localize the peak fractions and obtain differential counts. Leptotene–zygotene spermatocyte, pachytene spermatocyte, round spermatid, and elongated spermatid nuclei may be identified on the hemacytometer when viewed under phase contrast. After pooling selected fractions, Triton X-100 is added to a concentration of 0.2% to aid in pelleting the nuclei and enhance recovery. Nuclei are concentrated by centrifugation (500 g, 10 minutes).

B.   Centrifugal Elutriation

Centrifugal elutriation appears to be the method of choice for separation of testicular cells by velocity sedimentation (Grabske et al., 1975). For nuclei, however, aggregation is too great to permit separation by elutriation. We are presently trying to overcome this problem by proper choice of suspension medium.

The elutriator rotor (Model JE-6, Beckman Instruments, Palo Alto, California) is driven by a Beckman J21-B centrifuge on which the speed range has been reduced to 0–6000 rpm to permit finer adjustment. The set-up of the system shown in Fig. 5 differs slightly from that described by Glick et al., (1971). There is only one separation chamber in the rotor. The cells are passed through the peristaltic pump eliminating the need for the more cumbersome loading chamber (Grabske et al., 1975). The buffer chamber, syringe barrel for loading cell suspensions, pump and most of the tubing are kept in a refrigerated chamber at 4 $\pm$ 1°C. Regulation of the buffer temperature is important for reproducibility of the separations since viscosity, which is a factor in Eq. (1), varies with temperature.

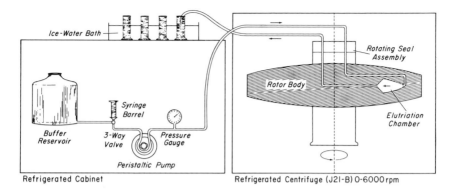

FIG. 5.   Experimental set-up for centrifugal elutriation.

The elutriator rotor and associated tubing are first filled with the desired buffer. We employ PBS with the addition of 0.5% BSA to preserve cell integrity and reduce cell adhesion. When more than 3×10^8 cells are separated in a run, 5 mM NDA is also included in the buffer. Cell suspensions prepared by a method employing trypsin are filtered through a 25-μm nylon screen just prior to loading into the elutriator. The sample is diluted, if necessary, to a cell concentration of less than 3×10^7/ml. The rotor speed and buffer flow rate are adjusted. The three-way valve is then turned to pump the cell suspension from the syringe barrel into the rotor. When the cells enter the elutriator chamber, the collection of the first fraction is begun. When the cell suspension is loaded into the rotor, the valve is turned to pump buffer through the chamber again. The flow rate and rotor speed are maintained until very few cells are being elutriated. The the flow rate and/or the rotor speed are adjusted to increase S (Eq. 2), the maximum sedimentation rate of cells which will be elutriated. At this point, the collection of a second fraction is begun and continued until elutriation of this next class of cells from the chamber is essentially complete. These steps are repeated for as many fractions as are desired. Then the remaining cells are washed from the chamber with a high flow rate while stopping the rotor.

Because of the wide range of sedimentation rates of testicular cells (from $0.7 \, \text{mm/hr} \cdot g$ to $\geq 10 \, \text{mm/hr} \cdot g$), the large cells tend to be packed at the bottom of the chamber while elutriating smaller cells. This problem is overcome by loading the rotor at an intermediate setting and passing all the small cells through in fraction 1. After sequential fractions of the larger cells are collected, fraction 1 is reintroduced into the chamber and fractions of the smaller cells are collected. The protocol for this procedure is given in Table III.

When only one fraction is desired, a simplified procedure is used. Cells are loaded into the elutriator under conditions such that cells having sedi-

TABLE III

CONDITIONS FOR SEPARATION OF RAT TESTICULAR CELLS BY CENTRIFUGAL
ELUTRIATION[a]

Fraction	Rotor speed (rpm)	Flow rate (ml/min)	Volume collected (ml)	Major cell types (percentage composition)[b]
1[c]	3000	13.5	180	—
2	3000	31.5	150	Spermatids 1–8 (59%) Spermatids 9–13 (25%)
3	3000	41.4	150	Spermatids 1–8 (81%)
4	2000	23.2	150	Spermatids 1–8 (47%)
5	2000	28.2	150	Spermatids 1–8[d] (52%) Pachytene (19%)
6	2000	40	150	Pachytene (86%)
W[e]	0	—	—	—
7[f]	5000	8.75	250	Subcellular debris
8	5000	22.5	150	Spermatids 11–19 (69%) Residual bodies (28%)
9	3360	22.5	150	Spermatids 9–13 (55%) Residual bodies (35%)
W[e]	0	—	—	—

[a] Temperature setting: 4°C; buffer: 0.5% BSA, 5 mM NDA in PBS; cells loaded: 2.5×10^9; load volume: 100 ml.
[b] Cytoplasmic fragments that did not contain large ribonucleoprotein aggregates were not scored (Platz et al., 1975).
[c] Cell suspension loaded during collection of fraction 1.
[d] Almost all are multinucleate cells.
[e] Wash out of chamber.
[f] Fraction 1 loaded during collection of fraction 7.

mentation rates lower than that of the desired cell type are elutriated from the chamber. The flow rate is then increased just enough to elutriate the desired cell type, leaving the more rapidly sedimenting cells in the chamber. Such a separation can be completed within 20 minutes (Grabske et al., 1975).

As many as 3×10^9 cells have been separated in a single run with no problems of cell aggregation. The only effect of such high cell numbers in the elutriation chamber is a decrease in the apparent sedimentation rates of the cells. This is likely a result of the fact that the cells fill an appreciable fraction of the chamber. Based on the packed volume of cells and sedimentable debris after 500 g centrifugation, 2.5×10^9 rat testis cells would completely fill the 4.5-ml chamber. Although the debris does not remain in the chamber, the presence of so many cells reduces the effective cross-sectional area available for fluid flow, thereby increasing the fluid velocity. This increase in fluid velocity causes cells with sedimentation rates greater than the S calculated from Eq. 2 to be elutriated from the chamber.

Typical fractions obtained by elutriation of hamster testicular cells are shown in Fig. 6. The fact that the separation is primarily based on size is apparent.

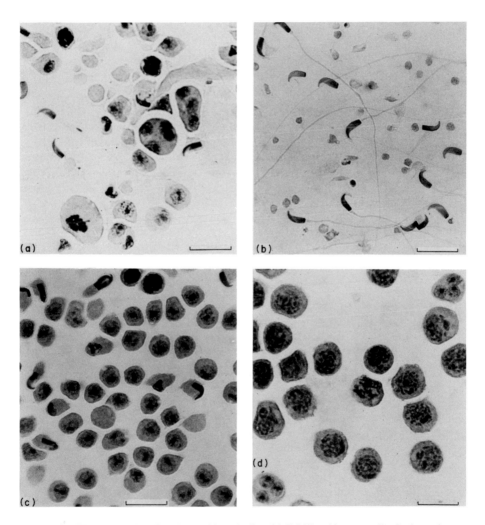

FIG. 6.   Hamster testis cells stained with periodic acid–Schiff and hematoxylin. Scale markers correspond to 25 μm. (a) Unseparated cell suspension. Fractions separated by centrifugal elutriation enriched in (b) spermatoza, (c) round spermatids, and (d) pachytene spermatocytes. Reprinted with permission from Grabske *et al.* (1975).

VI. Equilibrium Density Gradient Centrifugation

A. Cells

Equilibrium density gradient centrifugation can also be used to separate testicular cells. Although various gradient media may be employed, we have chosen Renografin because of its high density, low cost, low viscosity, availability in commercially prepared sterile solutions, and preservation of cell viability and integrity (Grdina et al., 1973).

Renografin-76 (E.R. Squibb and Sons) is a 76% w/v solution of an anionic iodinated derivative of benzoic acid. It contains the methylglucamine (66%) and sodium (10%) salts of N,N-diacetyl-3,5 diamino-2,4,6-triiodobenzoate (MW 613) dissolved in isotonic saline containing chelating buffers (citrate and EDTA). This solution is hypertonic with an osmolality of about 1700 mOsm.

In preliminary experiments the osmolality of the gradient solution was reduced by diluting Renografin-76 to 20% w/v with water followed by further dilution with an isotonic salt solution. However, the bands of cells obtained with this method were not as sharp as when the Renografin was merely diluted with isotonic salts yielding a more hypertonic gradient. In contrast to other cell types studied previously (Grdina et al., 1973), testicular cells are particularly sensitive to lysis upon rapid dilution from the hypertonic Renografin solution (Meistrich and Trostle, 1975). Since a several step washing procedure was required to remove the cells from the hypertonic Renografin, reexamination of the procedure of first diluting the Renografin with water to reduce the tonicity may be warranted for future use.

Although Renografin has no effect on intact cells except for decreasing their volume, it lyses isolated nuclei and cells with damaged membranes. This causes the release and precipitation of DNA which can trap cells in the gradient. To minimize this problem, crude DNase (100 μg/ml), as well as excess $CaCl_2$ to overcome the chelating strength of the buffers and activate the DNase, are included throughout the gradient.

In general, we have employed 16.5-ml gradients, consisting of 1.9 ml of a 24% w/v Renografin underlayer followed by linear 22% w/v to 8% w/v gradient of Renografin concentration. Gradients are generated as described by Grdina (1976). When purification of specific cell types are desired, shallower gradients are used to achieve better resolution. The density of these Renografin solutions at 4°C (gm/cm^3) is calculated from the refractive index measured at 24°C (N_{24}) by the equation: $\rho_4 = 3.5419N_{24} - 3.7190$.

In order to minimize the time in Renografin, the cells are loaded on top of chilled preformed gradients. Approximately 10^7 cells in 1 ml of Ringer's solution containing 20 μg of DNase per milliliter are layered on each gradient. The maximum number of cells that can be separated in each 17-ml gradient, without the formation of clumps, is 3×10^7.

Gradients were spun for 20 minutes at 10,000 rpm (13,000 g_{av}) at 4°C in an SW 27.1 rotor in a Beckman L5-50 ultracentrifuge. The rotor is decelerated with the brake off.

The method of unloading the centrifuge tube is an important factor affecting the quality of the separation. It appears that during centrifugation cells become stuck against the walls of the centrifuge tube, and during unloading, as the meniscus of the liquid passes along the walls of the tube, the cells elute off. When the gradients are unloaded from the top with a Buchler Auto-Densi Flow Collector, cell bands tend to be skewed toward greater densities and the pellet fraction is contaminated with other cells. On the other hand, when the gradients are unloaded by puncturing the bottom of the tube, cell bands tend to be skewed toward lower densities and the top fraction is contaminated with other cells. The most satisfactory procedure for use when bands of cells are visible in the tube appears to be puncturing the tube from the side with a hypodermic needle and withdrawal of the desired region in a syringe (Grdina, 1976).

B. Nuclei

The choice of a medium for the separation of nuclei by equilibrium density centrifugation is based on some different criteria than for separation of cells. Isotonicity is not required. The greater density of nuclei necessitates finding a medium of sufficient density to prevent the nuclei from pelleting.

We have employed metrizamide as the medium for separating different classes of elongated spermatid nuclei from rat testes. Metrizamide, 2-(3-acetamido-5-N-methylacetamido-2,4,6-triiodobenzamido)-2-deoxy-D-glucose (Gallard-Schlesinger Corp.), is a nonionic iodinated benzamide derivative with molecular weight 789. Whereas Renografin swells and lyses these nuclei, metrizamide has no effect on either the morphological properties or basic protein composition of the elongated spermatid nuclei. Metrizamide has been reported to be inert, noncytotoxic and show no irreversible inhibition of enzyme or protein assays.

Nuclei of spermatids in steps 12–19 of development are prepared by sonication and centrifugation through sucrose. Although there is not much change in shape during these stages, the intensity of staining with hematoxylin decreases significantly between steps 15 and 16, permitting differential counts of these nuclei. These preparations contain 33% step 12–15 nuclei and 67% steps 16–19 nuclei.

Solutions of 59% w/v and 69% w/v metrizamide are prepared in buffer containing 1 mM $MgCl_2$, 1 mM sodium phosphate, 0.25% Triton X-100, and 5 mM NDA at pH 6.5. Spermatid nuclei are pelleted and resuspended in a small volume of this buffer. Then 0.11-ml aliquots are added to 3.2 ml of each of the light and the dense metrizamide solutions. A linear 5-ml gradient containing between 10^7 and 2×10^8 nuclei uniformly dispersed throughout the gradient is generated from these two solutions as described by Grdina (1976). Gradients are spun at 20,000 rpm (37,400 g_{av}) for 20 minutes in an SW 50.1 rotor in a Beckman ultracentrifuge. Fractions are collected from the top of the gradient with the Auto-Densi Flow Collector or, alternatively, bands are removed by puncturing the side of the tube.

Bands of nuclei are formed at $\rho = 1.31$ gm/cm^3 and at $\rho = 1.34$ gm/cm^3. The lighter band contains nuclei in steps 12–15 of development with up to 93% purity. The denser band is more concentrated and contains step 16–19 nuclei with as high as 99% purity. Epididymal sperm nuclei also band at $\rho = 1.34$ gm/cm^3.

Although these studies are still preliminary and the conditions may have to be subsequently changed, excellent separation is clearly possible. Further experiments are in progress to maximize the numbers of nuclei loaded, to determine whether nuclei in the region between the two bands correspond to any specific intermediate stages, and to separate other types of testicular nuclei.

VII.   Effect of Preparation of Suspension on Separation

A.   Aggregation of Cells and Nuclei

Methods of preparation that minimize cell or nuclear aggregation provide more satisfactory material for separation. Aggregation limits the concentration and hence the number of cells or nuclei that can be separated. The most important step in minimizing cell aggregation is the treatment of the cell suspensions with trypsin. Trypsin treatment removes glycopeptides, which are specifically involved in cell adhesion, from the plasma membranes of mammalian cells.

Nuclear aggregation is minimized by avoiding pelleting of the nuclei after lysis of the cells. The presence of Triton X-100 and NDA in the buffer also appears to reduce aggregation of the nuclei. Sperm heads may be sonicated to break up clumps.

B.   Multinucleate Cells

Spermatogenic cells of the same generation arising from a single stem cell are connected by cytoplasmic bridges (Dym and Fawcett, 1971). Many cells at the same stage of development are connected in such syncytia, but whether an appreciable number of them actually coalesce within the testis to form multinucleate cells is a matter of disagreement. Johnson (1974) indicates that only 1 % to 5 % of testicular spermatids are present as multinucleates, but Bryan and Wolosewick (1973) believe the percentage to be much greater.

In trypsin-prepared cell suspensions, 13 % of the round spermatids are multinucleates (Meistrich et al., 1973a). Alternatively, if nuclei are counted, 27 % of the round spermatid nuclei are found in multinucleate cells. Similar results are obtained with the mechanical or EDTA–trypsin method. On the other hand, the percentage of multinucleate spermatogonia is reduced from 34 % with the trypsin preparation to about 5 % by employing the EDTA-trypsin method. A more gentle procedure, involving enzymic digestion prior to any mechanical mincing or dispersion, reduced the overall number of nuclei present in multinucleates by half (Bellvé and Romrell, 1974). This observation indicates that some of the multinucleates in cell suspensions are produced by handling the tissue, and that their number can be reduced.

Multinucleate cells have volumes which are multiples of the single cell volume (Loir and Lanneau, 1974; Meistrich and Koehler, 1975). However, they possess the same density as the corresponding uninucleates since they band at the same position upon equilibrium centrifugation (Meistrich and Trostle, 1975). The formation of multinucleates, therefore, has no effect on separation by equilibrium density centrifugation, but it limits the purity of larger cell types obtained by velocity sedimentation. For example, a binucleate cell with twice the usual cell volume sediments 1.6 times as fast as a uninucleate cell of the same type (Eq. 1). Binucleate round spermatids cosediment with secondary spermatocytes and tri- and tetranucleate-round spermatids with late pachytene primary spermatocytes (Table IV). Reduction in the numbers of multinucleates would obviously result in a significant improvement in purity of secondary and pachytene spermatocytes.

C.   Damage to Elongating Spermatids

During the stages of elongation, the mammalian spermatid nucleus is embedded in invaginations of the Sertoli cell and the cytoplasm is extended along the flagellum. During the final stages of spermiogenesis, the nucleus and cytoplasm are engulfed in separate regions of the Sertoli cell, connected only by a thin extension of cytoplasm. These factors are responsible for the susceptibility of elongated spermatids to loss of a discrete quantity of their cytoplasm during preparation of cell suspensions. Several observations indicate that following cytoplasmic detachment, the plasma membranes must be immediately resealed. The nucleated portion generally excludes trypan blue. The ultrastructure of the cytoplasmic portion and its organelles is normal (Barcellona and Meistrich, 1976). Both the nuclear and cytoplasmic pieces are resistant to trypsin proteolysis.

TABLE IV

PERCENTAGE OF UNINUCLEATE AND MULTINUCLEATE CELLS IN
FRACTIONS B AND D OF MOUSE TESTIS CELLS SEPARATED BY
VELOCITY SEDIMENTATION AT UNIT GRAVITY

	B	D
Uninucleate cells		
Spermatogonia and early primary		
spermatocytes	0	5
Pachytene spermatocytes	75	3
Secondary spermatocytes	0	18
Spermatids	1	6
Nongerminal Cells	7	2
Unknowns and degenerating	1	3
Multinucleate cells		
Spermatogonia and early primary		
spermatocytes	4	4
Spermatids		
Binucleate	1	55
Tri- and tetranucleates	9	1
Cytoplasmic fragments	1	5

The cytoplasmic fragments detached from elongating spermatids should not all be referred to as residual bodies. Most of these fragments represent the cytoplasm of steps 11–15 spermatids, while the residual body has been defined as the cytoplasm of the step 16 spermatid which exhibits densely basophilic ribonucleoprotein aggregates (Firlit and Davis, 1965).

The integrity of the flagellum or its attachment to the nucleus is also delicate, especially in the earlier stages of development. The loss of the flagellum from these cells appears to be primarily a result of mechanical factors. In contrast to epididymal spermatozoa (Millette *et al.,* 1973), the attachment of the flagellum to the nucleus of testicular sperm is relatively trypsin resistant.

Damage to maturing spermatids follows a definite trend. Round spermatids always possess their full cytoplasmic complement, but no flagella. Spermatids in steps 9–13 usually lose their flagella and a discrete amount of their cytoplasm leaving only a small cytoplasmic remnant adjacent to the nucleus. Some of these spermatids do retain all their cytoplasm, but still lose their flagella. A variable number retain their flagella either with or without their cytoplasm remaining. The more mature (steps 14–16) spermatids are almost all devoid of cytoplasm. Some possess intact flagella; some are just sperm heads.

This differential loss of cytoplasm and flagella aids in the separation of different classes of spermatids. We obtain, in order of decreasing sedimentation rates, spermatids with full cytoplasm but no flagella (mostly steps 1–8), spermatids with partial cytoplasm but no flagella (mostly steps 9–13), spermatids with full cytoplasm and flagella (steps 9–16), free spermatid nuclei (steps 11–16), and spermatids with flagella but no cytoplasm (mostly steps 14–16). Similarly, in order of increasing density we obtain spermatids with full cytoplasm (mostly steps 1–8), spermatids with partial cytoplasm (mostly steps 11–13), and spermatids with no cytoplasm (mostly steps 14–16). Elimination of this differential damage to spermatids would not improve the separation, but would result in greater cell viability.

The trypsin preparation method appears to be best for the separation of late spermatids. The EDTA–trypsin method results in a greater loss of flagella from the elongated spermatids (Meistrich and Trostle, 1975). The mechanical method (Lam *et al.*, 1970) produces even more damage. Steps 11–15 spermatids are mostly just nuclei (peak at 1.1 mm/hr, days 19–23). With the trypsin method, however (Meistrich, 1972), many steps 11–13 spermatids have a partial cytoplasmic complement (peak at 2.0 mm/hr, days 19–21) and many steps 14–15 spermatids have intact flagella (peak at 0.8 mm/hr, days 22–24). A more gentle mechanical preparation method (Sanborn *et al.*, 1975) does yield a large number of elongated spermatids with intact flagella.

D.   Spermatogonia

The spermatogonia appear to be quite susceptible to damage. Mechanical methods of preparation damage almost all these cells. More intact spermatogonia are obtained with trypsin, but some of them appear to have a reduced amount of cytoplasm. The use of EDTA alone (Davis and Schuetz, 1975) or EDTA plus trypsin (Meistrich and Trostle, 1975) results in a greater yield of spermatogonia. These cells display a more homogeneous sedimentation rate profile (Fig. 7A). Unfortunately, in the mouse these cells cosediment with the more numerous round spermatids and the purity of spermatogonia obtained with the Staput is still only 10%. EDTA–trypsin also causes some spermatogonia (Fig. 7B), and young primary spermatocytes to have reduced densities. This density shift lowers the effectiveness of density gradient centrifugation for separation of these cells.

VIII.   Discussion

A.   Comparison of Methods

An investigator desiring to separate one or more testicular cell types in high purity must consider which method is most suitable for the particular purpose. The important factors to weigh are (a) whether intact cells are

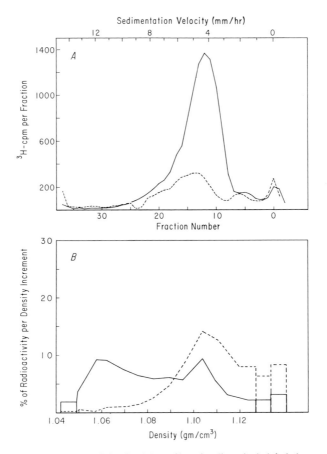

Fig. 7. Sedimentation (A) and density (B) profiles of radioactively labeled mouse spermatogonia and preleptotene spermatocytes. Mice were sacrificed 1 hour after injection of 1 μCi per gram body weight [³H]-TdR. Dashed line indicates preparation by trypsin method; solid line, by EDTA–trypsin method. Staput separations were run as described by Meistrich (1972). Renografin density gradients were centrifuged as described by Meistrich and Trostle (1975).

required or nuclei would suffice, (b) the purity desired and which contaminants are most serious, (c) the quantity desired, (d) the time available for separation; and (e) degradation or loss of macromolecules of interest.

First we shall consider the individual separation methods. A comparison of these procedures is presented in Table V.

Of the methods for velocity sedimentation separation of cells, centrifugal elutriation appears to be the method of choice. With elutriation, separation equivalent to that obtained with the Staput can be achieved in shorter times, with greater numbers of cells, smaller volumes in which cells are collected, and less BSA consumed. The shorter times are important when another step of purification is needed or if biochemical extraction of the separated fractions requires fresh material. Although we had initially reported that the purity of specific cell types was not as high as with unit gravity sedimentation (Grabske *et al.*, 1975), additional experience with the technique actually indicates that we can obtain pachytene spermatocytes and round spermatids in higher purity by elutriation (Tables III and VI). We believe that with proper choice of sedimentation limits on the various fractions, the purity of other cell types can be increased as well. The only disadvantage is the high initial cost of the equipment. The technique of gradient centrifugation (Pretlow *et al.*, 1974) also has the advantage of speed, but has several drawbacks. The number of cells separated is less than with the Staput because of the smaller size of the centrifuge tubes. Wall effects should, at least theoretically, reduce the resolution of this technique. The use of zonal rotors, however, could overcome some of these objections (Boone *et al.*, 1968). Comparison of separations obtained by gradient centrifugation (Pretlow *et al.*, 1974) with elutriation (Grabske *et al.*, 1975) indicates that, with the exception of spermatozoa, greater enrichments of cell types from hamster testes are obtained by elutriation.

In the case of velocity sedimentation of nuclei, similar separations can be obtained by the Staput method (Meistrich and Eng, 1972) and by gradient centrifugation (Utakoji *et al.*, 1968). The separation of round and elongated mouse spermatid nuclei with the Staput has been improved by the use of MP buffer in place of PBS. The difference in sedimentation rates of round and elongated spermatid nuclei is even greater in the rat. The use of zonal centrifugation to rapidly achieve separation of 10^9 rat testis nuclei has been described (Grimes, 1973). Differential counts of the nuclei were not performed precluding comparisons with other methods.

Several different media have been tested for equilibrium density centrifugation of testicular cells (Table V). BSA gradients have been used for separation of testicular cells from fish (Moav *et al.*, 1974), immature rats (Davis, 1975), and adult mice (M. L. Meistrich, unpublished observations). BSA preserves cell viability and minimizes cell clumping permitting the separation of large numbers of cells (Shortman, 1968). Except for the difficulties in preparation and the lack of commerically available dialyzed isotonic preparations, BSA would be the medium choice. Renografin or colloidal silica appear to be the most convenient alternatives. Metrizamide is also convenient and has been used for separating large numbers (10^9) of amphibian testicular cells (Risley *et al.*, 1975). Ficoll has also been tested, but did not yield a good separation (Pretlow *et al.*, 1974). Ficoll possesses some of the same disadvantages as BSA and causes more cell clumping and loss of cell viability.

Two different media have been used for density separation of nuclei

TABLE V

COMPARISON OF METHODS FOR SEPARATING TESTICULAR CELLS AND NUCLEI

Principle	Cells or nuclei	Technique	Gradient medium (concentrations)	Disadvantages of gradient medium	Maximum number loaded	Separation time employed
Velocity sedimentation	Cells	Staput	BSA (1–4% w/v)	Expensive for large runs	1.5×10^9 (1.5×10^7/ml)	4 Hr
	Cells	Elutriator	No gradient required	—	$\geq 3 \times 10^9$	20–90 min depending on number of fractions
	Cells	Gradient centrifugation[a]	Ficoll (2.7–5.5% w/w)	Viscous	3×10^7	21 Min
	Nuclei	Staput	Sucrose (1.3–10%)	—	5×10^8 (5×10^6/ml)	4–16 Hr
	Nuclei	Gradient centrifugation[b]	Sucrose (59–71% w/w)	Viscous	$\sim 10^8$	1 Hr

42

Equilibrium density centrifugation	Cells	Renografin (8–24% w/v)	Hypertonic	3×10^7 (17 ml gradient)	20 Min
	Cells	BSA	Expensive; viscous; difficult to prepare; maximum density 1.11 gm/cm^3	6×10^{8c} (13 ml gradient)	45 Min[c]
	Cells	Colloidal silica[d] (5–22% w/v)	Forms gel; toxic to cells in absence of protective agent	1.2×10^7 (28 ml gradient)	23 Min
	Nuclei	Metrizamide (57–67% w/v)	Expensive	$\sim 10^8$ (5 ml gradient)	20 Min
	Nuclei	Colloidal silica[e] (6–28% w/v)	Forms gel; coats nucleus	$> 10^7$ (75 ml gradient)	1 Hr

[a] Pretlow et al. (1974).
[b] Utakoji et al. (1968), Utakoji (1970), Kumaroo et al. (1975).
[c] For lymphoid cells, from Shortman (1968).
[d] Loir and Lanneau (1975a).
[e] Loir and Wyrobek (1972).

TABLE

MAXIMUM PURITIES OF VARIOUS TESTICULAR CELLS AND

Cell type[a]	Unseparated cell suspension[b] Mouse	Velocity sedimentation of cells (Staput) Mouse	Velocity sedimentation of cells (elutriation) Mouse	Equilibrium density centrifugation Mouse	Velocity sedimentation of nuclei Mouse	Rat
Spermatogonia[d]	1.2	10	4.7	19	—	—
Early primary spermatocytes[e]	2.0	6	4	9	—	—
Pachytene spermatocytes	3.9	74	80	28 (32)[g]	76	—
Secondary spermatocytes	0.8	14	21	4	—	—
Meiotic cells	0.3	4	3	1.4 (18)	—	—
Spermatids 1–8	23	81 (85)	70 (79)	66 (77)	40	—
Spermatids 9–10	2.2	3 (12)	4 (17)	7	—	—
Spermatids 11–13[h]	14	18 (74)	13 (61) } (100)	55 (87) }	70 }	70 (100) }
Spermatids 14–16[h]	13	37 (81)	—	63 (65)		
Cytoplasmic fragments + residual bodies	37	74	78	92	—	98[i]
Sertoli	0.4	29	1	1.4	—	—
Leydig	0.7	17	12	7	—	—

[a] Uninucleate and multinucleate cells are included together under each category.

[b] Prepared with trypsin. Suspension also contains 0.4 % macrophages, 1 % erythrocytes, and 2.5 % cells that were unidentifiable or degenerating.

[c] Data calculated from Loir and Wyrobek (1972).

[d] Including preleptotene spermatocytes.

[e] Leptotene, zygotene, and early pachytene.

(Table V). Colloidal silica gradients have been used to separate elongating spermatid nuclei from mouse (Loir and Wyrobek, 1972) and cricket testes (Kaye and McMaster-Kaye, 1975) on the basis of increased density with progressive nuclear condensation. The basic proteins of the cricket nuclei showed no alterations following this separation. We have employed metrizamide, but these studies are too preliminary to determine whether metrizamide or colloidal silica will prove to be a superior medium. Although separation of testicular nuclei was achieved by sucrose gradient centrifugation, equilibrium was not reached (Kumaroo et al., 1975). The similarity of the separation with that obtained by velocity sedimentation methods, indicated that nuclei were being separated primarily by size and not by density.

Of all the methods described, we find that velocity separation of cells by centrifugal elutriation has the most widespread utility. Velocity sedimentation of cells provides the greatest enrichment of the greatest variety of cell types (Table VI) and can handle the largest number of cells.

VI

NUCLEI OBTAINED FROM ADULT MOUSE OR RAT TESTES

Sonication Rat	Velocity sedimentation of cells + density separation of cells Mouse	Velocity sedimentation of cells + velocity sedimentation of nuclei		Velocity sedimentation of cells + density separation of nuclei Mouse	Sonication + density separation of nuclei Rat	Velocity sedimentation of cells + sonication Rat
		Mouse	Rat			
—	29	—	—	—	—	—
—	9	—	—	—	—	—
—	71, 47^f	95	92	—	—	—
—		—	—	—	—	—
—	34 (49)	—	—	—	—	—
—	90 (92)	96	93	—	—	—
—		—	—	70	—	—
33	70 (92)	—	79 (100)	~98	93	97
66	—	—		98	99	~85
—	97	—	—	—	—	—
—		—	—	—	—	—
—	33	—	—	—	—	—

^{f} 71 % refers to late pachytene spermatocytes; 47 % to mid-pachytene cells.
^{g} Values in parentheses are calculated after exclusion of cytoplasmic fragments and residual bodies.
^{h} Steps 11–13 and steps 14–16 refer to mouse spermiogenesis. Corresponding stages of rat spermiogenesis are steps 12–15 and steps 16–19, respectively.
^{i} In nuclear suspensions, the ribonucleoprotein aggregates contained within the residual bodies are scored.

However, where high purity is desired a sequential combination of different methods is required. Rerunning a sample through a second sequence of a given separation generally would not result in much improvement of the purity. This is particularly true of velocity sedimentation of cells with the Staput or elutriation. The separation of cells according to their sedimentation rate is virtually complete. Contamination is a result of different cell types having sedimentation rates that are very similar and some variation in sedimentation rates between individual cells of a given morphological type.

To obtain higher purities, two-step (or two-dimensional) cell and nuclear separations are employed. The purities obtained by such two-step separations are given in Table VI. We are presently using velocity sedimentation of cells by centrifugal elutriation as the first step in most procedures. The further purification of cells can be performed by equilibrium density gradient centrifugation. In our studies involving nuclear components such as protein or

DNA, we perform the second step of separation on the isolated nuclei. Staput separation of nuclei prepared from fractions of cells obtained by elutriation, yields pachytene spermatocytes and round spermatids in 95% purity. We have also shown by measurement of contaminating histones that the elongated spermatid nuclei prepared by sonication can be further purified by velocity sedimentation on the Staput apparatus (Platz *et al.*, 1975).

B.   Choice of Methods

Methods for separating each class of spermatogenic cell will be discussed in this section. The choice of fractions of separated cells to use for each cell type is based on the physical parameters given in Fig. 8.

Purification of spermatogonia and early primary spermatocytes from adult animals has proved to be difficult. Davis and Schuetz (1975), however, report the purification of spermatogonia, preleptotene, leptotene, and zygotene spermatocytes from testes of immature (10–25 days old) rats by velocity sedimentation. In our separations of immature mouse (16–18 days old) testes, we have found a large number of cells, which were difficult to identify by our methods, contaminating the various fractions. The cells are immature forms of either Sertoli cells or of other testicular cell types. We have not yet been able to obtain spermatogonia or early primary spermatocytes from immature mice in similar purities to those reported for the rat. Nevertheless, immature animals or animals treated to produce incomplete spermatogenesis must be used when these cell types are desired in high purity.

Enrichment of mid-pachytene spermatocytes is best achieved by velocity sedimentation followed by equilibrium density separation of cells. The fraction of cells sedimenting at 7.6 mm/hr contains 15% mid-pachytene spermatocytes contaminated by binucleate spermatids and secondary spermatocytes. Equilibrium density centrifugation results in enrichment of mid-pachytene cells to 47%. Contamination by spermatids is still present and could probably be reduced by a second density centrifugation run.

Late pachytene spermatocytes are obtained in relatively high purity (80%) by velocity sedimentation. Methods of preparation of cell suspensions which first remove interstitial material and lower the number of multinucleate cells enhance the purity of this cell type (Bellvé and Romrell, 1974). Further purification of pachytene nuclei can be achieved by the Staput method.

Secondary spermatocytes are difficult to purify. The Staput fraction most enriched in these cells is contaminated by multinucleate spermatids (55%) and by spermatogonia and early primary spermatocytes (11%) (Table IV). Further purification by equilibrium density centrifugation of cells would separate secondary spermatocytes from spermatogonia and early primary

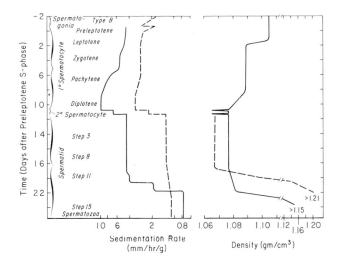

FIG. 8.   Changes in physical properties of mouse spermatogenic cells and nuclei with differentiation. Sedimentation rates of cells and nuclei at unit gravity in BSA gradients are taken from Meistrich (1972) and Meistrich and Eng (1972). Densities of cells in Renografin are from Meistrich and Trostle (1975). Densities of nuclei in colloidal silica gradients are from Loir and Wyrobek (1972).

spermatocytes, but not from spermatids. Velocity sedimentation of nuclei would separate them from spermatids, but not from spermatogonia and early primary spermatocytes. Therefore, three steps of purificiation would be required to obtain secondary spermatocyte nuclei in high purity. An alternative approach would be to eliminate the spermatogonia and early primary spermatocytes by X-irradiation of the testes about 1 week prior to sacrifice.

Cells in the metaphase and anaphase of meiotic division can be separated by velocity sedimentation followed by equilibrium density centrifugation. Besides cytoplasmic fragments, the major contaminants are again multinucleate spermatids which could be minimized by different preparation methods.

Round spermatids are enriched to 79% by velocity sedimentation. Further enrichment is obtained by either density separation of cells (90%) or velocity sedimentation of nuclei (96%). We have found no separation of round spermatids at different steps of maturation.

Separation of spermatids in early stages of elongation (steps 9–10) appears to be best obtained by selection of a fraction of cells sedimenting slightly slower than round spermatids, preparing nuclei and performing an equilibrium density separation of nuclei in colloidal silica or possibly in metrizamide.

It does not appear possible to obtain highly enriched fractions of these cells in intact form.

Nuclei of elongated spermatids (step 12 and later) are readily prepared from rat testes on a large scale by sonication. These nuclei may be further separated into at least two fractions; a light fraction consisting almost exclusively of nuclei within steps 12–15 and a dense fraction within steps 16–19. An alternative procedure is to first separate spermatids by velocity sedimentation of cells and then to employ sonication to destroy everything but the elongated spermatid nuclei (Grimes *et al.*, 1975). This method has yielded more highly enriched steps 12–15 spermatids (Table VI). The later spermatids are not so homogeneous, and the purity of this fraction is variable because it is dependent on the percentage of steps 12–15 spermatids with flagella. In the mouse, selective elimination of spermatids in steps 12–14 can rapidly be achieved by the use of trypsin and DNase; however, the use of these hydrolytic enzymes on isolated nuclei limits the value of these preparations for further biochemical analysis.

The purification of intact elongated spermatids presents a problem because of the loss of cytoplasm and flagella from some of these cells during preparation of cell suspensions. However, the detached cytoplasmic fragments cosediment with the elongated spermatids possessing only a partial amount of cytoplasm. A fraction (fraction I, Table I) can therefore be obtained which consists of 97% step 9–16 spermatids and their detached cytoplasm. Further separation of this fraction by equilibrium density centrifugation yields one fraction consisting of 97% cytoplasmic fragments and another of 70% steps 11–13 spermatids which have a small amount of remaining cytoplasm. Isolation of true residual bodies, i.e., cytoplasmic fragments from step 16 spermatids, appears to be difficult.

IX.   Applications

Techniques for separation of testicular cells and nuclei are useful for a wide variety of studies dealing with spermatogenesis. In this section, I shall describe the use of the separation methods for analytical studies, list applications of the techniques, and present one example of information on basic nuclear protein synthesis during spermatogenesis we have obtained using cell separation.

Although the major emphasis has been placed on developing separation methods for large-scale preparations of homogeneous cell populations, these methods can also be used as analytical tools. If cells are radioactively labeled, the cell type containing the radioactivity can be determined by either autoradiography or cell separation. In the cell separation studies, all fractions are collected and the radioactivity in each fraction is measured by liquid scintillation counting and plotted versus fraction number. Peaks of radioactivity correspond to the sedimentation rates of specific testicular cell or nuclear types. Thus, although the individual fractions are not pure, the correspondence of the radioactivity peak with the distribution of a cell type indicates that that cell type is labeled. If the label were in a contaminating cell type, it would also appear more prominently in the fraction most enriched in that contaminating cell type. Several studies employing the Staput technique as an analytical tool are listed in Table VII. Although equivalent studies could be performed by autoradiography, cell separation has several advantages. Results are obtained in a quantitative form within 1 day and are representative of the average of the entire testis. Autoradiograms often require more than 1 month of exposure, a trained observer is needed for analysis, and obtaining quantitative values necessitates grain counting, which is tedious.

Cell separation has been applied to the biochemical analysis of spermatogenesis. The applications are listed in Table VIII.

We are currently using a variety of cell-separation methods described here for studies of nuclear protein changes during rat spermatogenesis. An example of the differences in basic nuclear protein content and synthesis in different separated testicular cell types (Platz et al., 1975) is presented in

TABLE VII

ANALYTICAL STUDIES OF SPERMATOGENESIS EMPLOYING CELL SEPARATION

Species	Radiochemicals injected	Results	References
Mouse	$[^3H]$-Thymidine	Kinetics of spermatogenesis influenced by temperature and genetic factors, but not by steroid hormones	Meistrich et al. (1973b), Bruce et al. (1973), Meistrich et al. (in preparation)
Rat	$[^3H]$-Thymidine	Quantitative analysis of the effect of hormones on progression of spermatogonia through spermatogenesis in hypophysectomized rats	Go et al. (1971b), Vernon et al. (1975)
Mouse	$^{109}CdCl_2$ $^{203}HgCl_2$, CH_3 $^{203}HgCl$	Heavy metals bind primarily to late spermatids and spermatogonia	Lee and Dixon (1973, 1975)
Ram	Various $[^3H]$-amino acids	Highly stage-specific synthesis of different nucleoproteins during spermiogenesis	Loir and Lanneau (1975c)

TABLE VIII

BIOCHEMICAL STUDIES OF SPERMATOGENESIS EMPLOYING CELL SEPARATION

Species	Separation methods	Results	References
Mouse	Staput	Synthesis of highly basic nuclear protein in spermatids at about step 11	Lam and Bruce (1971)
Rat	Staput, sonication, elutriator	Changes in basic and nonhistone proteins during spermatogenesis. Stage specific synthesis of basic proteins during pachytene and during spermatid elongation	Platz et al. (1975) Grimes et al. (1975)
Ram	Staput	Replacement of histones by more highly basic proteins during spermatid elongation	Loir and Lanneau (1975b)
Mouse	Staput	Histone synthesis in pachytene spermatocytes	Goldberg et al. (1974), Bellvé et al. (1975)
Trout	Staput	Histone acetylation in premeiotic cells and spermatids; histone phosphorylation and methylation predominantly in premeiotic cells	Louie and Dixon (1972), Candido and Dixon (1972), Honda et al. (1975).
Cricket	Staput	New histones associated with elongating spermatids	Kaye and McMaster-Kaye (1974)
Rat	Staput	First appearance of carnitine acetyltransferase in early primary spermatocytes	Vernon et al. (1971)
Mouse	Staput	Changes in activities of glycosyltransferases during spermatogenesis and their presence in Golgi of specific cell types	Letts et al. (1974a,b)
Mouse	Elutriator	Lactate dehydrogenase-X synthesis in spermatocytes and spermatids	Meistrich et al. (1976b)
Hamster	Gradient centrifugation of nuclei	Analysis of pachytene nuclear RNA	Utakoji (1970)
Rat	Elutriator	Nuclear androgen exchange activity in elongated spermatids	Sanborn et al. (1975)
Mouse	Elutriator, Staput	Abnormal DNA content of spermatids of sterile hybrid mice	Meistrich et al. (1975b)

Fig. 9. These absorbance and radioactivity profiles of acid-soluble proteins separated by polyacrylamide gel electrophoresis from Staput fractions of cells separated from rats 24 hours after injection of [^{3}H]-arginine show several striking features. The biosynthesis and presence of two rapidly

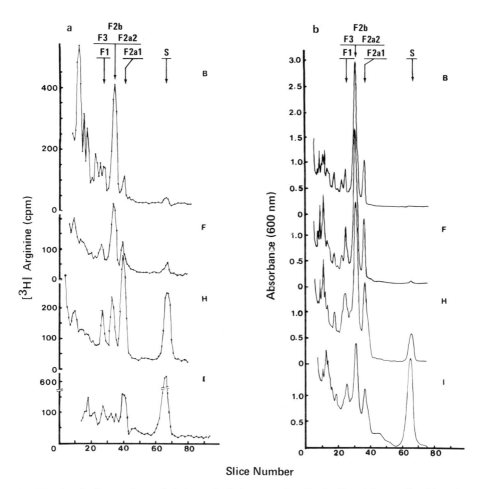

FIG. 9.   Basic proteins and their synthesis in rat testis cells. Radioactivity profiles (a) and densitometric scans (b) of acid-soluble proteins separated by polyacrylamide gel electrophoresis are shown. Various fractions obtained by Staput separation are enriched in (B) late pachytene spermatocytes, (F) round spermatids, (H) early elongated spermatids, and (I) late elongated spermatids. Proteins were radioactively labeled by intratesticular injection of [3H]-arginine 1 day prior to sacrifice. Positions of major histones and spermatidal proteins (S) are indicated by arrows. Reprinted with permission from Platz et al. (1975).

migrating basic proteins (labeled S) is localized to elongated spermatids (fractions H and I). The biosynthesis of a basic protein which migrates with histone F2b occurs in pachytene spermatocytes (fraction B). Another basic protein, which migrates with histone F2al, is synthesized in early elongated spermatids (fraction H). We have subsequently shown, using centrifugal elutriation to separate cells, that the latter, which is a unique protein designated TP2, is synthesized in steps 12–15 spermatids (Grimes et al., 1975).

X.   Conclusions

Velocity sedimentation, equilibrium density centrifugation, and ultrasonic disruption have been used to obtain enrichment of all classes of cells and nuclei from adult rodent testes. Several types of nuclei, including pachytene spermatocytes, round spermatids, early elongated spermatids, and late elongated spermatids can be obtained in purities in excess of 95 % by using the appropriate combinations of methods. As many as 3×10^9 cells in the initial suspension may be separated. Little or no damage at either the morphological or biochemical level occurs during separation.

Using these cell separation methods, we can exploit spermatogenesis as a model of cell differentiation and begin to map the sequence of macromolecular events involved in sperm development. The availability of purified populations of specific testicular cell types will permit the elucidation of other important events of spermatogenesis, such as genetic recombination, meiosis, spermatid morphogenesis, and nuclear inactivation and condensation.

Acknowledgments

The author wishes to thank Patricia K. Trostle, Vincent W. S. Eng, Dale W. Koehler, and Betty O. Reid for technical assistance and contributions to the methods; Dr. W. R. Bruce for providing guidance and laboratory facilities during the development of some of these methods; and Drs. D. J. Grdina, R. E. Meyn, and R. D. Platz for critical reading of the manuscript. This work was supported in part by U.S. Public Health Service Grant CA-17364.

References

Anderson, N. G., and Wilbur, K. M. (1952). *J. Gen. Physiol.* **35**, 781.
Barcellona, W. J., and Meistrich, M. L. (1976). Submitted for publication.
Bellvé, A. R., and Romrell, L. J. (1974). *J. Cell Biol.* **63**, 19a.
Bellvé, A. R., Bahtanager, Y. M., and Romrell, L. J. (1975). *J. Cell Biol.* **67**, 26a (abstr.).
Boone, C. W., Harell, G. S., and Bond, H. E. (1968), *J. Cell. Biol.* **36**, 369.
Bruce, W. R., Furrer, R., Goldberg, R. B., Meistrich, M. L., and Mintz, B. (1973). *Genet. Res.* **22**, 155.
Bryan, J. H. D., and Wolosewick, J. J. (1973). *Z. Zellforsch. Mikrosk. Anat.* **138**, 155.
Candido, E. P. M., and Dixon, G. H. (1972). *J. Biol. Chem.* **247**, 5506.
Davis, J. C. (1975). *J. Cell Biol.* **67**, 86a (abstr.).
Davis, J. C., and Schuetz, A. W. (1975). *Exp. Cell Res.* **91**, 79.
Dym, M., and Fawcett, D. W. (1971). *Biol. Reprod.* **4**, 195.
Firlit, C. F., and Davis, J. R. (1965). *Q. J. Microsc. Sci.* **106**, 93.
Galena, H. J., and Ternei C. (1974). *J. Endocrinol.* **60**, 269.
Glick, D., von Redlich, E., Juhos, E. Th., and McEwen, C. R. (1971). *Exp. Cell. Res.* **65**, 23.
Go, V. L. W., Vernon, R. G., and Fritz, I. B. (1971a). *Can. J. Biochem.* **49**, 753.
Go, V. L. W., Vernon, R. G., and Fritz, I. B. (1971b). *Can. J. Biochem.* **49**, 768.

Goldberg, R. B., Geremia, R., and Bruce, W. R. (1974). *Proc. Can. Fed. Biol.* **17**, 48 (abstr.).
Grabske, R. J., Lake, S., Gledhill, B. L., and Meistrich, M. L. (1975). *J. Cell Physiol.* **86**, 177.
Grdina, D. J. (1976). *In* "Methods in Cell Biology" (D. M. Prescott, ed.), Vol. 14, p. 213. Academic Press, New York.
Grdina, D. J., Milas, L., Hewitt, R. R., and Withers, H. R. (1973). *Exp. Cell Res.* **81**, 250.
Grimes, S. R., Jr. (1973). Ph.D. Thesis, University of North Carolina, Chapel Hill.
Grimes, S. R., Jr., Platz, R. D., Meistrich, M. L., and Hnilica, L. S. (1975). *Biochem. Biophys. Res. Commun.* **67**, 182.
Grimes, S. R., Jr., Platz, R. D., Meistrich, M. L., and Hnilica, L. S. (1976). Submitted for publication.
Honda, B. M., Candido, E. P. M., and Dixon, G. H. (1975). *J. Biol. Chem.* **250**, 8686.
Johnson, D. R. (1974), *Cell Tissue Res.* **150**, 323.
Kaye, J. S., and McMaster-Kaye, R. (1974). *Chromosoma* **46**, 397.
Kaye, J. S., and McMaster-Kaye, R. (1975). *J. Cell Biol.* **67**, 204a (abstr.).
Kumaroo, K. K., Jahnke, G., and Irvin, J. L. (1975). *Arch. Biochem. Biophys.* **168**, 413.
Lam, D. M. K., Furrer, R., and Bruce, W. R. (1970). *Proc. Nat. Acad. Sci. U.S.A.* **65**, 192.
Lam, D. M. K., and Bruce, W. R. (1971). *J. Cell Physiol.* **78**, 13.
Lee, I. P., and Dixon, R. L. (1972). *Toxicol. Appl. Pharmacol.* **23**, 20.
Lee, I. P., and Dixon, R. L. (1973). *J. Pharmacol. Exp. Ther.* **187**, 641.
Lee, I. P., and Dixon, R. L. (1975). *J. Pharmacol. Exp. Ther.* **194**, 171.
Letts, P. J., Meistrich, M. L., Bruce, W. R., and Schachter, H. (1974a). *Biochim. Biophys. Acta* **343**, 192.
Letts, P. J., Pinteric, L., and Schachter, H. (1974b). *Biochim. Biophys. Acta* **372**, 304.
Loir, M., and Lanneau, M. (1974). *Exp. Cell Res.* **83**, 319.
Loir, M., and Lanneau, M. (1975a). *Exp. Cell. Res.* **92**, 499.
Loir, M., and Lanneau, M. (1975b). *Exp. Cell Res.* **92**, 509.
Loir, M., and Lanneau, M. (1975c). *Experientia* **31**, 401.
Loir, M., and Wyrobek, A. (1972). *Exp. Cell Res.* **75**, 262.
Louie, A. J. and Dixon, G. H. (1972). *J. Biol. Chem.* **247**, 5498.
Marushige, Y., and Marushige, K. (1975). *J. Biol. Chem.* **250**, 39.
Meistrich, M. L. (1972). *J. Cell Physiol.* **80**, 299.
Meistrich, M. L., and Eng, V. W. S. (1972). *Exp. Cell. Res.* **70**, 237.
Meistrich, M. L., and Koehler, D. W. (1975). *Radiat. Res.* **64**, 564 (abstr.).
Meistrich, M. L., and Trostle, P. K. (1975). *Exp. Cell Res.* **92**, 231.
Meistrich, M. L., Bruce, W. R., and Clermont, Y. (1973a). *Exp. Cell Res.* **79**, 213.
Meistrich, M. L., Eng, V. W. S., and Loir, M. (1973b). *Cell Tissue Kinet.* **6**, 379.
Meistrich, M. L., Reid, B. O., and Barcellona, W. J. (1975a). *J. Cell Biol.* **64**, 211.
Meistrich, M. L., Lake, S., Steinmetz, L. L., and Gledhill, B. L. (1975b). *J. Cell Biol.* **67**, 279a (abstr.).
Meistrich, M. L., Reid, B. O., and Barcellona, W. J. (1976a). *Exp. Cell Res.* **99**, 72.
Meistrich, M. L., Trostle, P. K., and Erickson, R. P. (1976b). *J. Cell Biol.* **70**, 59a (abstr.).
Miller, R. G., and Phillips, R. A. (1969). *J. Cell Physiol.* **73**, 191.
Millette, C. F., Spear, P. G., Gall, W. E., and Edelman, G. M. (1973). *J. Cell Biol.* **58**, 662.
Moav, B., Goldberg, A., and Avivi, Y. (1974). *Exp. Cell Res.* **83**, 37.
Nyquist, S. E., Acuff, K., and Mollenhauer, H. H. (1973). *Biol. Reprod.* **8**, 119.
Peterson, E. A., and Evans, W. E. (1967). *Nature (London)* **214**, 824.
Platz, R. D., Grimes, S. R., Meistrich, M. L., and Hnilica, L. S. (1975). *J. Biol. Chem.* **250**, 5791.
Poste, G. (1971). *Exp. Cell Res.* **65**, 359.
Pretlow, T. G., Scalise, M. M., and Weir, E. E. (1974). *Am. J. Pathol.* **74**, 83.
Rickwood, D., and Birnie, G. D. (1975). *FEBS Lett.* **50**, 102.

Risley, M. S., Eckhardt, R. A., and Eppig, J. J. (1975). *J. Cell Biol.* **67**, 363a (abstr.).
Sanborn, B. M., Steinberger, A., Meistrich, M. L., and Steinberger, E. (1975). *J. Steroid. Biochem.* **6**, 1459.
Shortman, K. (1968). *Aust. J. Exp. Biol. Med. Sci.* **46**, 375.
Steinberger, A. (1975). *In* "Methods in Enzymology" (B. W. O'Malley and J. G. Hardman, eds.), Vol. 39, Part D, pp. 283–296. Academic Press, New York.
Utakoji, T. (1970). *In* "Methods in Cell Physiology" (D. M. Prescott, ed.), Vol. 4, pp. 1–17. Academic Press, New York.
Utakoji, T., Murumatsu, M., and Sugano, H. (1968). *Exp. Cell Res.* **53**, 447.
Vernon, R. G., Go, V. L. W., and Fritz, I. B. (1971). *Can. J. Biochem.* **49**, 761.
Vernon, R. G., Go, V. L. W., and Fritz, I. B. (1975). *J. Reprod. Fertil.* **42**, 77.
Wyrobek, A. J., Meistrich, M. L., Furrer, R., and Bruce, W. R. (1976). *Biophys. J.* **16**, 811.

Chapter 3

Separation of Mammalian Spermatids

M. LOIR AND M. LANNEAU

*I.N.R.A.– Laboratoire de Physiologie de la Reproduction,
Nouzilly, France*

I. Introduction

Spermiogenesis, the process by which spermatids differentiate into spermatozoa, provides a unique system for the study of cell differentiation in higher animals. Although in fishes fairly pure spermatid populations are readily obtained after gonadotropin injection (Marushige and Dixon, 1969), in mammals the separation of pure spermatid populations is a prerequisite for the investigation of biochemical events during spermiogenesis because of the diversity of cell types present in the mammalian testis.

Mammalian spermatids are among the smallest testis cells. Nonround spermatids undergo an important decrease in cell size as they mature. The volume of the spermatozoa is 18 (ram) to 28 (mouse) times lower than that of a round spermatid. This reduction is due first to the gradual decrease of cytoplasmic and nuclear volumes and finally to the loss of cytoplasm at the time of spermiation. In addition, these cells undergo a dramatic increase in nuclear density (Loir and Wyrobek, 1972). These two changes in the cell parameters allow the separation of spermatids from testis cell mixtures into relatively pure populations either on the basis of cell size or buoyant density. The first parameter was exploited five years ago by Lam *et al.* (1970), who used the Staput technique of sedimentation at unit gravity (Miller and Phillips, 1969). Separations on the basis of the second parameter were achieved more recently by Meistrich and Trostle (1975) and Loir and Lanneau (1975b).

One of the two procedures described here (the Staput technique) is also used for the separation of spermatogonia and spermatocytes and is described by Meistrich in this volume, Chapter 20. In this chapter, emphasis will be given to aspects or modifications specifically related to the separation of spermatids.

II. Separation According to Cell Size

The sedimentation velocity (SV) of a cell moving through a uniform stable medium of density close to 1.00 gm/cm^3 under the influence of a constant gravitational or centrifugal field depends primarily on cell size. However, it must be remembered that it depends also on the density and shape of the cell and on the viscosity of the medium.

A. Saline Solutions and Media

In Earle's saline solution and phosphate-buffered saline (PBS), the spermatids of rodents and domestic animals retain a normal ultrastructure for several hours. Because it is easier to prepare and use, PBS is routinely used. However, when they are dispersed and/or incubated at 32°C in Eagle's minimal essential medium, ram and mouse testis cells incorporate [^{14}C] leucine into proteins at a linear rate during a longer time than when PBS is present. Recently, Romrell *et al.*, (1976) have demonstrated that, in the mouse, germ cells have higher respiratory quotients when dispersed in an enriched Krebs–Ringer bicarbonate buffer solution than in PBS. Thus, when optimum preservation of the spermatids is necessary, a culture medium should be used throughout the separation procedure.

In the rat, the seminiferous tube fluid should be slightly hypertonic

(Tuck *et al.,* 1970). In the ram, although it should be isotonic with plasma (Setchell and Waites, 1971), the osmotic balance for round spermatids is 312–318 mOsm/kg H_2O. For these reasons we routinely adjust the osmotic pressure of PBS to 315–325 mOsm/Kg H_2O with 0.014 *M* KCL because the K^+ concentration is especially high in the seminiferous tube fluid (Setchell and Waites, 1971). Osmolarity is measured by freezing point depression.

The pH of solutions is adjusted to 7.15–7.30.

B.   Cell Suspensions

1.   METHOD OF PREPARATION

The first separations of testis cells by the Staput method used mechanically prepared suspensions (Lam *et al.,* 1970; Go *et al.,* 1971). Such cell suspensions contain free nuclei, many damaged cells and other debris which contaminate the spermatid populations after separation.

The highest yield of viable, clean spermatids is obtained by a procedure involving the use of trypsin (Meistrich, 1972; Loir and Lanneau, 1974). After cell dispersion, the enzymic solution is removed by low-speed centrifugation after a 3- to 4-fold dilution with either 8% fetal calf serum or 6% bovine serum albumin (BSA). This step depresses the enzyme action and has the additional advantage of protecting cell integrity. Soybean trypsin inhibitor has been used in the rat (Platz *et al.,* 1975). In the ram, the electrophoregrams of the basic nucleoproteins of spermatids (Loir and Lanneau, 1975c) are identical whether lima bean trypsin inhibitor is present or not after trypsinization. Germ cells are resuspended at a cell concentration of 5 to 20 × 10^6/ml in a volume and at a concentration of BSA or Ficoll that depends on the conditions of the separation to be carried out. The suspension is cooled to 1°–4°C over a period of 45 minutes.

Two modifications of this method involving either the use of EDTA (Meistrich and Trostle, 1975) or collagenase (Romrell *et al.,* 1976) have been used to improve the homogeneity of spermatogonia and spermatocyte populations. However, these do not improve the separation of spermatids, and the first method provides lower yields of spermatids.

2.   CYTOLOGY OF SPERMATIDS

In the mammalian testis, many spermatids are connected by cytoplasmic bridges (Dym and Fawcett, 1971; Bryan and Wolosewick, 1973). In the mouse, rat, and domestic animals, all the multinucleate spermatids present in suspensions are of the multinucleate-individual type recently called symplast by Romrell *et al.,* (1976). These authors have demonstrated that symplasts result mainly from the widening of cytoplasmic bridges that allows the adjoining cells to flow together. During the dispersion of the testis cells, some cytoplasmic bridges between connected spermatids are broken. Romrell *et al.* (1976) have suggested that in the mouse, the plasma membrane of the disconnected cells is quickly repaired.

The nonround spermatids have an unexpected behavior in the presence of trypsin. Many of these cells which *in situ* have a postnuclear cytoplasm, lose it during the preparation of suspensions. Detached cytoplasmic lobes (CL) are of variable size, and spermatids have differing amounts of cytoplasm or no cytoplasm at all (Figs. 1 and 2), this depending on the intensity of stirring during trypsinization. Electron microscopy not only confirms the origin of CL, but also provides evidence that such spermatids and CL have a normal ultrastructure and a continuous plasma membrane (Figs. 3–5). In this case again it may be assumed that the plasma membrane is quickly closed after the detachment of the CL.

C.   Cell Separation at Unit Gravity

1.   SEDIMENTATION CHAMBERS

Sedimentation chambers, hereafter called the "Staput", similar to that described by Lam *et al.* (1970), of 12.5 cm diameter are used. Preparative-scale separations may be carried out either by using two such chambers simultaneously (up to 7×10^8 testis cells, corresponding to about 1.6×10^8 spermatids) or a larger chamber (28 cm in diameter; 20×10^8 testis cells; Platz *et al.,* 1975). Separations are carried out at $1°–4°C$.

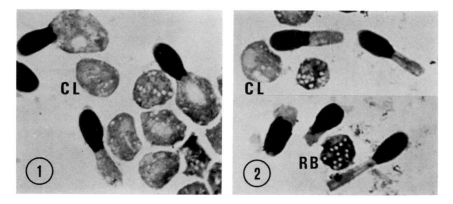

FIGS. 1 and 2.   Photomicrographs of ram spermatids after Staput separation. Toluidine blue-stained smears. (1) Spermatids S12–13 with a large or very small amount of cytoplasm and cytoplasmic lobes. Cell band III. (2) spermatids S9–12 with only the manchette, residual bodies (RB), and a cytoplasmic lobe (CL). Cell band II.

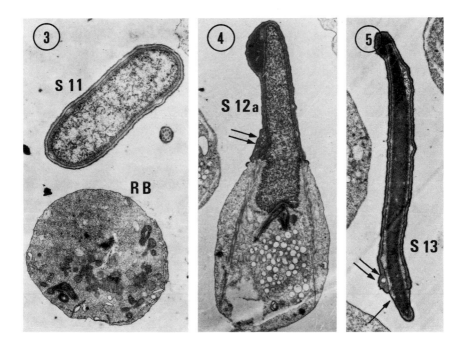

FIGS. 3–5. Electron micrographs of ram spermatids after sedimentation in a Ficoll gradient. Staput fractions were fixed for 40 minutes in 2.5% glutaraldehyde in PBS then postfixed for 2 hours in 2% OsO_4, 7.5% w/v sucrose solution in PBS. After a 20-minute wash the pellets were preembedded in 2% agar, dehydrated, and embedded in Spurr's medium. The plasma membrane is continuous around the reduced postnuclear cytoplasm of the S12a spermatid, around the maturing residual body (RB) and the posterior part of the free S13 nucleus (arrow). The acrosomes show alterations (double arrow. (3) × 7700; (4) × 5700; (5) × 11,800.

2. GRADIENTS

Ficoll and BSA are used as gradient material. Because BSA is more expensive than Ficoll, the respective advantages of these two materials have been tested. BSA supports the separation of 4 times more cells than Ficoll, and a shorter sedimentation time is necessary to have identical cell migrations. For similar distances of migration the distribution of the various spermatid classes is somewhat more broadened in Ficoll than in BSA (see Table I). Although cell viability at the end of sedimentation differs little whether Ficoll or BSA is used, the viability of round spermatids stored in the gradient medium at 4°C over 20 hours is 90% with BSA vs 76% with Ficoll. After sedimentation in Ficoll the spermatids show ultrastructural changes, while those sedimented in BSA deviate little from the cells fixed *in situ* (Figs. 6–9) (Loir and Lanneau, 1974).

Nonlinear gradients are used instead of linear gradients because they support the separation of 4 times more cells. They are generated as described by Miller and Phillips (1969) with 0.5, 1, 2.75% Ficoll solutions or 0.5, 2, 4% BSA solutions.

3. COLLECTION OF SPERMATID POPULATIONS

Usually cells are allowed to sediment over 3 to 4 hours. The standardization of the sedimentation time on the basis of the distance run by a well defined band is preferable to the use of a fixed time because the SV of germ cells can vary between animals and throughout the year (de Reviers *et al.,* 1976). Sedimentation times of 3 to 4 hours appear to provide a satisfactory compromise between duration and resolution. Longer sedimentation times should increase the homogeneity of spermatid populations, but in fact this is not evident from 5-hour sedimentations, and in any case the broadened distribution of intermediate spermatids (see Section II,C,5) cancels the advantage of a hypothetical increase in resolution.

When the distribution of spermatids as a function of SV or when the incorporation of labeled precursors are under investigation, 25 to 30 fractions of 11.5 ml are collected (small Staput) and treated separately. When an equilibrium density separation follows the Staput separation or when the biochemical techniques to be used need more cells than are present in an 11.5-ml fraction, four or five spermatid populations are collected. Once the volumes of gradient corresponding to the desired populations have been determined, the standardization of the sedimentation time by the migration of a cell band allows identical spermatid populations from all the separations throughout an experiment to be obtained readily.

4. ANALYSIS OF THE FRACTIONS AND POPULATIONS

a. Cellular Composition. Several methods are available for determining the cell distribution in the fractions containing the spermatids, CL and residual bodies (RB).

In the fastest method, the cells in every fraction are counted in a hema-

---

FIGS. 6–9.   Electron micrographs of ram round spermatids fixed, (6) *in situ,* (7) after trypsinization in PBS, following 5-hour Staput separation (8) in a BSA gradient, and (9) in a Ficoll gradient. Electron microscopy preparation: see Figs. 3–5. After sedimentation in Ficoll the nuclear membrane and the endoplasmic reticulum are absent and multivesicular and multigranular bodies (arrow) have developed. After trypsinization some acrosomes show slight modifications (double arrow). (6) × 11,500; (7) × 11,700; (8) × 12,800; (9) × 10,300.

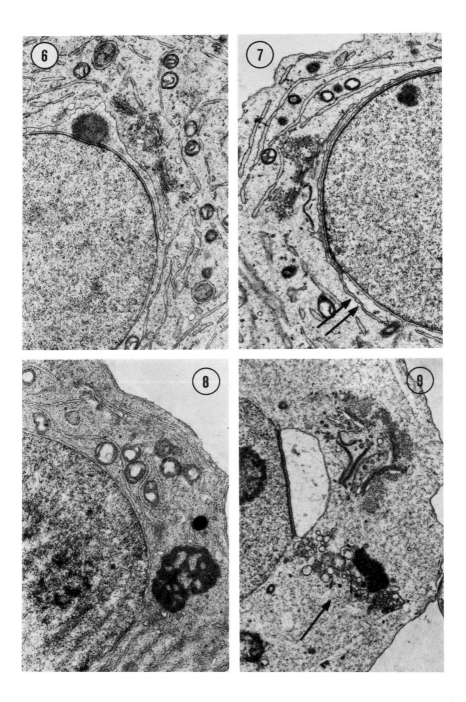

cytometer using a phase contrast microscope. However, at the low magnification compatible with hemacytometers, CL and late spermatids with a large amount of cytoplasm are not easily distinguished from one another. This can be overcome by additional counts at high magnification. A more accurate determination of the distribution of the nucleate cells is obtained by counting nuclei prepared in each fraction with detergents, which simultaneously destroy CL and RB. The nuclei of elongated mammalian spermatids are resistant to trypsin and DNase. This property allows the confirmation of the distribution of these cells within a gradient (Loir and Lanneau, 1974).

The cytological technique is the more convenient to analyze fractions and especially to identify the low concentrations of contaminating cells. Various methods involving smearing or plastic-embedding and staining, either with periodic acid–Schiff/hematoxylin (Meistrich et al., 1973) or with toluidine blue (Figs. 1 and 2) (Loir and Lanneau, 1974; Romrell et al., 1976), yield good material for cytological examination.

Another method uses labeling with tritiated thymidine. The comparison of autoradiographic analyses with the peaks of radioactivity in the velocity sedimentation profiles (Fig. 10) provides additional information on the distribution of spermatids and contaminating cells (Meistrich, 1972; Loir and Lanneau, 1974).

   b. Cell Integrity.   The viability of round spermatids in the fractions can be tested with the trypan blue dye exclusion method, whereas that of nonround spermatids is difficult to determine with the same method. With the above-mentioned cytological techniques, degenerative changes can be observed at the light microscope level. Observation of the mobility of the tail provides a functional evaluation of the nonround spermatid integrity. Electron microscopy allows the detection of ultrastructural alterations in comparison to spermatids fixed in situ.

5.   Distribution of Spermatids in Relation to SV

The distributions of the various mononucleate ram, mouse, and rat spermatids in the fractions following a Staput sedimentation are shown in Tables I–III. They appear to be similar in the three species if we consider the nucleate cells.

The round spermatids migrate as one band, the modal SV of which is between 3.5 and 5 mm/hour according to the species and to the conditions of the separation. A round spermatid population can be collected with a purity equal to or greater than 80 % in the mouse and in domestic animals. In the rat, the purity is only about 70 %, this being due to a higher contamination by spermatogonia, early primary spermatocytes, and nongerminal cells in this species. In the ram, as the volume of round spermatids increases then decreases slightly, the low SV part of the band is enriched with young

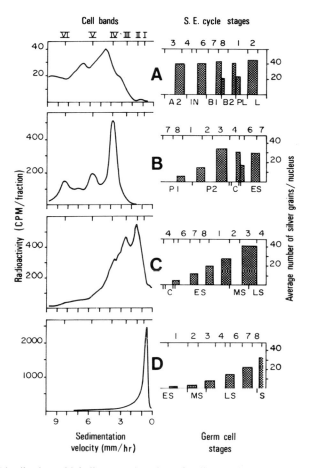

FIG. 10. Distribution of labeling as a function of sedimentation velocity of nuclei of ram spermatogenic cells (autoradiography). (A) 20 hours, (B) 17 days, (C) 23 days, and (D) 30 days after injection of [^{3}H]-thymidine. It appears that the spermatid bands IV and III are slightly contaminated by spermatogonia and leptotene spermatocytes. Abbreviations: A$_2$, IN, B1, B2 = spermatogonia; PL = preleptotene stage; L = leptotene; P1, P2 = pachytene; C = secondary spermatocytes; ES = round spermatids; MS = elongating spermatids; LS = elongated spermatids; S = testicular spermatozoa (Loir and Lanneau, 1974).

and old round spermatids and the high SV part is enriched with intermediate round spermatids.

The oldest spermatids (testicular spermatozoa, S), which have the lowest SV, have narrow sedimentation profiles, especially in BSA gradients (Figs. 10 and 11). By contrast, the distribution of the intermediate spermatids

TABLE I

Distribution of Mononucleate Ram Spermatids at Various Stages of Maturation in Spermatid Populations after 4 Hours of Sedimentation in a Nonlinear Ficoll Gradient (6 Experiments) and after 3.5 Hours of Sedimentation in a Nonlinear Bovine Serum Albumin (BSA) Gradient (5 Experiments)

Spermatid population	Gradient material	Sedimentation velocity limits (mm/hr)	Spermatid stages[a]					
			1–8a	8b–9a	9b–11	12	13–S	CL[b] + RB
1	Ficoll	0.45–1.20	—	1 ± 1	8 ± 2	10 ± 4	80 ± 7	—
	BSA	0.50–1.30	—	—	1 ± 1	13 ± 5	84 ± 5	52 ± 9
2	Ficoll	1.20–2.35	7 ± 2	9 ± 2	25 ± 3	26 ± 3	31 ± 6	—
	BSA	1.30–2.55	3 ± 3	17 ± 8	25 ± 4	36 ± 3	16 ± 6	58 ± 4
3	Ficoll	2.35–3.50	45 ± 11	20 ± 8	16 ± 4	9 ± 2	8 ± 2	—
	BSA	2.55–3.90	50 ± 13	30 ± 11	14 ± 3	3 ± 1	1 ± 1	43 ± 8
4	Ficoll	3.50–4.80	84 ± 4	9 ± 2	2 ± 1	1 ± 1	1 ± 1	—
	BSA	3.90–5.30	91 ± 3	5 ± 9	1 ± 1	1 ± 1	—	5 ± 3

[a] Values are given as percentage of nucleate cells and late spermatid nuclei. Blood cells were not counted. Average ± SD. In some experiments differential counts were made after collection of the cells and after preparation of nuclei; usually the two counts differed only slightly.

[b] (CL), Cytoplasmic lobes; (RB) residual bodies. Values are given as percentage of total particles (nucleate cells plus nuclei plus CL and RB). Average ± SD.

TABLE II

DISTRIBUTION OF MONONUCLEATE MOUSE SPERMATIDS IN FRACTIONS AFTER 4 HOURS OF
SEPARATION IN A LINEAR BOVINE SERUM ALBUMIN (BSA) GRADIENT (4 EXPERIMENTS)[a]

Fraction	Mean SV (mm/hr)	Spermatid stages[b]				
		1–8	9–10	11–13	14–16	CL[b] + RB
K	0.75	0	1 ± 1 (2)	8 ± 7 (17)	37 ± 11 (81)	55 ± 18
J	1.40	0	1 ± 1 (3)	7 ± 3 (19)	20 ± 6 (54)	63 ± 9
I	2.10	1 ± 0 (4)	3 ± 2 (11)	18 ± 5 (69)	2 ± 2 (8)	74 ± 5
H	3.00	10 ± 2 (31)	3 ± 2 (9)	15 ± 5 (47)	0	68 ± 9
G	4.10	77 ± 4 (83)	2 ± 1	8 ± 2 (8 5)	0	7 ± 2
F	5.00	79 ± 7 (83)	1 ± 1	4 ± 2	0	5 ± 1

[a] After Meistrich et al. (1973).
[b] Values are given as percentage of total particles ± SD. Calculated values as percentage of nucleate cells plus nuclei are given in parentheses.

TABLE III

DISTRIBUTION OF MONONUCLEATE RAT SPERMATIDS IN POPULATIONS AFTER 4 HOURS
SEDIMENTATION IN A NONLINEAR BOVINE SERUM ALBUMIN (BSA) GRADIENT[a]

Spermatid population	SV limits (mm/hr)	Spermatid stages[b]				
		1–8	9–11	12–17	18–19	CL[c] + RB
1	0.35–1.4	0	1	43	55	70
2	1.4–3.3	8	10	52	11	72
3	3.3–5.0	61	5	9	1	49
4	5.0–6.6	70	9	2	0	8

[a] C. Auriault and M. Loir, unpublished results.
[b,c] See footnotes a and b, respectively, of Table I.

(elongating and elongated spermatids) is somewhat broadened. Detailed analyses carried out in the mouse (Meistrich et al., 1973) and in the ram (Loir and Lanneau, 1974) (Fig. 11) have demonstrated that this is a result of different states of integrity. The sedimentation rate of spermatids at a specific maturation stage is a function of the amount of cytoplasm and of the presence or the absence of the manchette, while the flagellum acts as a drag to decrease the sedimentation rate. The sedimentation profile of a class of spermatids in a similar state of integrity is narrow. For example,

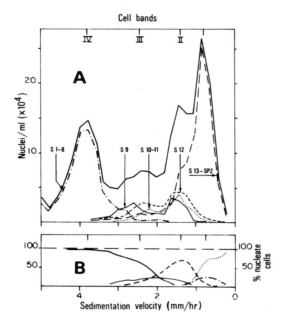

FIG. 11.   Distribution of ram spermatids as a function of sedimentation velocity after 5 hours of sedimentation in a Ficoll gradient. (A) Distribution of spermatids at various stages of maturation. For counting, the nuclei were prepared in every fraction with cetrimide. (B) Distribution of spermatids; ———, with a large cytoplasmic lobe; ———, with a small one; ------, with the manchette only; .—.—., free nuclei; ······, free nuclei with flagella. Toluidine blue-stained smears (Loir and Lanneau, 1974).

this is apparent for ram spermatids S9–S12 which have lost their cytoplasm but not the manchette (Fig. 11).

After separation, some acrosomes show alterations of variable magnitude (Figs. 4 and 5). Observations in our laboratory (in the ram) indicate that the acrosome is very sensitive to variations in the cell environment resulting from the isolation of cells. Morphological abnormalities appear more rapidly in PBS or Krebs–Ringer solution than in Eagle's medium and develop with time even at low temperature.

The CL and RB migrate concomitantly with nonround spermatids (Tables I–III). Rodents and domestic animals differ with regard to these particles since they are more numerous in rodents. According to Romrell et al. (1976), in the mouse, RB should be so numerous in some fractions that a population could be collected with a purity greater than 85%. In the rat, the maximal percentage of RB is 60% in a fraction with a SV around 2.4 mm/hour (Platz et al., 1975). It should be noted that high contaminations of spermatid populations by CL and RB can be completely suppressed by detergents when only the nuclei are needed.

Nonround spermatid populations are also contaminated by erythrocytes (SV: 1 to 2 mm/hr) and small white blood cells. These sources of contamination can be eliminated before castration by perfusion of the testis either via the thoracic aorta in rodents (Romrell et al., 1976) or via the testicular artery in domestic animals.

D. Centrifugal Elutriation

Centrifugal elutriation was studied by Grabske et al. (1975) to separate testis cells in the mouse and the hamster. This method is closely related in principle to sedimentation at unit gravity. It is based on velocity sedimentation, but it involves forces at greater than unit gravity and so allows faster sedimentation.

The suspension of testis cells is prepared as for a Staput separation. Up to 20×10^8 cells can be processed. The buffer used throughout the separation is 0.5% BSA in PBS. Elutriator separations take less than half the time required for equivalent separations with the Staput method. If only one cell population is wanted, it can be obtained in less than 30 minutes. The integrity of cells is not decreased by elutriation, and recovery is 100%. However, the homogeneity of the spermatid populations is somewhat lower or, in the best case (testicular spermatozoa), it is identical to values obtained with the Staput.

III. Separation According to Buoyant Density

Spermatid separation on the basis of sedimentation velocities provides populations enriched in specific spermatid stages that are useful for biochemical studies (Loir and Lanneau, 1975a, c; Platz et al., 1975) despite the fact that they are not pure. However, further purification may be necessary when an acute resolution of phenomena is needed.

If nuclei are of primary interest, high resolution can be achieved with isopycnic centrifugation of isolated nuclei (Loir and Wyrobek, 1972). Because in this method nuclei are isolated a long time before eventual biochemical investigations are carried out and they are processed in conditions that modify their extractability, it is more convenient to work with entire spermatids. Isopycnic centrifugation of mixtures of testis cells does not provide more pure spermatid populations than Staput separation (Meistrich and Trostle, 1975). On the other hand, isopycnic centrifugation of fractions obtained by the Staput method achieves significant enrichment of specific spermatid classes. Two methods have been reported. One involves the use of Renografin as gradient material (Meistrich and Trostle, 1975) to separate various mouse testis cells and is described elsewhere in this volume (Chapter 2). The other uses colloidal silica gradients and has been used for the separation of ram spermatids (Loir and Lanneau, 1975b). Because true cell densities can be measured with silica gradients, application of this method is preparative as well as analytical. It is described here at a level within the means of most cellular biology laboratories.

A. Spermatid Suspensions

Nuclear density increases progressively throughout spermiogenesis. Because the cytoplasm density is low and does not change (see Section III, F,2), the more abundant the cytoplasm, the lower is the buoyant density of nonround spermatids at a specific stage, after separation by the Staput method. In order to narrow the distribution of spermatids at the same maturation stage in the gradients, stirring during trypsinization must be vigorous enough to cause most of the nonround spermatids to lose their cytoplasm (see Fig. 12). It is noteworthy that if the detachment of the postnuclear cytoplasm favors the preparation of fairly pure populations of spermatids, it results in a mixing of the CL at different stages of maturation.

Ram testis cells are separated at unit gravity in nonlinear BSA or Ficoll gradients. When the band of round spermatids has run 14.5 mm, five populations are collected (Table IV). They are washed once in PBS and resuspended at a concentration of 0.75 to 7×10^6 particles/ml in 4.2 ml of 8% w/v Ficoll in PBS ($d_4°C = 1.028$ gm/cm^3; 325 mOsM/kg H$_2$O; 7.1 centipoises at 4°C).

B. Gradient Material

Since the density of mature mammalian spermatozoa is around 1.17 gm/cm^3, the material must be dense enough to produce gradients covering the density range 1.04–1.17 gm/cm^3. Ficoll, Renografin, and colloidal silica meet this requirement. Ficoll has not been used because of its high viscosity. Meistrich and Trostle (1975) chose Renografin, but they confirmed that its greatest disadvantage is its hypertonicity. Colloidal silica has the advantage of low viscosity and low cost and allows the formation of isotonic gradients with a constant physiological pH. However, alone it is toxic to some cells, but this can be overcome by the addition of protective molecules (Pertoft and Laurent, 1969).

TABLE IV

CONCENTRATION AND YIELD OF VARIOUS RAM SPERMATID CLASSES (CL) (upper part) AND OF CYTOPLASMIC LOBES (CL) PLUS RESIDUAL BODIES (RB) (lower part) IN STAPUT FRACTIONS AND IN DEFINED DENSITY INTERVALS AFTER ISOPYCNIC CENTRIFUGATION IN COLLOIDAL SILICA GRADIENTS[a]

Staput fraction	SV limits (mm/hr)	Staput gradient material	Particles types	Concentration[b]		Density intervals (gm/cm³)	Yield[c]	No. of expts.
				In Staput fractions	In density intervals			
I	0.50–1.25	Ficoll	13-S	80 = 8	97 ± 0.5	1.100–1.165	60	4
	0.55–1.30	BSA	13-S	84 = 5	—	—	—	8
II, III	1.25–2.60	Ficoll	8b–12	63 ± 10	84 ± 2.5	1.050–1.090	62	5
	1.30–2.85	BSA	8b–12	73 ± 6.5	85 ± 3.5	1.050–1.090	72	7
IV	3.05–3.80	Ficoll	1–8 (1–4,8b)	80 ± 5	93 ± 4.5	1.040–1.050	61	3
	3.40–4.20	BSA	1–8 (1–4,8b)	86 ± 4	—	—	—	8
V	3.80–4.60	Ficoll	1–8(3–8a)	83 ± 4	—	—	—	5
	4.20–5.10	BSA	1–8(3–8a)	91 ± 4	—	—	—	8
I	—	Ficoll, BSA	CL + RB	56 ± 6	92 ± 4	1.040–1.060	—	4
II	—	Ficoll, BSA	CL + RB	57 ± 2.5	81 ± 12	1.040–1.060	—	7
III	—	Ficoll, BSA	CL + RB	53 ± 5	61 ± 24	1.040–1.060	—	5

[a] After Loir and Lanneau (1975b).
[b] Spermatid concentrations are expressed as percentage of nucleate cells and CL + RB concentrations as percentage of total particles. Average ± SD.
[c] Yield is expressed as the number of spermatids of a class recovered in the density interval as percentage of the number of the same spermatids layered onto the gradient.

C.   Stock Solutions for Gradient Formation

1.   PHYSICAL PROPERTIES

Osmolarity is measured by freezing point depression. To determine the precise density (d), first of the stock solutions, then of the fractions, standard curves of d vs percentage of silica sols and refractive index (N) are drawn from accurate determinations of d and N of the commercial silica sols and of various dilutions of these sols with PBS. Measurements of N are made at 4°C with a Zeiss refractometer.

Two colloidal silica sols ($d = 1.30$ gm/cm^3) were compared: HS40 Ludox and AS40 Ludox purchased from E. I. Du Pont de Nemours Co., Seppic Paris. The stabilizing counterion is sodium in HS40 sol and ammonium in AS40 sol. Once the pH is adjusted to 7.25 with 5 N HCl, the osmolarities are 370 and 180 mOsm/kg H$_2$O, respectively. Stock solutions are prepared in the cold just before use by dilution with PBS at various concentrations in order to adjust the osmolarity to the required values. The low-density (L) solutions (1.036 gm/cm^3 at 4°C) contain 12% v/v Ludox diluted with normal PBS for both sols, and the osmolarity is adjusted to 325 mOsm/kg H$_2$O with 0.012 M and 0.024 M KCl. The high-density (H) solutions (1.165 gm/cm^3) contain 55% v/v Ludox diluted with normal PBS for HS40 sol (340 mOsm/kg H$_2$O) and with 1.95 × concentrated PBS for AS40 sol (325 mOsm/kg H$_2$O). The final dilution of PBS is 0.9 in the two L stock solutions. It is only 0.45 in the H HS40 stock solution vs 0.87 in the H AS40 stock solution. Despite this difference, cell viability does not differ in the two H solutions. The buoyant densities of cells, however, are consistently lower in AS40 gradients. For example, the density of erythrocytes is 1.106 ± 0.0017 gm/cm^3 in HS40 gradients and 1.092 ± 0.0007 in AS40. The former value is nearer to that determined by Legge and Shortman (1968) with physiological BSA gradients. The low values obtained with AS40 Ludox might result from the high ammonium concentration in the gradients (0.020 to 0.090 M). For this reason the HS40 stock solutions are routinely used with a third one (I) of intermediate density (1.069 gm/cm^3) containing 21% v/v HS40 diluted with normal PBS and adjusted to 325 mOsm/kg H$_2$O with KCl. The viscosities at 4°C are, respectively, 1.8, 2.3, and 5.6 centipoises. Although the osmolarity of the H solution could be decreased to 325 mOsm/kg H$_2$O, it is kept at 340, since we have shown that the density of spermatozoa is increased by only 0.0017 gm/cm^3 if it is measured at 340 mOsm/kg H$_2$O instead of 300 (Loir et al., 1976).

2.   REDUCTION OF THE CYTOTOXICITY OF SILICA

The viability of spermatids stored in HS40 solutions decreases with time. We have improved the cell viability, first by reducing the time of contact of the cells with silica to a maximum of 45 minutes. This is achieved by using

Ficoll as layering medium, by reducing the centrifugation time to 23 minutes (the cells come to their isopycnic positions in 18 minutes), by collecting simultaneously the two gradients processed in one centrifugation and by immediate dilution of the fractions with PBS. Second, we have investigated the possibility of using protective molecules. We have tried to use poly-2-vinylpyridine N-oxide (Nash et al., 1966). Unfortunately this compound is incompatible with silica and, when used to treat the cells before the separation procedure, its protective effect is counteracted and canceled by the high osmolarity in solution. According to Pertoft (1969), polyethylene glycol (PEG) added to the gradient medium preserves the cell integrity through its coating action. However, we were unable to mix useful concentrations of PEG 4000 with HS40 to obtain liquid physiological solutions. Either they gel or they have high pH and osmolarity. The addition of 4% fetal calf serum to the stock solutions increases the spermatid viability but shortens the gelation time too much to be usable. Pertoft et al. (1968) demonstrated that a sulfhydryl reagent, e.g., reduced glutathione, preserves the cell functions. Reduced glutathione was not found to be very efficient, but we have evidence that 0.2% cysteine has a significant protective effect. For this reason it is routinely added to the HS40 stock solutions.

Recently, Ludox AM30 ($d = 1.21$ gm/cm^3; stabilizing counterion sodium) has been used successfully for isopycnic centrifugation (Morgenthaler et al., 1975). Because in this sol, aluminate ions replace silica at the surface of the colloidal particles, the cytotoxicity should theoretically be lower than that of AS and HS sols. After adjusting the pH to 7.25, the osmolarity of Ludox AM is 103 mOsm/kg H$_2$O. By dialysis against PBS plus 0.012 M KCl, the osmolarity is increased to 325 mOsm/kg H$_2$O. After preliminary assays the cytotoxicity of Ludox AM is effectively somewhat lower than that of Ludox HS or AS.

D. Formation, Centrifugation, and Collection of Gradients

1. SEPARATION PROCEDURE

Linear and nonlinear 28-ml gradients are generated from L and H or from L and I solutions using a peristaltic pumping system (Leif, 1968). The H or I solution is pumped into the L (mixing) chamber at one-half the rate that the gradient medium is pumped into the centrifuge tube in the case of linear gradients and at a lower rate in the case of nonlinear gradients. Gradients formed from L and I solutions are convenient for the separation of the Staput population IV (see Fig. 16), while gradients formed from L and H solutions are necessary for the other Staput populations. The use of nonlinear instead of linear gradients is of interest only for Staput population III.

A 4-ml cell suspension containing 3 to $28 + 10^6$ particles is layered onto each of two gradients. They are immediately centrifuged at 4°C and 4000 g in a Heraeus IV KS centrifuge for 23 minutes including the acceleration time (5 minutes). Care is taken to accelerate the rotor very slowly to 1000 rpm and deceleration occurs with the brake off. Twenty-seven fractions of 1.2 ml are collected in 14 minutes from the top of the gradient in tubes containing 3.8 ml of PBS. The dense chase solution (pure HS40 sol) is pumped into the bottom of the tubes. To allow the detection of the end of the gradient, the first milliliter contains 0.1 % toluidine blue. The fractions are concentrated to 0.2 ml by low-speed centrifugation. The density of every fraction was determined in some gradients. It appears that it varies regularly with very low distortion in successive fractions and that the measurement of d in every fifth fraction provides a valuable determination of the variation of d throughout the gradient. Routinely, every fifth fraction is collected undiluted and checked for d.

2.   CENTRIFUGE TUBES AND COLLECTING DEVICE

When equilibrium density centrifugation is used for cell separation, the resolution and the cell recovery can be reduced by the streaming and wall effect artifacts and by the method of collection. To obtain optimum results we use improvements in technique proposed by Leif et al. (1972) and by Morton (1973). Concentration of cells at the interface between sample and gradient is avoided with 8 % Ficoll, which is 4 times more viscous than L solution. Collisions of the cells with the tube wall are minimized by the use of an altuglas tube reducer fitted at the top of the 50-ml glass centrifuge tube. The 4.2-ml sample chamber of the reducer has a mean diameter 0.65 times that of the tube. To minimize band distortion during the collection of fractions, they are collected through a conical altuglas tube adapter, placed on the surface of the sample in order to reduce the time before the first fraction is collected. Collection is done through a 100 mm-long polyethylene tube. The high resolution provided by this separation procedure has been checked with ram ejaculated spermatozoa; the sharp narrow visible band of spermatozoa is collected without significant distortion (Loir et al., 1976).

E.   Analysis of the Fractions

1.   CELL VIABILITY

In some experiments, viability was determined by the trypan blue dye exclusion test, but this test was not routinely used. For round spermatids we have checked that less refractile cells are dying cells since they are stain-able with trypan blue. For nonround spermatids we consider the morphology, appearance, and mobility of the tail.

The fractions below 1.040 gm/cm³ contain mainly dead and swollen spermatids and CL. They are not used. In fractions with higher densities the viability of round spermatids if between 79 and 92%, and the nonround spermatids have a somewhat higher viability, a good appearance, and slight motility.

2. CELL DISTRIBUTION IN THE FRACTIONS

Distribution was determined by counting in a hemacytometer using a phase contrast microscope (see Section II, C, 4). Typical distributions of spermatids, CL, and RB are shown in Fig. 14, 15, and 16.

Density intervals can be determined in which round (Sl-8a), elongating (S 8b-12), and elongated (S 13-S) spermatids are found at a maximal concentration consistently throughout the different separations (Table IV). In these density intervals, the three classes of spermatids are obtained with purities 12–21% higher than in Staput populations. The S 13-S spermatid population obtained in the density interval 1.100–1.165 gm/cm³ is uncontam-

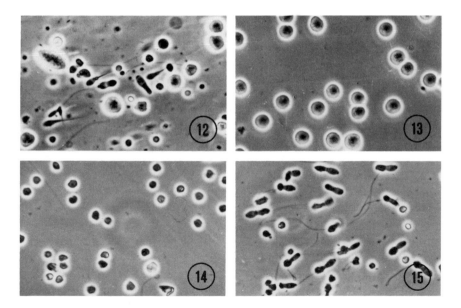

FIGS. 12–15. Phase contrast photomicrographs of ram: (12) testis cells after trypsinization; (13) round spermatids after Staput separation in BSA gradient; (14) cytoplasmic lobes and residual bodies after isopycnic centrifugation, fraction II-7 (see Fig. 16); and (15) spermatids S11–12 after isopycnic centrifugation, fraction II-13.

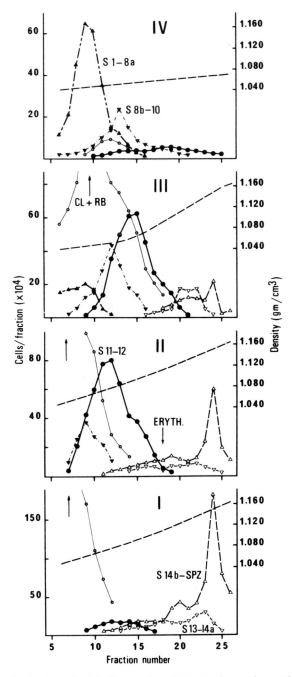

FIG. 16. Distribution in colloidal silica gradients following isopycnic centrifugation of ram spermatids of the Staput populations I to IV. Variation of density in the fractions. Fraction 4 corresponds to the top of the gradients. Number of spermatids (× 10⁶) layered onto the gradients: IV, 4.4; III, 8.5; II, 12; I, 12. Eryth: erythrocytes. [From Loir and Lanneau (1975b)].

inated by CL or RB. On the other hand, the S 8b-12 spermatid populations obtained in the density interval 1.050–1.090 gm/cm^3 are almost as contaminated by CL and RB as in Staput populations. Once again this contamination is not serious when only spermatid nuclei are needed. Fairly pure populations (81 to 92%) of mixed CL and RB can be obtained from Staput populations I and II (Table IV), but no pure RB population can be obtained.

3. CELL RECOVERY AND YIELD

The recovery of cells loaded onto the gradients (density higher than 1.040 gm/cm^3) is, respectively, 90, 84, 80, and 75% for the Staput populations I to IV. These values result mainly from the fact that most of the dead cells are swollen and have a density below 1.040 gm/cm^3 and so are not taken into account.

The yield (Table IV) of the three classes of spermatids in their respective density intervals is between 60 and 72%, and the numbers of spermatids collected in these intervals vary from 7.5 to 9.7 × 10^6 when 28 × 10^6 particles (corresponding to about 12 × 10^6 nucleate cells) are layered onto the gradient.

F. Cell Density

1. SIGNIFICANCE OF THE MEASURED DENSITY VALUES

The gradient medium has physiological characteristics. Centrifugation is short and at a relatively low force so that physical stress should be avoided. If the shape of the gradient used for one Staput population is changed, the distribution of the cells as a function of d remains identical. When the two peak fractions of an initial separation of caput epididymal spermatozoa are rerun on new gradients, they reband as before at the same density (Loir et al., 1976). As already mentioned, the erythrocyte density obtained with our technique is near that obtained with physiological BSA gradients. On the other hand, the measured density of the ejaculated spermatozoa is 1.160 gm/cm^3 (Loir et al., 1976) for a calculated value of 1.150 gm/cm^3 (Benedict et al., 1967). All these observations suggest that the measured densities are likely to be the true cell densities.

The reproducibility of the procedure was tested by duplicate separations of ejaculated spermatozoa, the density of every fraction being measured. The mean difference between the two obtained values is 0.002 ± 0.0015 gm/cm^3 (average ± SD; 12 pairs of gradients).

2. DENSITIES OF SPERMATIDS, CYTOPLASMIC LOBES, AND RESIDUAL BODIES

The buoyant density is lowest in round spermatids and cytoplasmic lobes. It is 1.045 ± 0.0010 gm/cm^3 and 1.045 ± 0.0037 gm/cm^3, respectively, regardless of the SVs. In nonround spermatids it increases to 1.152 ± 0.0007

gm/cm^3 (testicular spermatozoa). As expected, the density of a given class of nonround spermatids increases as the SV decreases, i.e., when the amount of cytoplasm diminishes. On the other hand, in a given Staput population it increases with the stage of maturation of the spermatids. RB have a somewhat higher density (1.058 ± 0.0051 gm/cm^3) than CL, this possibly reflecting the high density of the "Sphere chromatophile."

IV.   Discussion

The separation of mammalian spermatids on the basis of cell size allows a single class of spermatids, the round spermatids, to be obtained with a purity equal to or greater than 70%. The old spermatid class can also be obtained with satisfactory purity if only the nucleate cells are taken into account, since this population is mixed with about equal quantities of CL plus RB. The populations of intermediate spermatids obtained are not homogeneous and are contaminated with high numbers of CL plus RB. Separation on the basis of buoyant density, of the spermatids already separated according to cell size, allows the purity of the round spermatid and old spermatid populations to be improved, and cytoplasmic contaminants in the latter to be removed. At the same time, populations of the various intermediate spermatids can be obtained, with a significantly improved homogeneity.

After Staput separation and centrifugal elutriation, the viability of the spermatids is high, especially if BSA is used. It is lower after isopycnic centrifugation in HS40 silica gradients, particularly for round spermatids and CL. Ludox AM30 is somewhat less toxic for mammalian spermatids than Ludox HS40. Furthermore, according to Morgenthaler et al. (1975) this sol is easily mixed with PEG to provide nontoxic solutions, so it is likely that higher spermatid viabilities could be obtained by using AM30 sol instead of HS40 sol.

Staput separations of 1 to 20×10^8 testis cells provide enough spermatids to carry out biochemical investigations. After isopycnic centrifugation of spermatids in two silica gradients we obtain spermatid populations of fewer than 2×10^7 cells. Usually, they are still sufficient for biochemical studies, We have not checked the maximum number of particles we can load onto our 28-ml gradients, but according to Pertoft et al. (1968), it probably should not be much higher than 28×10^6, the maximum number we load. So if larger spermatid populations are needed after isopycnic centrifugation on silica gradients, a suitable way to scale up the separation procedure might be by use of the continuous-flow zonal centrifugation method. This method has recently been used successfully by Morgenthaler et al. (1975)

with Ludox AM30 gradients. Finally, to decrease the duration of the separation procedure, it could be of interest to separate testis cells by centrifugal elutriation.

ACKNOWLEDGMENTS

The author's work was supported in part by the Délégation Générale à la Recherche Scientifique et Technique (72–7 0098 and 74–7 0038). We wish to thank Dr. Meredith Lemon for valuable assistance with the translation of the manuscript and Mr. J. L. Courtens for his help in the preparation of sections for electron microscopy.

REFERENCES

Benedict, R. C., Schumaker, V. N., and Davies, R. E. (1967). *J. Reprod. Fertil.* **13**, 237–249.
Bienkowski, R., and Hoffman, L. (1974). *J. Cell Biol.* **63**, 26a.
Bryan, J. H. D., and Wolosewick, J. J. (1973). *Z. Zellforsch, Mikrosk Anat.* **138**, 155–169
de Reviers, M. T., Loir, M., and Pelletier, J. (1976). *J. Reprod. Fertil.* **46**, 203–209.
Dym, M., and Fawcett, D. M. (1971). *Biol. Reprod.* **4**, 195–215.
Go, V. L. W., Vernon, R. G., and Fritz, I. B. (1971). *Can. J. Biochem.* **49**, 753–760.
Grabske, R. J., Lake, S., Gledhill, B. L., and Meistrich, M. L. (1975). *J. Cell Physiol.* **86**, 177–190.
Lam, D. M. K., Furrer, R., and Bruce, W. R. (1970). *Proc. Natl. Acad. Sci. U.S.A.* **65**, 192–199.
Legge, D. G., and Shortman, K. (1968). *Br. J. Haematol.* **14**, 323–335.
Leif, R. C. (1968). *Anal. Biochem.* **25**, 271–182.
Leif, R. C., Kneece, W. C., Warters, R. L., Grinvalsky, H., and Thomas, R. A. (1972). *Anal. Biochem.* **45**, 357–373.
Loir, M., and Lanneau, M. (1974). *Exp. Cell Res.* **83**, 319–327.
Loir, M., and Lanneau, M. (1975a). *Experientia* **31**, 401–402.
Loir, M., and Lanneau, M. (1975b). *Exp. Cell Res.* **92**, 499–505.
Loir, M., and Lanneau, M. (1975c). *Exp. Cell Res.* **92**, 509–512.
Loir, M., and Wyrobek, A. (1972). *Exp. Cell Res.* **75**, 261–265.
Loir, M., Dacheux, J. L., and Lanneau, M. *J. Reprod. Fertil.* (submitted for publication).
Marushige, K., and Dixon, G. H. (1969). *Dev. Biol.* **19**, 397–414.
Meistrich, M. L. (1972). *J. Cell Physiol.* **80**, 299–312.
Meistrich, M. L., and Trostle, P. K. (1975). *Exp. Cell Res.* **92**, 231–244.
Meistrich, M. L., Bruce, W. R., and Clermont, Y. (1973). *Exp. Cell Res.* **79**, 213–227.
Miller, R. G., and Phillips, R. A. (1969). *J. Cell Physiol.* **73**, 191–202.
Morgenthaler, J. J., Mardsen, M. P. F., and Price, C. A. (1975). *Arch. Biochem. Biophys.* **168**, 289–301.
Morton, B. (1973). *Anal. Biochem.* **52**, 421–429.
Nash, T., Allison, A. C., and Harington, J. S. (1966). *Nature (London)* **210**, 259–261.
Pertoft, H. (1969). *Exp. Cell Res.* **57**, 338–350.
Pertoft, H., and Laurent, T. C. (1969). *Prog. Sep. Puri.* **2**, 71–90.
Pertoft, H., Bäck, O., and Lindahl-Kiessling, K. (1968). *Exp. Cell Res.* **50**, 355–368.
Platz, R. D., Grimes, S. R., Meistrich, M. L., and Hnilica, L. S. (1975). *J. Biol. Chem.* **250**, 5791–5800.
Romrell, L. J., Bellvé, A. R., and Fawcett, D. W. (1976). *Dev. Biol.* **49**, 119–131.
Setchell, B. P., and Waites, G. M. H. (1971). *J. Reprod. Fertil. Suppl.* **13**, 15–27.
Tuck, R. R., Setchell, B. P., Waites, G. M. H., and Young, J. A. (1970). *Pfluegers Arch.* **318**, 225–243.

Chapter 4

Isolation of Skeletal Muscle Nuclei

LeROY KUEHL

Department of Biochemistry,
University of Utah,
Salt Lake City, Utah

I. Introduction

A number of procedures for preparing nuclei from striated muscle have appeared in the literature (Edelman *et al.*, 1965; Severin *et al.*, 1965; Zaalishvili *et al.*, 1966; Marchok and Wolff, 1968; Franke and Schinko, 1969; Sobel and Kaufman, 1970; Silakova *et al.,* 1971; Kuehl, 1975a), and a new procedure for isolating muscle nuclei will be described later in this chapter. The main features of each of the above methods are summarized in Table I. We have tried nearly all of these procedures in our own laboratory; those which have proved most satisfactory will be described in detail in a subsequent section.

Most of the published methods for preparing muscle nuclei are unsatisfactory, either with respect to yield or degree of cytoplasmic contamination (Table I). A particularly troublesome aspect of those methods which give a very low yield is the possibility that the nuclei obtained are not really derived from muscle fibers, but rather from some other cell type (e.g., from the

TABLE I

SUMMARY OF METHODS FOR ISOLATION OF SKELETAL MUSCLE NUCLEI

Reference	Source of muscle	Homogenization procedure[a]	Purification procedure	Approx. yield	Amount of cytoplasmic contamination[b]	Remarks
Edelman et al. (1965)	Rat, femoral musculature	P.E. 0.022 inch clearance in 0.32 M sucrose, 1 mM MgCl$_2$, 0.8 mM phosphate, pH 6.8	Cent. from 2.15 M sucrose, 1mM MgCl$_2$, 3.5 mM phosphate, pH 6.8	<1%[c]	Low	Isolated nuclei were extensively characterized
Severin et al. (1965)	Rabbit, femoral musculature	P.E. [?] in 2.2 M sucrose 3.4 mM CaCl$_2$	Cent. from 2.2 M sucrose, 3.4 mM CaCl$_2$	<1%[c]	Low	—
Zaalishvili and Gachechiladze (1966)	Rabbit, femoral musculature	Blender, low speed then P.E. 0.45 mm clearance in 0.32 M sucrose, 5 mM MgCl$_2$, 10 mM histidine, pH 7	Cent. from 2.2 M sucrose	—	Low	—
Marchok and Wolff (1968)	Chick, embryonic and 8-day hatched leg muscle	Blender, moderate speed in 2.2 M sucrose, 3 mM MgCl$_2$	Cent. from 2.2 M sucrose, 3 mM MgCl$_2$	20–30%[d] (embryos), 10–15% (hatched)	Low[d]	—
Franke and Schinko (1969)	Rat, femoral musculature	Blender, high speed in 0.4 M sucrose, 4% gum arabic, 4 mM n-octanol, 10 mM Tris, pH 7.0	Cent. from 2.5 M sucrose	<1%[c]	Moderate[c]	Morphology of isolated nuclei well preserved

	Tissue	Homogenization	Centrifugation	Yield	Contamination	Comments
Sobel and Kaufman (1970)	Rat, soleus muscle	Blender, high speed in 0.32 M sucrose, 25 mM KCl, 3 mM MgCl$_2$, 1 mM HSCH$_2$CH$_2$OH	Cent. from 2.2 M sucrose, 2.5 mM KCl, 1.2 mM MgCl$_2$	15%[c]	High[c]	—
Silakova et al. (1971)	?	P.E. 0.45 mm clearance in 0.25 M sucrose 1.8 mM CaCl$_2$	Cent. from 2.2 M sucrose, 1.8 mM CaCl$_2$	1%[c]	Low[c]	—
Kuehl (1975a)	Rat, femoral musculature	P.E. 0.022 inch clearance, then P.E. 0.010 inch clearance in 0.25 M sucrose, 0.5% Triton X–100, 1 mM MgCl$_2$	Cent. through 2.4 M sucrose, 5 mM MgCl$_2$, 1% BSA	10%	Low	Isolated nuclei were extensively characterized
Kuehl, this chapter	Rat, femoral musculature	Blender, moderate speed in 0.25 M sucrose, 2.5 mM MgCl$_2$ 25 mM TES, pH 7.5 then P.E. 0.010 inch clearance in 0.25 M sucrose, 2.5 mM MgCl$_2$, 0.5% Triton X–100, 25 mM TES, pH 7.5		15%	Low	—
—	—		—		—	—

[a] Abbreviations: P.E. = Potter–Elvehjem homogenizer; TES = N-tris(hydroxymethyl)methyl-2-aminoethane sulfonic acid; BSA = bovine serum albumin; Cent. = centrifugation.
[b] As judged microscopically.
[c] As determined in our laboratory.
[d] When this method was applied to muscle from adult rats, approximately 5% yields of moderately contaminated nuclei were obtained.

endothelial cells that line the capillary network permeating the muscle or from connective tissue or fat cells).

Preparation of nuclei from skeletal muscle presents several problems not generally encountered in isolation of nuclei from other tissues:

1. The concentration of nuclei in striated muscle is low relative to that in other tissues.

2. Muscle fibers are tough, and homogenization procedures vigorous enough to cause their rupture frequently destroy the nuclei also.

3. Purification of the nuclei on sucrose gradients may be unsatisfactory owing to the presence, in muscle homogenates, of cytoplasmic material with a density similar to that of nuclei.

4. Muscle nuclei may disintegrate under conditions where nuclei from other tissues are stable.

II.   Concentration of Nuclei in Skeletal Muscle

Table II records the number of nuclei per gram of wet weight of various tissues of the rat. The low concentration of nuclei in skeletal muscle is partially compensated for by the fact that this tissue is present in relatively large amounts. Thus, 25 gm of muscle can easily be obtained from the hindquarters of an adult rat, compared with approximately 8 gm of liver, 3 gm of kidneys, 1 gm of heart and less than 1 gm of thymus and spleen. Nevertheless, the investigator who is accustomed to isolating nuclei from liver may be unprepared for the small pellets that will be obtained from skeletal muscle.

TABLE II

CONCENTRATION OF NUCLEI IN VARIOUS RAT TISSUES

Tissue	Nuclei per gram wet tissue $\times 10^{-9}$	References
Thymus	3.8	
Spleen	2.0	
Lung	0.91	Thompson *et al.* (1953), Schneider
Kidney	0.57	and Klug (1946)
Liver	0.28	
Heart	0.22	
Brain	0.18	Schneider and Klug (1946)
Skeletal muscle	0.05	Kuehl (1975a)

[a]Values given were calculated from data in the indicated references. Brain nuclei were assumed to contain 0.67 pg of DNA phosphorus per nucleus.

III. Homogenization

Effective release of nuclei from muscle fibers is not achieved in a Potter-Elvehjem homogenizer in the absence of detergent, and yields of nuclei obtained by methods employing such a procedure are low (1 % or less). Addition of the nonionic detergent, Triton X-100 (Rohm and Haas), to the homogenizing medium greatly increases the proportion of nuclei that may be released from muscle fibers in a Potter-Elvehjem homogenizer, and this observation has served as the basis for an isolation method developed in our laboratory (Kuehl, 1975a).

An alternative homogenization procedure which has proven satisfactory makes use of a high-speed, VirTis-type homogenizer. We used this technique a number of years ago (Kuehl, 1964) to release nuclei from plant cells, which, like muscle fibers, are difficult to rupture without destroying the nuclei. The method that we developed was later used in modified form by Franke and Schinko (1969) to prepared nuclei from rat muscle. Zaalishvili and Gachechiladze (1966), Marchok and Wolff (1968), and Sobel and Kaufman (1970) have also employed a high-speed homogenizer to isolate muscle nuclei. Although good release of nuclei from the muscle fibers can be achieved by this procedure, it also presents several problems: (1) Nuclei are damaged during homogenization; at higher speeds, the damage becomes appreciable and the yield of nuclei declines. (2) Material having a density similar to that of nuclei is generated during homogenization, particularly at higher speeds.

IV. Purification on Sucrose Gradients

Centrifugation through dense sucrose (2.1–2.5 M) has usually been used to separate muscle nuclei from cytoplasmic components. This procedure is quite satisfactory unless dense cytoplasmic material has been generated during homogenization, in which case the nuclear pellets may be appreciably contaminated. Such dense cytoplasmic elements are produced upon vigorous homogenization of muscle. This material sometimes displays the banding pattern characteristic of myofibrils, but frequently does not. In the latter case, the dense elements may represent myofibrils that have been extensively altered during homogenization. Little dense material is found in homogenates prepared in the Potter-Elvehjem homogenizer, presumably because the conditions of homogenization are much gentler. Also we have not observed such material in homogenates prepared in the presence of Triton X-100. Contamination by dense cytoplasmic particles can be reduced, but not eliminated, by increasing the density of the sucrose through which the nuclei are sedimented but only at the expense of lowered yields of nuclei. Some of the dense cytoplasmic material will pass through a layer of 2.5 M sucrose, whereas appreciable numbers of nuclei are lost at sucrose concentrations above 2.3 to 2.4 M.

In purifying muscle nuclei by centrifugation through sucrose, it is important to avoid overlaying a dense sucrose solution with a suspension of impure nuclei in dilute sucrose, for in this case centrifugation causes a rapid movement of the cytoplasmic components to the interface, where they form a tight pad that serves as an effective barrier to nuclei subsequently reaching the interface. To prevent this, the crude nuclear suspension should be made up to a sucrose concentration of 2.0 M or greater.

V. Stability of Nuclei

We have on occasion observed disintegration of muscle nuclei during purification or storage (Kuehl, 1975a), a phenomenon that has also been noted by Edelman et al. (1965). In our experience the relative stability of different preparations of nuclei has been extremely variable and we have, as yet, been unable to assign a cause to this variability. We have found that the nuclei may be stabilized by inclusion of 1 % bovine serum albumin in the solutions to which the nuclei are exposed and have published a procedure for isolating muscle nuclei that makes use of this observation (Kuehl, 1975a). Other investigators have been able to obtain satisfactory preparations of skeletal muscle nuclei without taking special precautions to stabilize the nuclei, and we, too, have often obtained satisfactory preparations in the absence of serum albumin.

VI. Detailed Procedures for Isolating Muscle Nuclei

The procedures described below have been worked out using muscle tissue from adult albino rats (weight 250–350 gm). Tissue from younger animals gives somewhat higher yields. Yield and purity of the nuclei do not appear to be altered by fasting the animals prior to sacrifice.

In the filtration procedures, a 10-mesh sieve is interchangeable with coarse cheesecloth (grade 10, 20 × 12 threads/inch); a 20-mesh sieve, with one thickness of fine cheesecloth (grade 60, 32 × 28 meshes/inch); an 80-mesh sieve, with three thicknesses of fine cheesecloth; and a 325-mesh sieve, with Miracloth (Chicopee Mills, Inc., N. Y.). Slightly better yields are obtained with the metal sieves, since they do not absorb liquid.

All steps of the isolation procedure are done at a temperature of $0°$ to $4°C$.

A. Method 1

This technique was developed in our laboratory (Kuehl, 1975a), where it has been in routine use for several years. The tissue is homogenized in a Potter-Elvehjem homogenizer in the presence of Triton X-100, and the nuclei are purified by centrifugation through 2.4 M sucrose. Nuclei are stabilized by the addition of 1 % bovine serum albumin to some of the solutions used in the isolation. Nuclei prepared by this method have been characterized microscopically, by chemical and enzymic assay procedures, and by fractionation and autoradiographic studies (Kuehl, 1975a,b). Nuclei are obtained in approximately 10 % yield, cytoplasmic contamination is low, and most of the nuclei appear undamaged when examined with a light microscope.

1. SOLUTIONS

Solution A: 0.25 M sucrose, 1 % bovine serum albumin, 5 mM MgCl$_2$
Solution B: 0.25 M sucrose, 0.5 % Triton X-100, 1 mM MgCl$_2$
Solution C: 2.3 M sucrose, 1 % bovine serum albumin, 5 mM MgCl$_2$
Solution D: 2.4 M sucrose, 1 % bovine serum albumin, 5 mM MgCl$_2$

2. PROCEDURE

Step 1. A 20-gm portion of muscle is passed three times through a Latapie mincer with a plate containing perforations 1 mm in diameter. To facilitate mincing, 5–10 ml of solution A are added to the muscle as it is being passed through the mincer.

Step 2. The mince is mixed with 100 ml of solution B and homogenized by 50 up-and-down strokes in a Potter-Elvehjem homogenizer having a clearance of 0.022 inch between the plastic pestle and glass homogenizing vessel. The homogenizer is run at 600 rpm.

Step 3. The homogenate is filtered through one layer of fine cheesecloth, and the filtrate is saved.

Step 4. The filter residue from step 3 is suspended in 50 ml of solution B and rehomogenized, this time in a homogenizer with a 0.010-inch clearance.

Step 5. The homogenate is filtered through one layer of cheesecloth; the residue is discarded; the filtrate is combined with the filtrate from step 3.

Step 6. The combined filtrates (step 5) are passed through one layer of Miracloth, and the residue on the filter is washed with a small amount of solution B.

Step 7. The filtrate from step 6 is centrifuged for 5 minutes at 800 g; the supernatant is discarded.

Step 8. The sediment is resuspended in 75 ml of solution B and centrifuged as in step 7 to give a crude nuclear pellet.

Step 9. The nuclear pellet from step 8 is suspended in 75 ml of solution C; 25-ml portions of this suspension are layered over 13-ml portions of solution D and centrifuged for 75 minutes at 65,000 g_{av} in a swinging-bucket rotor (Beckman type SW27). The supernatant is carefully aspirated, and the walls of the tube are wiped with a tissue.

Step 10. The pellets of purified nuclei are suspended in medium A.

B.   Method 2

This is a variation of method 1, which has not yet been published. An initial treatment in a VirTis homogenizer is followed by a second homogenization in a Potter–Elvehjem homogenizer in the presence of Triton X-100. The nuclei are then purified by centrifugation through 2.3 M sucrose.

This method is more convenient and rapid than method 1 and gives a better yield of nuclei. However, cytoplasmic contamination (as judged microscopically) is somewhat greater, and a higher proportion of the nuclei have been damaged during isolation.

1. SOLUTIONS

Solution A: 0.25 M sucrose, 2.5 mM MgCl$_2$, 25 mM N-tris (hydroxymethyl) methyl-2-aminoethane sulfonic acid (TES) buffer, pH 7.5 (pH 7.1 at 22°C)

Solution B: 0.25 M sucrose, 2.5 mM MgCl$_2$, 3% w/v Triton X-100, 25 mM TES buffer, pH 7.5

Solution C: 2.3 M sucrose, 2.5 mM MgCl$_2$, 25 mM TES buffer, pH 7.5

2. PROCEDURE

Step 1. A 25-gm portion of muscle is finely minced with a pair of scissors.

Step 2. The mince is suspended in 100 ml of solution A and homogenized for 30 seconds in a VirTis Model 23 homogenizer at setting 10 (approximately 10,000 rpm).

Step 3. A 25-ml portion of solution B is added to the homogenate, and the resulting mixture is rehomogenized by 50 up-and-down strokes in a Potter–Elvehjem homogenizer having a clearance of 0.010 inch between the pestle and the homogenizing vessel. The homogenizer is run at 600 rpm.

Step 4. The homogenate is filtered through one layer of coarse cheesecloth, then through an 80-mesh sieve, and, finally, through a 325-mesh sieve.

Step 5. The filtrate is centrifuged for 10 minutes at 800 g, the supernatant is discarded, and the pellets are drained briefly.

Step 6. The pellets are suspended in 56 ml of solution C; 28-ml portions of this suspension are layered over 10-ml portions of solution C and centrifuged for 75 minutes at 82,000 g_{ave} in a swinging-bucket rotor (Beckman type SW 27). The supernatant is carefully aspirated, and the walls of the tube are wiped with a tissue.

Step 7. The pellets of purified nuclei are suspended in solution A.

C.   Method 3

This is a slight modification of the method of Sobel and Kaufman (1970). The tissue is homogenized in a high-speed blender, and the nuclei are purified by centrifugation from 2.2 M sucrose. Use of detergent is avoided.

Nuclei are obtained in approximately 15% yield; cytoplasmic contamina tion is appreciably greater than with method 1 or 2, and an appreciable proportion of the nuclei have been damaged during isolation. Cytoplasmic contamination can be reduced somewhat by sedimenting the nuclei through a 2.3 or 2.4 M sucrose solution, but only at the expense of yield.

1.   SOLUTIONS

Solution A: 0.32 M sucrose, 25 mM KCl, 3 mM MgCl$_2$
Solution B: 2.4 M sucrose, 1.0 mM MgCl$_2$

2.   PROCEDURE

Step 1. A 25-gm portion of muscle is finely minced with a pair of scissors.

Step 2. The mince is suspended in 600 ml of solution A and homogenized five times in 30-second bursts in a VirTis Model 45 homogenizer with a speed setting of 5 (approximately 20,000 rpm).

Step 3. The homogenate is filtered through one layer of coarse cheesecloth, then through an 80-mesh sieve, and finally through a 325-mesh sieve.

Step 4. The filtrate is centrifuged for 10 minutes at 750 g and the supernatant is discarded.

Step 5. The pellets are suspended in a small volume of solution A and diluted exactly 1:10 with solution B. The resulting suspension is centrifuged for 75 minutes at 82,000 g_{ave} in a swinging-bucket rotor (Beckman type SW27). The supernatant is carefully aspirated and the walls of the tube are wiped with a tissue.

Step 6. The purified nuclei are suspended in solution A.

Method 1 is recommended when it is important that cytoplasmic contamination and/or breakage of nuclei be minimized. Method 2 is more rapid and convenient and gives a higher yield of nuclei, but cytoplasmic contamination is somewhat greater and a higher proportion of the nuclei are damaged. Method 3 should be employed when the use of detergent is objectionable.

88 LEROY KUEHL

REFERENCES

Edelman, J. C., Edelman, P. M., Knigge, K. M., and Schwartz, I. L. (1965). *J. Cell Biol.* **27**, 365–378.

Franke, W. W., and Schinko, W. (1969). *J. Cell Biol.* **42**, 326–331.

Kuehl, L. (1964). *Z. Naturforsch., Teil B* **19**, 525–532.

Kuehl, L. (1975a). *Exp. Cell Res.* **91**, 441–448.

Kuehl, L. (1975b). *Exp. Cell Res.* **92**, 221–230.

Marchok, A. C., and Wolff, J. A. (1968). *Biochim. Biophys. Acta* **155**, 378–393.

Schneider, W. C., and Klug, H. L. (1946). *Cancer Res.* **6**, 691–694.

Severin, S. E., Tseitlin, L. A., and Telepneva, V. I. (1965). *Dokl. Akad. Nauk SSSR* **160**, 953–955.

Silakova, A. I., Polishchuk, S. N., and Konoplitskaya, K. L. (1971). *Dokl. Akad. Nauk SSSR* **198**, 1468–1470.

Sobel, B. E., and Kaufman, S. (1970). *Arch. Biochem. Biophys.* **137**, 469–476.

Thomson, R. Y., Heagy, F. C., Hutchison, W. C., and Davidson, J. N. (1953). *Biochem. J.* **53**, 460–474.

Zaalishvili, M. M., and Gachechiladze, N. A. (1966). *Soobshch. Akad. Nauk Gruz. SSR* **41**, 443–449.

Chapter 5

Isolation of Neuronal Nuclei from Rat Brain Cortex, Rat Cerebellum, and Pigeon Forebrain

CLIVE C. KUENZLE, ALOIS KNÜSEL, AND
DANIEL SCHÜMPERLI

*Department of Pharmacology and Biochemistry,
School of Veterinary Medicine, University of Zürich,
Zürich, Switzerland*

I. Introduction

The cellular composition of the brain is heterogeneous. It comprises neurons, various types of glia, ependymal cells. and capillary endothelia. Accordingly, if one attempts to isolate neuronal nuclei from brain, one is faced with the problem of separating them from nuclei of other origins. One possibility is to obtain first a mixed population of nuclei, which are then separated according to cell parentage (Austoker *et al.*, 1972; Løvtrup-Rein and McEwen, 1966). Such procedures have the advantage of yielding simultaneously nuclei of different lineage, if necessary in large quantities (Austoker *et al.*, 1972), but suffer from ambiguities in the assignment of individual nuclei to parental cell type (Kato and Kurokawa, 1966; Knüsel *et al.*, 1973) An alternative scheme circumventing this problem uses intermediate preparations of easily identifiable cells, which upon gentle disruption yield nuclei of defined origin. The drawback of this technique is that it is not applicable to all cell types, and that it does not lend itself to large-scale preparations. However, using this approach we have succeeded in isolating from rat brain cortex neuronal nuclei of high purity and in good yield (Knüsel *et al.*, 1973). The present report reviews this previous method and extends it to the isolation of neuronal nuclei from rat cerebellum (unpublished) and pigeon forebrain (Schümperli, 1976).

II. Materials and Methods

A. Materials

Polyvinylpyrrolidone (PVP), type K 30, with a nominal average molecular weight of 40,000, and bovine serum albumin (BSA), "fraction V," were from Fluka AG, Buchs SG, Switzerland. Nylon bolting cloth was from Schweizerische Seidengazefabrik AG, Zürich, Switzerland. All other chemicals were of analytical grade and were purchased from either Fluka or E. Merck AG, Darmstadt, West Germany.

B. Methods

Flow diagrams of the three procedures for nuclear preparation from rat cortex, rat cerebellum, and pigeon forebrain are outlined in Scheme 1. They consist of two parts. The first, based on the method of Johnson and Sellinger (1971), Sellinger and Azcurra (1974), and Sellinger *et al.* (1971, 1974), affords pure neuronal perikarya. These are obtained by disrupting the brain tissue, without the use of digestive enzymes, by sequential passages through nylon bolting cloth of decreasing pore sizes, and separating the cell types by isopycnic centrifugation. In the second part of the procedure, the intermediate perikarya are carefully homogenized, and the released neuronal nuclei are purified by differential velocity and isopycnic centrifugations. All details of preparation are given in Scheme 1.

III. Results and Discussion

The procedures for isolating neuronal nuclei as outlined in Scheme 1 take only 5–6 hours to perform. Thus, preparation and biochemical analyses can usually be carried out on the same day.

The nuclear recoveries are somewhat variable, and the following average counts ($\pm$SD) are obtained per animal sacrificed: 6.6×10^6 ($\pm 3.3 \times 10^6$) from rat cortex, 50×10^6 ($\pm 23 \times 10^6$; approximately 98% granule cells and 2% Purkinje cells) from rat cerebellum, and 7.9×10^6 ($\pm 6.0 \times 10^6$) from pigeon forebrain. The yields of rat cortex nuclei can be increased by omitting the densest layer (2.3 M sucrose) from the final sucrose gradient. This, however, occurs at the expense of purity, since under standard conditions nuclei carrying larger cytoplasmic remnants (see below and Fig. 1) will be retained at the 2.2/2.3 M sucrose interface.

In rats the yields are also strongly age-dependent. The values listed above refer to animals 28–30 days old. Substantially better yields are obtained with younger animals (approximately 3-fold at age 7 days), but recoveries decline rapidly after 30 days. Nevertheless, it is still possible to isolate nuclei from 60-day-old rats.

The procedures can be scaled down without loss of relative yield to accommodate a minimum of 3 rat cortices, 2 rat cerebella, or a single pigeon hemisphere. This simply requires that all volumes be reduced proportionately. However, scaling up is not possible unless several runs are performed in tandem such as to prevent overloading of the nylon meshes during tissue disruption and of the sucrose gradients during centrifugation.

The purities of the nuclear preparations with respect to neuronal origin are expected to parallel closely the composition of the parent perikaryal fractions. The latter, if derived from rat cortex and rat cerebellum, consist of at least 90% neuronal elements (Knüsel et al., 1973; C. Kuenzle, unpublished; Sellinger et al., 1971, 1974), and the same therefore applies to the nuclear preparations. The perikaryal and nuclear fractions from pigeon forebrain are less pure because of consistent contamination by capillary fragments and nucleated erythrocytes. Contributions by these elements can be kept to a minimum if utmost care is devoted to the removal of meninges prior to brain dissociation. Nevertheless, the final preparation contains, along with at least 80% neuronal nuclei, 10–15% endothelial nuclei in capillary fragments and 5–10% erythrocytes (Schümperli, 1976).

Contamination of the rat cortex nuclei by other subcellular components seems negligible as judged by electron (Fig. 1), phase-contrast (Knüsel et al., 1973) and interference-contrast microscopy (Fig. 2, left), as well as by analysis of marker enzymes (Knüsel et al., 1973). The most consistent impurities are small cytoplasmic remnants and occasional clumped mitochondria. In preparations from rat cerebellum (Fig. 2, middle), cytoplasmic contamination may be slightly more pronounced owing to the sporadic occurrence of intact Purkinje cell bodies. Pigeon forebrain nuclei (Fig. 2, right) are the most difficult to purify (Schümperli, 1976), and larger cytoplasmic fragments are often seen to adhere to their membranes.

RAT CEREBELLUM

decapitate 36 rats (28-30 days old)

RAT CORTEX

decapitate 18 rats (28-30 days old)

PIGEON FOREBRAIN

decapitate 8 pigeons (any age)

remove brain quickly from skull

place brain structure of interest in 0.32 M sucrose/2 mM MgCl$_2$/1 mM K-phosphate, pH 6.5 and wait until it sinks

prepare cortex by manual dissection

blot onto filter paper

carefully remove all meninges with forceps, remoisten brain in above medium

mince in 15 ml 7.5% (w/v) PVP/1% (w/v) BSA/10 mM CaCl$_2$

ease mince into truncated plastic syringe (20 ml), use plunger to gently push mince 2x through one layer of nylon bolting cloth (670 μm pore size) fused to the syringe by pressing onto a heating plate

push 2x through one layer of nylon cloth (335 μm pore size)

keep on ice for 10 min

(RAT CEREBELLUM / RAT CORTEX)

push 3x through one inner (335 μm pore size) and one outer (112 μm pore size) layer of nylon cloth

keep on ice for 10 min

push 3x through one inner (335 μm pore size) and one outer (75 μm pore size) layer of nylon cloth

adjust vol. to 132 ml with 7.5% PVP/1% BSA/10 mM CaCl$_2$, apply onto 6 gradients, centrifuge in Spinco SW 27 rotor at 42,000 g_{av} x 30 min

discard by aspiration, clean walls with absorbent tissues

sample (22 ml)

1.0 M sucrose/1% BSA (5 ml)
1.75 M sucrose/1% BSA (10 ml)

myelin
glia, neurons
capillaries

neurons (pellet)

(PIGEON FOREBRAIN)

push 1x through one layer of nylon cloth (335 μm pore size)

adjust vol. to 50 ml with 7.5% PVP/1% BSA/10 mM CaCl$_2$, divide into two equal portions

push each 2x through one inner (335 μm pore size) and one outer (112 μm pore size) layer of nylon cloth using separate syringes for each of the two 25-ml portions

keep on ice for 10 min

push 1x through one inner (335 μm pore size) and one outer (112 μm pore size) layer of nylon cloth using separate syringes for each of the two 25 ml-portions; pool

push 2x through one inner (335 μm pore size) and one outer (75 μm pore size) layer of nylon cloth

adjust vol. to 120 ml with 7.5% PVP/1% BSA/10 mM CaCl$_2$, apply onto 6 gradients, centrifuge in Spinco SW 27 rotor at 42,000 g_{av} x 30 min

discard by aspiration

sample (20 ml)

1.0 M sucrose/1% BSA (4 ml)
1.8 M sucrose/1% BSA (7 ml)
2.0 M sucrose/1% BSA (5 ml)

myelin
glia, neurons
capillaries
neurons

pellet = dark, naked nuclei and erythrocytes

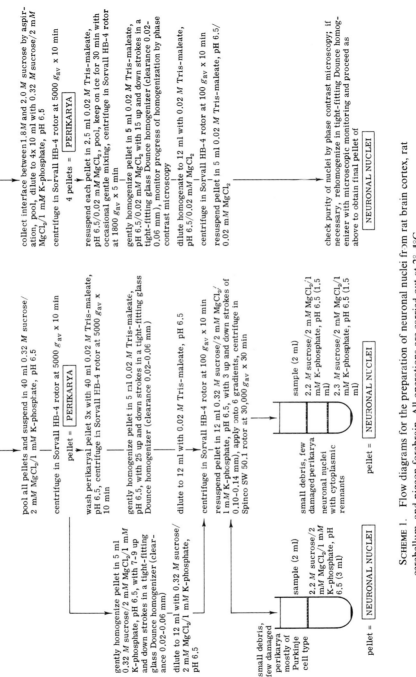

SCHEME 1. Flow diagrams for the preparation of neuronal nuclei from rat brain cortex, rat cerebellum, and pigeon forebrain. All operations are carried out at 2°–4°C.

93

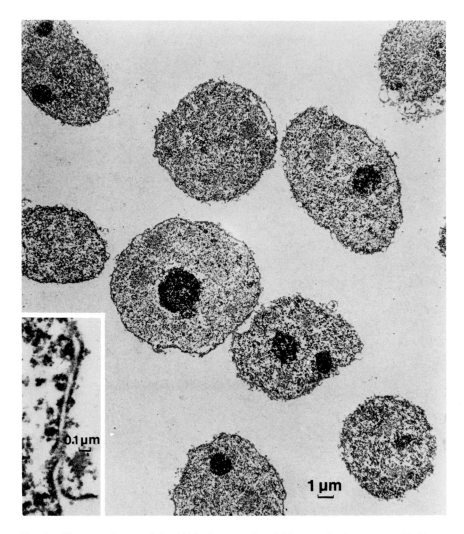

FIG. 1. Electron micrograph (× 4400) of neuronal nuclei from rat brain cortex purified by a final centrifugation through 2.2 *M* and 2.3 *M* sucrose. The section shown in the inset (× 35,000) is representative of the structural preservation of both the inner and the outer membrane and shows small cytoplasmic fragments attached. Cytoplasmic contamination is somewhat more pronounced if the layer of 2.3 *M* sucrose is omitted in the final centrifugation. Reproduced by courtesy of the Rockefeller University Press, New York.

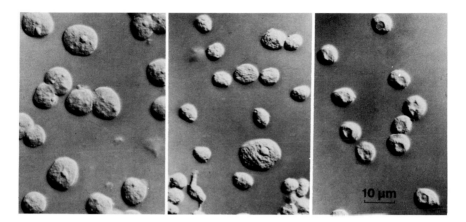

FIG. 2.   Interference-contrast micrographs (× 900) of neuronal nuclei from rat brain cortex (left), rat cerebellum (middle), and pigeon forebrain (right). In the cerebellar preparation one large Purkinje cell nucleus is surrounded by smaller nuclei originating from granule cells. Note the characteristic depression in the nuclei from pigeon forebrain.

IV.   Conclusion

The methods presented above afford neuronal nuclei from rat brain cortex, rat cerebellum, and pigeon forebrain within 5–6 hours from sacrifice. The yields are generally good and amount to 6, 50, and 8 million nuclei per animal, respectively. The fact that the scheme uses intermediate preparations of easily identifiable neurons raises confidence in the assignment of at least 80–90 % of the recovered nuclei to neuronal origin. Contamination by nonnuclear organelles is minimal.

ACKNOWLEDGMENTS

We thank Dr. G. S. Kistler for electron micrographs (Fig. 1) and Dr. A. Bregnard for interference-contrast micrographs (Fig. 2). This work was supported in part by the Swiss National Science Foundation, Grant 3.709.72.

REFERENCES

Austoker, J., Cox, D., and Mathias, A. P. (1972). *Biochem. J.* **129**, 1139–1155.
Johnson, D. E., and Sellinger, O. Z. (1971). *J. Neurochem.* **18**, 1445–1460.
Kato, T., and Kurokawa, M. (1966). *Proc. Jpn. Acad.* **42**, 527–532.

Knüsel, A., Lehner, B., Kuenzle, C. C., and Kistler, G. S. (1973). *J. Cell Biol.* **59**, 762–765.

Løvtrup-Rein, H., and McEwen, B. S. (1966). *J. Cell Biol.* **30**, 405–415.

Schümperli, D. (1976). Thesis, School of Veterinary Medicine, University of Zürich.

Sellinger, O. Z., and Azcurra, J. M. (1974). *Res. Methods Neurochem.* **2**, 3–38.

Sellinger, O. Z., Azcurra, J. M., Johnson, D. E., Ohlsson, W. G., and Lodin, Z. (1971). *Nature (London), New Biol.* **230**, 253–256.

Sellinger, O. Z., Legrand, J., Clos, J., and Ohlsson, W. G. (1974). *J. Neurochem.* **23**, 1137–1144.

Chapter 6

Isolation of Metaphase Chromosomes from Synchronized Chinese Hamster Cells

MASAKATSU HORIKAWA AND TAKASHI SAKAMOTO[1]

Division of Radiation Biology, Faculty of
Pharmaceutical Sciences, Kanazawa University,
Kanazawa, Japan

I. Introduction

In recent years, many investigators have reported the isolation of metaphase chromosomes from mouse ascites tumor (Cantor and Hearst, 1966), mouse lymphocytic leukemia cells (Chorazy *et al.,* 1963), and cultured mammalian cells (Somers *et al.,* 1963; Huberman and Attardi, 1966; Maio and Schildkraut, 1966; Salzman *et al.,* 1966; Burkholder and Mukherjee, 1970). Isolated metaphase chromosomes with excellent preservation of morphology and biological activities are very useful for analyzing the fine structure

[1] *Present address:* Research Laboratories, NISSUI Pharmaceutical Co., Ltd., 1805 Inari-cho, Souka-shi, Saitama 340, Japan.

and physicochemical properties of chromosomes. On the other hand, studies on the fractionation of isolated chromosomes and the uptake of fractionated chromosomes by recipient cells are very important in the genetic mapping of mammalian cell chromosomes. Recently, the incorporation and replication of transferred foreign metaphase chromosomes in recipient cells have been shown by Sekiguchi *et al.* (1973), and both replication and expression of chromosomal genes after the uptake of isolated mammalian metaphase chromosomes by recipient cells have been demonstrated by many workers (McBride and Ozer, 1973; Willecke and Ruddle, 1975; Burch and McBride, 1975; Sekiguchi *et al.,* 1975).

For these experiments, however, a large mitotic cell population of cultured cells is needed to isolate a large number of highly purified metaphase chromosomes, especially for further fractionation. In this chapter we shall describe a method for obtaining a large mitotic cell population from cultured Chinese hamster Don cells (CH-Don-13 clone) by two treatments with the combination of Colcemid and harvesting techniques described by Watanabe and Horikawa (1973), by which a relatively large number of metaphase chromosomes can be isolated. The mitotic cell population obtained by this method shows no significant biological or cytological damage, and the chromosomes isolated are morphologically intact.

II.   Cell Line, Medium, and General Techniques

A.   Cell Line and Medium

Chinese hamster Don cells were kindly supplied by Dr. H. Kato, National Institute of Genetics, Misima. The cells were cultured in a medium containing 90% Eagle's MEM supplemented with 10^{-3} M sodium pyruvate, 2×10^{-4} M L-serine, 2×10^{-3} M L-glutamine, and 10% bovine serum. CH-Don-13 cells obtained by colonial cloning from the original Chinese hamster Don cells were used in the present studies. The modal chromosome number was 22. In this medium at 37°C, the average doubling time of these cells was about 22 hours as determined by cell growth studies.

B.   Determination of Cell Number and Mitotic Index

The cells were counted in a hemacytometer of the Bürker type. The mitotic index was determined in cell preparations fixed with absolute methanol and stained with Giemsa solution.

III. Optimum Concentration of Colcemid and Duration of Treatment for Accumulating Mitotic Cells

As in the previous study (Watanabe and Horikawa, 1973), 1×10^6 cells in a monolayer in 200-ml square culture bottles were treated with various concentrations of Colcemid (Demecolcine, Nakarai Chemicals, Ltd. Kyoto) at 37°C for various lengths of time, and the total number of cells and the number of mitotic cells per bottle were determined periodically during Colcemid treatment.

Figure la shows the percentage of mitotic cells per 200-ml square culture bottle, following the treatment of 1×10^6 CH-Don-13 cells with various concentrations of Colcemid for 6 hours. Treatment with more than 0.025 μg of Colcemid per milliliter seemed to be effective for accumulating mitotic cells. No marked decrease was observed in the percentage of mitotic cells, when treated with 0.1 μg of Colcemid per milliliter, as had been observed with HeLa S3 cells (Watanabe and Horikawa, 1973). Figure 1b shows the percentage of mitotic cells per bottle, following the treatment of 1×10^6 cells with 0.025 μg of Colcemid per milliliter for various lengths of time. As can be seen in this figure, the percentage of mitotic cells per bottle reached a maximum when treated with 0.025 μg of Colcemid per milliliter for 6 hours and decreased when treatment with this concentration of Colcemid continued

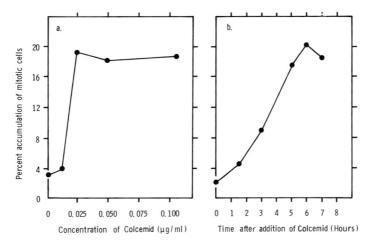

FIG. 1. Percent accumulation of mitotic cells per bottle after treatment of 1×10^6 CH-Don-13 cells (a) with various concentrations of Colcemid for 6 hours, and (b) with 0.025 μg of Colcemid per milliliter for various lengths of time. From Horikawa and Sakamoto (1974).

for more than 6 hours. These results indicate that treatment of cells with 0.025 µg of Colcemid per milliliter for 6 hours is best for collecting mitotic cells in a CH-Don-13 population, as in HeLa S3 cells (Watanabe and Horikawa, 1973).

IV.   Collection of a Large Highly Purified Mitotic Cell Population by Repeated Treatments with a Combination of Colcemid and Harvesting Techniques

For collecting a mitotic cell population by harvesting alone or by single and repeated treatments with a combination of Colcemid and harvesting techniques (Watanabe and Horikawa, 1973), 2×10^6 CH-Don-13 cells were cultured in Roux culture bottles, as shown in Fig. 2a. After 2 days of culture, cells were treated with the medium containing 0.025 µg of Colcemid per milliliter at 37°C for 6 hours. After one gentle washing with Eagle's MEM to remove the dead cells and the cellular debris, the mitotic cells in each bottle were collected by gentle pipetting (population I) and chilled at 0° to 4°C (population IC).

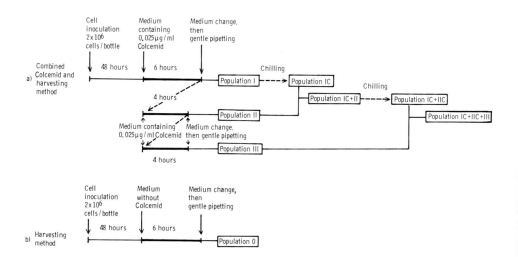

Fig. 2.   Schematic diagram of experimental procedures for obtaining synchronized mitotic cell populations from CH-Don-13 cells. From Horikawa and Sakamoto (1974).

In order to obtain a larger purified mitotic cell population from a given cell population, the remaining CH-Don-13 cells were again treated with 0.025 μg of Colcemid per milliliter at 37°C for 4 hours. After one gentle washing with Eagle's MEM, the mitotic cells were collected by gentle pipetting (population II) and combined with the chilled mitotic cell population collected previously (population IC + II). In some cases, the remaining cell population was once again treated with 0.025 μg of Colcemid per milliliter for 4 hours, and the mitotic cells were collected (population III) and combined with the previously collected two chilled mitotic cell populations (population IC + IIC + III). In the control cultures (Fig. 2b), mitotic cells were collected by harvesting alone without pretreatment with Colcemid; that is, the medium was changed to normal medium after 2 days of incubation of 2×10^6 cells. The culture was incubated for an additional 6 hours, and the medium was again renewed. Then mitotic cells were collected by gentle pipetting of culture bottles (population 0).

Table I shows the differences in the number of mitotic cells and the mitotic index in cell populations (population 0 and I) obtained by the harvesting method and by the combination of Colcemid and harvesting. As shown in Table I, the combined method is much better than the harvesting method with regard to the number of mitotic cells obtained and their mitotic index. Furthermore, as shown in Table II, it was found that the mitotic index of each cell population (population II or III) collected by second or third treatment with the combination of Colcemid and harvesting techniques was still high (87–91.7%), and the total number of mitotic cells in populations IC + II and IC + IIC + III obtained after the second and third treatment with the combined method of Colcemid and harvesting was about 2.0 and 2.5 times larger, respectively, than population I obtained by treatment with a single combined method.

TABLE I

DIFFERENCES IN TOTAL NUMBER OF MITOTIC CH-DON-13 CELLS AND MITOTIC INDEX IN CELL POPULATIONS OBTAINED BY HARVESTING METHOD AND BY COMBINED COLCEMID AND HARVESTING METHOD[a]

Method	Population	Original number of cells used for synchronization	Total number of cells harvested	Total number of mitotic cells harvested	Mitotic index
Harvesting	0	6.0×10^6	6.3×10^3	5.1×10^3	80.4
		1.3×10^7	6.0×10^4	4.3×10^4	71.0
Combined	1	3.5×10^6	3.5×10^5	3.2×10^5	90.4
Colcemid and harvesting		4.0×10^6	4.8×10^5	4.4×10^5	91.0

[a]From Horikawa and Sakamoto (1974).

TABLE II

DIFFERENCES IN TOTAL NUMBER OF MITOTIC CH-DON-13 CELLS AND MITOTIC INDEX IN CELL
POPULATIONS OBTAINED BY SINGLE AND REPEATED TREATMENT WITH COMBINED COLCEMID AND
HARVESTING METHOD[a]

Method	Population	Original number of cells used for synchronization	Total number of cells harvested	Total number of mitotic cells harvested	Mitotic index
Combined	I	3.5×10^6	3.5×10^5	3.2×10^5	90.4
Colcemid		4.0×10^6	4.8×10^5	4.4×10^5	91.0
and	II	3.0×10^6	3.5×10^5	3.2×10^5	91.7
harvesting		3.5×10^6	4.9×10^5	4.4×10^5	89.7
	IC + II	—	7.0×10^5	6.4×10^5	—
		—	9.7×10^5	8.8×10^5	—
	III	2.7×10^6	2.0×10^5	1.8×10^5	89.4
		3.0×10^6	2.7×10^5	2.4×10^5	87.0
	IC + IIC + III	—	9.0×10^5	8.2×10^5	—
		—	12.4×10^5	11.2×10^5	—

[a] From Horikawa and Sakamoto (1974).

V.   Biological and Cytological Effects of Single and Repeated Treatments with Colcemid and of Chilling on Chinese Hamster Cells

For determining the effects of single and repeated treatments with Colcemid (0.025 μg/ml) and of chilling on the growth of CH-Don-13 cells, various mitotic cell populations collected after treatment were placed in 60-mm glass petri dishes containing normal growth medium. After various periods of incubation at 37 °C in a humidified 5% CO_2 incubator, the number of cells in each dish was determined. In addition, the biological activity of mitotic CH-Don-13 cells collected by harvesting alone or by harvesting after Colcemid treatment was determined by their colony-forming ability in 60-mm glass petri dishes incubated at 37 °C in a humidified 5% CO_2 incubator for 12 days. Similarly, the effect of chilling at 0 °to 4 °C on collected mitotic cells was determined by their colony-forming ability.

Figure 3b shows the increase in the total number of mitotic cells in populations I, IC + II or IC + IIC + III obtained after the first, second, or third treatment of 2×10^8 CH-Don-13 cells in 40 Roux culture bottles with the combination of Colcemid and harvesting techniques, by the procedures described above. However, we should also like to know the biological effect of single or repeated treatment with Colcemid and of chilling on CH-Don-13 cells. Figure 3a shows the plating efficiencies of mitotic cells

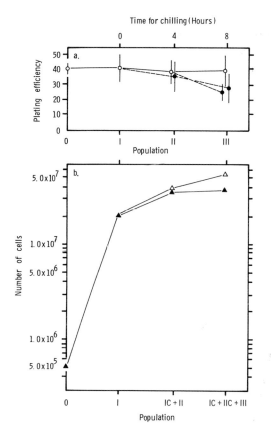

Fig. 3. (a) Plating efficiencies of populations 0, I, II, and III (O), and changes in plating efficiency of populations I and II by chilling (0°–4°C) (●) for various lengths of time. (b) Changes in total numbers of mitotic cells (△) and of viable mitotic cells (▲) in populations I, IC + II, and IC + IIC + III obtained after first, second, and third treatment of 2×10^8 CH-Don-13 cells with combination of Colcemid and harvesting techniques. From Horikawa and Sakamoto (1974).

(populations 0, I, II or III) collected by harvesting alone or by the first, second, or third treatment with the combined method. Simultaneously, the changes in the plating efficiencies of populations I or II by chilling at 0° to 4°C for various lengths of time are shown in Fig. 3a. As can be seen in this figure, there is no significant difference in plating efficiency among the populations 0, I, II and III. However, it can be concluded that chilling at 0° to 4°C for 4 or 8 hours decreases the plating efficiency of populations I and II. When we consider such a decrease in the plating efficiency caused by chilling, the total number of viable mitotic cells in populations IC + II and IC + IIC +

III may be calculated, as shown in Fig. 3b. If Fig. 3b is the correct case, collection of mitotic cells by the third treatment with the combined method has no meaning, because there is no significant difference in the number of viable mitotic cells between populations IC + II and IC + IIC + III.

On the other hand, the cytological effect of single or repeated treatment with 0.025 μg of Colcemid per milliliter and of chilling on CH-Don-13 cells was investigated by reincubation in normal culture medium of populations I, IC, II, IC + II or III, respectively. At 19 to 25 hours (roughly correspond-ing to one generation) after reincubation, cells of each population were collected, and chromosome numbers and chromosomal aberrations were analyzed. Cell preparations for chromosome study were obtained by an air-drying method and Giemsa staining.

As shown in Fig. 4, there are no significant changes in chromosome num-bers of populations I(b) and II(d), when compared with that of population 0(a) obtained by harvesting alone. In addition, no significant differences were observed in chromosomal aberrations among these populations. However, in populations IC(c) and IC + II(e), the number of cells with 27 to 36 chromosomes increased slightly, although no significant increase in chromosomal aberrations was observed in these cell populations. These results suggest that two treatments (for 6 hours and additional 4 hours) with 0.025 μg of Colcemid per milliliter for obtaining population IC + II cause no significant damage in the chromosome level of CH-Don-13 cells, whereas chilling at 0° to 4°C for 4 hours increases the number of chromosomes. On the other hand, in population III(f), the number of cells with tetraploid chromosomes and with 27 to 36 chromosomes increased, but few chromo-

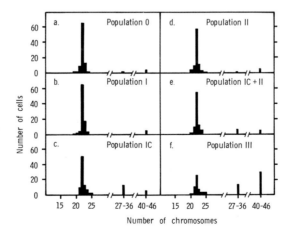

FIG. 4.   Distribution of chromosome numbers in various cell populations (populations 0, I, IC, II, IC + II and III) of CH-Don-13 cells. From Horikawa and Sakamoto (1974).

some aberrations were observed in this cell population. Therefore, it can be concluded that population IC + II obtained by two treatments with the combination of Colcemid and harvesting techniques is the most suitable for isolating large numbers of metaphase chromosomes.

Next, we examined the degree of cell progression through the cell cycle in normal culture medium in various mitotic cell populations collected by the harvesting method or by repeated treatments of the combined method. Figure 5 shows the changes in the relative cell numbers and the percentage of mitotic cells of populations 0, II and IC + II at various culture periods after harvesting. As can be seen in this figure, there is a good agreement in the synchrony of these three mitotic cell populations, although some delay was observed in the increase of cell numbers of population IC + II immediately after cultivation. From this figure, the total generation time can be determined as about 19 to 20 hours.

VI. Isolation of Metaphase Chromosomes from Mitotic Cell Population

The chromosomes from the mitotic cells (population IC + II) collected after two treatments with the combination of Colcemid (0.025 μg/ml) and harvesting techniques were isolated by essentially the same method as described by Maio and Schildkraut (1967, 1969) with the following slight

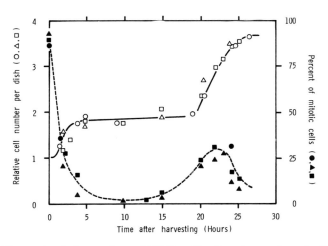

FIG. 5. Changes in relative numbers of CH-Don-13 cells and percentage of mitotic cells in populations 0 (○, ●), II (△, ▲), and IC + II (□, ■) at various culture times after harvesting. From Horikawa and Sakamoto (1974).

modification: during the process of removing nuclei and unbroken cells from the chromosomes and fine cellular debris in suspension, centrifugation was repeated in the horizontal head of a Kubota KR-180B centrifuge at 100 to 150 g increasing gradually, for 8 minutes. On the other hand, the chromosomes in the pooled supernatant fractions were sedimented by centrifugation in the horizontal head centrifuge for 20 minutes at 550 g, instead of 10 minutes at 2500 g, and the pellets were washed once by resuspension in 0.02 M Tris (pH 7.0) containing 0.1% saponin and recentrifugation at 550 g for 20 minutes, instead of 2500 g for 10 minutes.

Figures 6a and 6b show stained preparations of metaphase chromosomes isolated from the mitotic cell population IC + II of CH-Don-13 cells by this method. Comparison of these isolated chromosomal figures (Figs. 6a and 6b) and a typical preparation of diploid karyotype (Fig. 6c), obtained from a single CH-Don-13 cell by the air-drying method and Giemsa staining, showed that the isolated chromosomes were morphologically well preserved and that the various chromosome types found in the mitotic figures were represented with similar frequencies in the population of isolated chromosomes.

VII.   Discussion

Ever since cell fusion between cultured mouse cells was first discovered by Barski *et al.* (1960), the hybridization of cultured mammalian cells has been investigated by many workers as a potential tool for the study of genetic linkage in mammalian somatic cells. On the other hand, studies on the uptake of isolated chromosomes by recipient cells have been attempted by Yosida and Sekiguchi (1968), Burkholder and Mukherjee (1970), Kato *et al.* (1971), Sekiguchi *et al.* (1973), and others. These studies have provided a powerful tool for the study of the mechanism of transfer of genetic information in isolated chromosomes to recipient cells, as recently reported by many workers (McBride and Ozer, 1973; Willecke and Ruddle, 1975; Burch and McBride, 1975; Sekiguchi *et al.*, 1975), as well as for genetic mapping of mammalian cell chromosomes. Simultaneously, the isolated chromosomes with excellent preservation of morphology and biological activities are very useful materials for studying the fine structure and physicochemical prop-

FIG. 6.   Metaphase chromosomes of CH-Don-13 cells. (a and b) Metaphase chromosomes isolated from mitotic cell population IC + II of CH-Don-13 cells, stained with Giemsa. (c) Karyotype preparation showing chromosome complement of single CH-Don-13 cells, stained with Giemsa. From Horikawa and Sakamoto (1974).

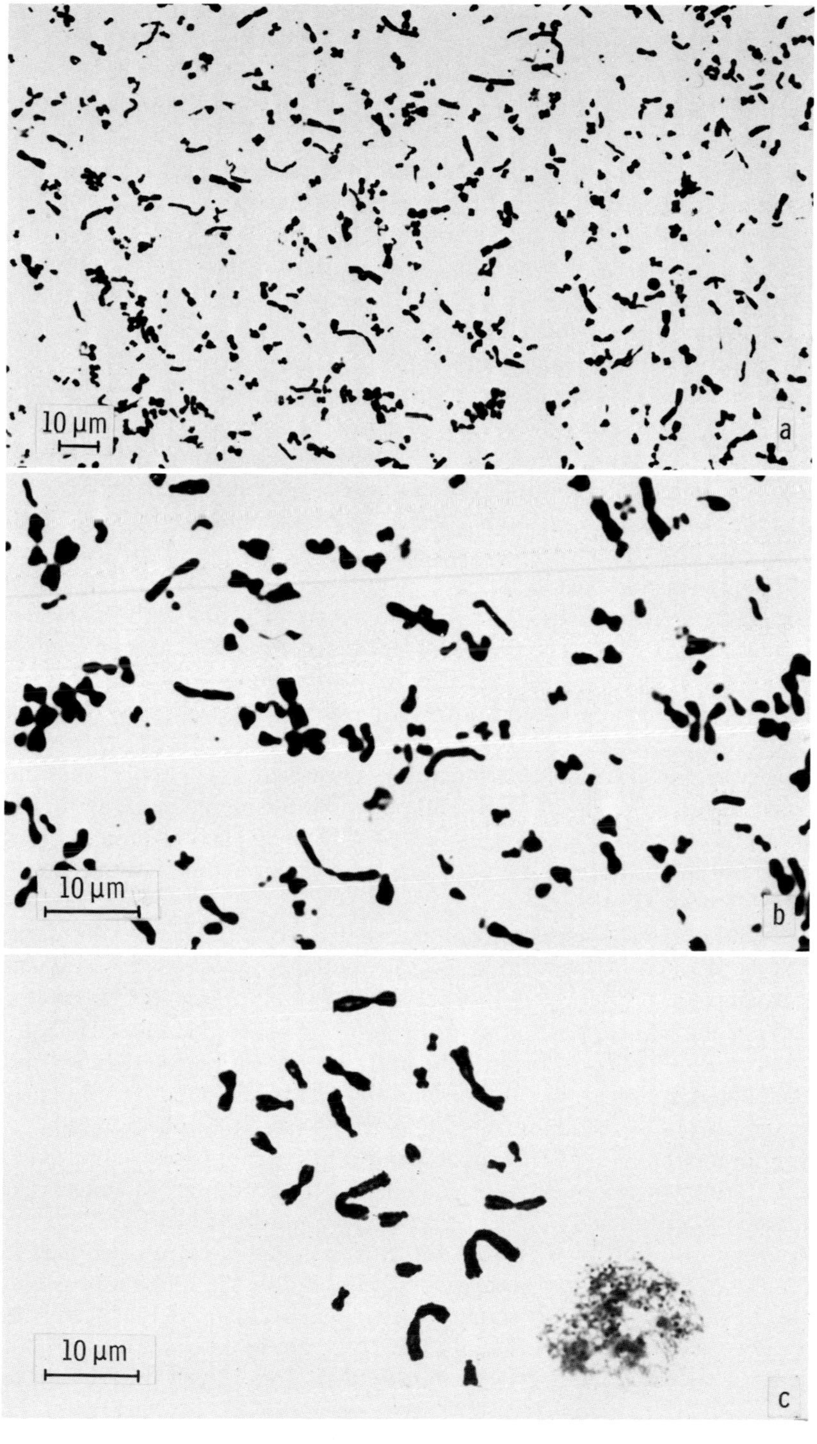
10 µm
a
10 µm
b
10 µm
c

erties of chromosomes (see Salzman *et al.,* 1966; Cantor and Hearst, 1966; Huberman and Attardi, 1966; Maio and Schildkraut, 1967, 1969; Mendelsohn *et al.,* 1968; Wray and Stubblefield, 1970).

The further purification and fractionation of isolated chromosomes require a large number of metaphase chromosomes, which should be isolated from a large mitotic cell population of cultured cells. We have established the optimum conditions for obtaining a large mitotic cell population from cultured Chinese hamster Don cells (CH-Don-13 clone) by two treatments with a combination of the Colcemid and harvesting techniques, recently established by Watanabe and Horikawa (1973) for HeLa S3 cells.

As indicated by plating efficiency (Fig. 3a), chromosome number (Fig. 4), chromosome aberrations (see text), and cell progression through the cell cycle (Fig. 5), two treatments (for 6 hours and additional 4 hours) with 0.025 μg of Colcemid per milliliter do not cause any significant biological or cytological damage to CH-Don-13 cells, whereas chilling at 0° to 4°C for 4 hours causes a slight decrease in plating efficiency and an increase in chromosomes in population I. These results show that the decrease in the plating efficiency of cells caused by chilling correlates well with the increase in number of chromosomes. In any case, the total number of viable mitotic cells in population IC + II obtained by two treatments with a combination of Colcemid and harvesting is about 70 times that obtained by the harvesting method alone, as seen in Fig. 3b, even though we consider the decrease in plating efficiency of cells to be due to chilling. However, it was found that three treatments with 0.025 μg of Colcemid per milliliter increase the number of cells with tetraploid chromosomes and with 27 to 36 chromosomes (Fig. 4f) and that chilling of population I for 8 hours decreases the plating efficiency of cells strikingly (Fig. 3a).

On the other hand, the chromosomes isolated from mitotic cell population IC + II by a slight modification of the method described by Maio and Schildkraut (1967, 1969) were morphologically well preserved. In addition, various chromosome types found in the mitotic figures of single CH-Don-13 cells obtained by an air-drying method and Giemsa staining were represented with similar frequencies in the population of these isolated chromosomes. Since the chromosome number of CH-Don-13 cells employed was 22, it was found that the total number of chromosomes isolated by this method was roughly estimated about 10% of that of the total number of mitotic cells used, when population IC + II (3 to 4.5 × 10^7 mitotic cells) obtained from 40 Roux culture bottles was used for chromosome isolation. Further improvement of this method for chromosome isolation will make it possible to isolate the metaphase chromosomes with a much higher yield. In any case, procedures established in this study for obtaining large quantities of metaphase chromosomes from a large mitotic CH-Don-13 cell population can serve as a meaningful system for further fractionation and biochemical analysis of isolated chromosomes.

REFERENCES

Barski, G., Sorieul, S., and Cornefert, F. (1960). *C. R. Hebd. Seances Acad. Sci.* **251**, 1825.
Burch, J. W., and McBride, O. W. (1975). *Proc. Natl. Acad. Sci. U.S.A.* **72**, 1797.
Burkholder, G. D., and Mukherjee, B. B. (1970). *Exp. Cell Res.* **61**, 413.
Cantor, K. P., and Hearst, J. E. (1966). *Proc. Natl. Acad. Sci. U.S.A.* **55**, 642.
Chorazy, M., Bendich, A., Borenfreund, E., and Hutchison, D. J. (1963). *J. Cell Biol.* **19**, 59.
Horikawa, M., and Sakamoto, T. (1974). *Jpn. J. Genet.* **49**, 197.
Huberman, I. A., and Attardi, G. (1966). *J. Cell Biol.* **31**, 95.
Kato, H., Sekiya, K., and Yosida, T. H. (1971). *Exp. Cell Res.* **65**, 454.
McBride, O. W., and Ozer, H. L. (1973). *Proc. Natl. Acad. Sci. U.S.A.* **70**, 1258.
Maio, J. J., and Schildkraut, C. L. (1966). *In* "Methods in Cell Physiology" (D. M. Prescott, ed.), Vol. 2, pp. 113–130. Academic Press, New York.
Maio, J. J., and Schildkraut, C. L. (1967). *J. Mol. Biol.* **24**, 29.
Maio, J. J., and Schildkraut, C. L. (1969). *J. Mol. Biol.* **40**, 203.
Mendelsohn, J., Moore, D. E., and Salzman, N. P. (1968). *J. Mol. Biol.* **32**, 101.
Salzman, N. P., Moore, D. E., and Mendelsohn, J. (1966). *Proc. Natl. Acad. Sci. U.S.A.* **56**, 1449.
Sekiguchi, T., Sekiguchi, F., and Yamada, M. A. (1973). *Exp. Cell Res.* **80**, 223.
Sekiguchi, T., Sekiguchi, F., Tachibana, T., Yamada, T., and Yoshida, M. (1975). *Exp. Cell Res.* **94**, 327.
Somers, C. E., Cole, A., and Hsu, T. C. (1963). *Exp. Cell Res., Suppl.* **9**, 220.
Watanabe, M., and Horikawa, M. (1973). *J. Radiat. Res.* **14**, 258.
Willecke, K., and Ruddle, F. H. (1975). *Proc. Natl. Acad. Sci. U.S.A.* **72**, 1792.
Wray, W., and Stubblefield, E. (1970). *Exp. Cell Res.* **59**, 469.
Yosida, T. H., and Sekiguchi, T. (1968). *Mol. Gen. Genet.* **103**, 253.

Chapter 7

Use of Metrizamide for Separation of Chromosomes

WAYNE WRAY

Department of Cell Biology,
Baylor College of Medicine,
Houston, Texas

I. Introduction

Isopycnic centrifugation is potentially one of the most powerful ways to fractionate the mammalian genome, but has been technically frustrating. High concentrations of salt, such as CsCl, rapidly dissociate metaphase chromosomes. This is not surprising in view of the data of MacGillivray *et al.* (1972) showing that high concentrations of salt dissociate nucleoproteins. Chromatin will band isopycnically in CsCl after fixation with formaldehyde (Hancock, 1970), but the reaction is irreversible and the material of little use thereafter. Obviously, formaldehyde-treated chromosomes are equally useless. We have used sucrose, Ficoll, and fructose as the support media for isopycnic banding of metaphase chromosomes, but at the density required to band the chromosomes ($\rho = 1.31$ gm/ml) the viscosity of the sugar solutions is so high as to make their use prohibitive. Chromatin has been shown to band in chloral hydrate gradients (Hossainy et al., 1973), but these gradients are very difficult to handle and are also quite viscous, which again make them unsuitable for banding metaphase chromosomes. In order to circumvent these problems, an elaborate system of nonaqueous gradients was developed by Stubblefield and Wray (1973) for buoying chromosomes. These gradients, which consist of carbon tetrachloride, ethylene dichloride, and diethylene glycol, accomplish their objective of isopycnically banding metaphase chromosomes ($\rho = 1.36$ gm/ml) in a low viscosity media. However, the technical problems in preparing these gradients, the solvent properties of these organic chemicals, especially their ability to extract certain proteins from the chromosomes, make this system amenable only to certain specified applications. The introduction of iodinated density gradient media for biological separations represents a significant advance over previous technology. The isopycnic centrifugation of chromatin in metrizamide (Rickwood et al., 1973; Birnie et al., 1973) solutions leads to the obvious extension of its use for isopycnic banding of metaphase chromosomes.

II.   General Methods

A.   Tissue Culture and Cell Lines

The experiments presented were performed using Chinese hamster ovary (CHO) and HeLa tissue culture cell lines. The cells were cultured in McCoy's medium 5A supplemented with 10% fetal calf serum. The pH of the medium was buffered by a bicarbonate system requiring 5% CO_2 in the atmosphere. The CHO cell line has a cell cycle of approximately 12 hours, and the HeLa cell cycle is about 22 hours. Cultures were subdivided every 24 hours.

B.   Cell Synchronization

In order to increase the number of cells in mitosis at the time of harvest, mitotic cells were accumulated with 0.06 μg of Colcemid per milliliter (Stubblefield, 1968) and collected by vigorous shaking of the culture flask. The mitotic cell populations obtained were routinely 94% to 97% metaphase cells.

C.   Metaphase Chromosome Isolation

Metaphase chromosomes were prepared by the method of Wray and Stubblefield (1970). The buffer solution for the isolation procedure consisted of 1.0 M hexylene glycol (2-methyl-2,4-pentanediol, Eastman Organic, Rochester, New York), 0.5 mM CaCl$_2$, and 0.1 mM (piperazine-N,N'-bis-2-ethanesulfonic acid) monosodium monohydrate (PIPES buffer), Calbiochem, San Diego, California) at pH 6.8. The only modification to the published procedure was that cells were broken by nitrogen cavitation using 250 psi pressure in a Parr pressure vessel (Parr Instrument Co., Moline, Illinois). Chromosomes were purified by sedimentation through a 10 % to 50 % sucrose–buffer gradient in a Sorvall SZ-14 zonal rotor at 1500 g for 30 minutes (Wray, 1973).

D.   Photography and Electron Microscopy

A Leitz Orthoplan microscope equipped with an Orthomat camera was used for phase contrast microscopy. Pictures were made on High Contrast Copy Film 5069 (Eastman Kodak, Rochester, New York) rated at an A.S.A. of 12, and developed in Acufine (Acufine, Inc., Chicago, Illinois) diluted with water 1 : 1 at 20 °C for 6 minutes. Examination of the metaphase chromosomes by electron microscopy was performed on a Phillips 200 electron microscope after critical point drying according to Anderson (1951).

E.   Biochemical Analysis and Polyacrylamide Slab Gel
Electrophoresis

DNA and protein were measured by the methods of Burton (1956) and Lowry *et al.* (1951). Chromosomal proteins were prepared for polyacrylamide gel electrophoresis by dissolving chromosomes in 3% sodium dodecyl sulfate (SDS) (BDH Chemicals, Ltd., Poole, England), 0.062 M Tris (Sigma, St. Louis, Missouri) pH 6.8, at 100 °C for 10 minutes. The amount of protein in the sample was assayed by spectrophotometric assay (Groves *et al.,* 1968) after which β-mercaptoethanol (Eastman Organic) was added and the samples were reheated. Chromosomal proteins were analyzed on 9.5% polyacrylamide slab gels using a Tris-glycine buffered SDS system (Laemmli, 1970). Gels were stained with 0.05% Coomassie blue (Colab, Glenwood, Illinois) in methanol : acetic acid : water 40 : 7.5 : 52.5 (v/v/v) and destained by diffusion in 7.5% acetic acid. Gels were photographed on Contrast Process Pan 4155 (Eastman Kodak) 4 × 5 inch sheet film and developed in D-11 for 5 minutes at 20 °C. Gels were scanned on a Helena Quick Scan Jr. (Helena Laboratories, Beaumont, Texas) which had been modified to accept slab gels.

F.   Properties of Metrizamide

Metrizamide (Nyegaard and Co., Oslo, Norway) is a triiodinated benz-amido derivative of glucose with the systematic name 2-(3-acetamido-5-N-methylacetamido-2,4,6-triodobenzamido)-2-deoxy-D-glucose. Its structural formula is shown in Fig. 1. Metrizamide, which has a molecular weight of 789.1, is readily soluble in aqueous media when it is added with stirring. Concentrations as high as 85% may be obtained, but temperatures higher than 55 °C should be avoided during dissolution. Figure 2 shows the correlation between the concentration of metrizamide solutions in the chromosome isolation buffer, the refractive index, and the specific density at 22 °C. Further information about the properties of metrizamide may be obtained in a booklet available from Nygaard and Co., Nycovien 2, Oslo 4, Norway, entitled "Metrizamide—A Gradient Medium for Centrifugation Studies."

G.   Recovery of Metrizamide

Because metrizamide is expensive, it may be desirable for many laboratories to recover it after use. This may be done rather simply by first passing the metrizamide solution through a mixed-bed Dowex resin, such as Bio-Rex RG 501-X8. This will remove any charged species of small molecular weight and allow the metrizamide, which is nonionic, to pass through the column. Then, metrizamide may be removed from the solution by the use of a Bio-Fiber 50 beaker dialyzer (Bio-Rad Laboratories) or conventional dialysis. Finally, lyophilization yields metrizamide which is as good as the initial product.

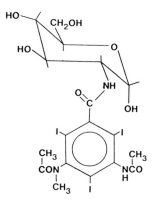

FIG. 1.   Structural formula of metrizamide.

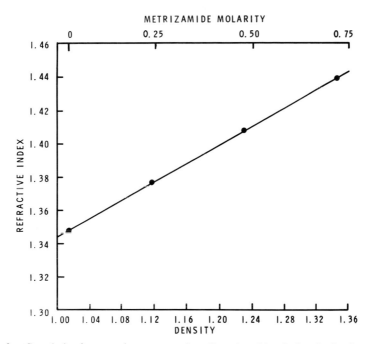

FIG. 2.   Correlation between the concentration of metrizamide solutions in the chromosome buffer, the refractive index, and the specific density at 22°C.

H.   Specific Problems and Solutions

Our major sources of difficulties during these studies were divided between problems inherent with working with metaphase chromosomes and problems associated with the isopycnic banding of the chromosomes. The particular problem endogenous to metaphase chromosome research is their very annoying tendency to stick to each other and to the sides of the centrifuge tubes. This made many early experiments difficult to interpret because chromosomes after centrifugation were plastered to the walls in a thin film down to the isopycnic density, at which point the chromosomes adhered in a heavy band to the walls of the tube. Upon fractionation, the chromosomes that dislodged were normally found in large clumps. Analysis of these gradients was almost impossible. These results were obtained whether the gradients were preformed or allowed to form by centrifugal force and had no relation to the speed or length of centrifugation time.

The first breakthrough came when chromosomes were mixed in 0.75 *M* metrizamide layered on the bottom of a preformed gradient and allowed to rise until they reached their buoyant density. This circumvented many of the wall effects associated with conventional centrifugation techniques. To further minimize wall effects and chromosomes aggregating, tubes have been treated with 1% Siliclad (Clay Adams), 1% dimethyldichlorosilane (PCR, Inc., Gainesville, Florida) in carbon tetrachloride, bovine serum albumin, or bovine submaxillary mucin, and chromosomes have been suspended in buffer containing 1% Siliclad, ethanol, isoamyl alcohol, and (2-naphthyl-6, 8-disulfonic acid) disodium salt (NDA; Eastman Organic). These have been used singularly and in various combinations. Time and speed of centrifugation also turned out to be very important as very short centrifugation times at relatively high speeds tend to minimize wall effects.

I.   Detailed Method for Isopycnic Banding of Chromosomes

The optimal conditions that we have established for isopycnic buoying of metaphase chromosomes in metrizamide are:

1. Treat 15-ml Corex centrifuge tubes for 5 minutes with 1% dimethyldichlorosilane in carbon tetrachloride.

2. Dry and bake the silicone on the tube at $140°–160°F$ for 24 hours.

3. Suspend chromosomes in 1 ml of chromosome isolation buffer containing 0.75 M metrizamide and 5 mM NDA and place the suspension in the Corex tube.

4. Layer on the chromosome suspension a preformed 10 ml 0.35 M to 0.75 M metrizamide-buffer gradient containing 5 mM NDA.

5. Centrifuge at 16,300 g (10,000 rpm) for 10 minutes in an HB-4 (swinging bucket) rotor in a Sorvall RC-2B centrifuge.

6. The gradient is fractionated from the top in 0.25-ml aliquots.

Under these conditions chromosomes buoy in a rather sharp and obvious band. There is minimal adherence to the walls of the tubes, and most chromosomes do not stick together. Thus, after the nagging technical problems have been eliminated, the metrizamide system for isopycnically buoying chromosomes seems to be ideal.

III.   Chromosome Characterization

A.   Morphology

The morphology of normal isolated chromosomes when examined with phase contrast light microscopy and electron microscopy (Fig. 3a) shows a structure with very good detail and excellent fine structure. Chromosomes exposed to metrizamide (Fig. 3b) in comparison cannot be distinguished from normal controls.

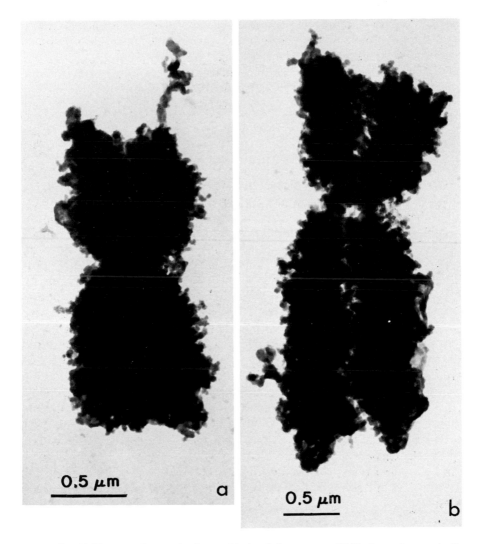

FIG. 3.   (a) Electron micrograph of control isolated chromosome. (b) Electron micrograph of isolated chromosome exposed to metrizamide.

B. Protein/DNA Relationship

The relative composition of isolated chromosomes is shown in Table I. The ratio of protein to DNA is about 2.2 : 1.0 for both control chromosomes and chromosomes exposed to metrizamide.

Figure 4 shows that when chromosomes have been doubly labeled with both [^{14}C]-thymidine and [^{3}H]-amino acid mix, the chromosomes have a buoyant density of $\rho = 1.24$ gm/ml, and the DNA and protein are located in a common peak.

C. Fidelity of Proteins on Chromosomes

Experiments were first carried out to demonstrate that the proteins associated with the chromosomes were indeed chromosomal and, moreover, that they were not redistributed among the chromosomes under the experimental conditions used. A control experiment to show that isolated chromosomes do not adventitiously adsorb cytoplasmic proteins was done as follows. Unlabeled chromosomes were isolated and purified through a 10%–40% (w/v) sucrose-buffer velocity gradient. Labeled cytosol was made by growing cells in the presence of a [^{3}H]-amino acid mixture for 12 hours. The cells were broken, and the nuclei and large organelles were removed by centrifugation at 16,300 g for 20 minutes. The labeled cytosol was mixed with the unlabeled chromosomes and incubated at room temperature for 30 minutes. The chromosomes were then purified through three successive 10%–20%–30%–40% (w/v) sucrose-buffer step gradients and prepared for autoradiography. The slides were developed after 6 months, and it was found that no radioactivity was associated with the chromosomes. Control chromosomes which had not been purified were heavily labeled.

It is important to demonstrate that the proteins that are associated with a particular chromosome are faithful to that chromosome and do not rearrange and transfer from chromosome to chromosome after isolation or in the presence of metrizamide. To show this, the following experiments were designed.

TABLE I

RELATIVE COMPOSITION OF ISOLATED CHROMOSOMES

Treatment	DNA	Protein	Protein/DNA
Control	100	180	1.80
Control	100	209	2.09
Control	100	220	2.20
Metrizamide	100	218	2.18

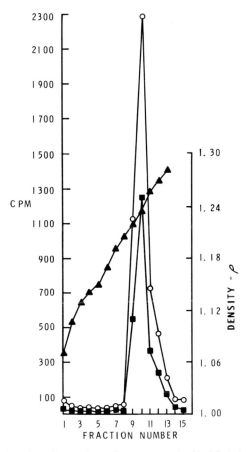

FIG. 4.   Isopycnic banding of metaphase chromosomes doubly labeled with [^{14}C]-thymidine (■) and [^{3}H]-amino acid mix (○). Tubes containing 5 ml of a preformed 0.35 M to 0.75 M metrizamide-buffer gradient were prepared and centrifuged as described in the text. (▲) ρ (gm/ml). From Wray (1976).

CHO cultures were grown in the presence of either [^{3}H]-thymidine, [^{3}H]-amino acid mix, or no label. Chromosomes from each were kept separate and independently isolated and purified. Four separate mixing experiments were then performed as diagrammed in Fig. 5. In two of the experiments [^{3}H]-thymidine-labeled chromosomes were mixed with unlabeled chromosomes. In the other two experiments [^{3}H]-amino acid-labeled chromosomes were mixed with unlabeled chromosomes. In one experiment in each of the above two sets metrizamide was present, in the other experiment in each set it was omitted. The rationale for the series of experiments is as follows: It is

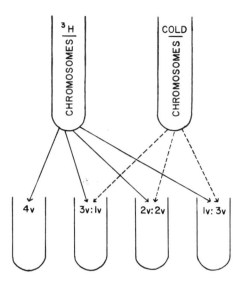

F$_{IG.}$ 5.   Flow chart of mixing experiments indicating the proportions of labeled and unlabeled chromosomes (v/v) present at each time point. For each time point in each experiment
the chromosomes were mixed and centrifuged onto a microscope slide with a cytocentrifuge
(Shandon). The slides were then prepared for autoradiography and allowed to develop for 1
month.

assumed that DNA present in a particular chromosome remains in that
chromosome so long as the chromosome maintains its morphologic integrity.
Therefore, when [³H]-thymidine-labeled chromosomes are mixed with
unlabeled chromosomes only the chromosomes which were originally
labeled will ever be labeled and thus show up as such by autoradiography.
When these labeled and unlabeled chromosomes are mixed on a volume/
volume basis the percent of labeled chromosomes present in the final mixture
should be proportional to the original mixing volumes (i.e., a linear fraction).
In these experiments the knowledge of exact numbers of labeled and unlabeled chromosomes per unit volume is not necessary. The number of labeled
and unlabeled chromosomes in a unit volume will determine the slope of the
line, but the only important point is that the graph of the line must be linear.
It is shown in Fig. 6 that, when [³H]-thymidine labeled chromosomes are
mixed with unlabeled chromosomes both in the presence and in the absence
of metrizamide, there is a linear function when the percentage of labeled
chromosomes on a slide is plotted against the mixing proportions (v/v)
labeled:unlabeled. The only way that a linear relationship may be obtained
is if there is no exchange of label (i.e., chromosomal components) among
chromosomes. Therefore when [³H]-amino acid chromosomes were mixed

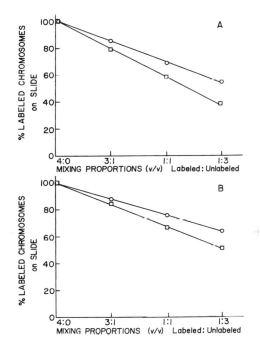

FIG. 6.   Experiments demonstrating the fidelity of DNA and proteins in chromosomes Four separate mixing experiments were conducted as described in Fig. 5 and the text. Panel A: In two of the experiments, [³H]-thymidine labeled chromosomes were mixed with unlabeled chromosomes. Panel B: In the other two experiments, ³H-amino acid labeled chromosomes were mixed with unlabeled chromosomes. In one experiment in each of the above two sets metrizamide was present (□), in the other experiment in each set it was omitted (○). At each of the mixing proportions in each experiment 1000 chromosomes were analyzed to obtain the percentage of labeled chromosomes on the slide. Clumps of chromosomes and uninterpretable grain patterns were omitted from analysis. In these experiments the chromosomes were either very heavily labeled or unlabeled. There were no chromosomes which had only a few grains associated with them. It is estimated that if 1.0% of the DNA or proteins transferred from one chromosome to another, they would be easily seen by this technique. The knowledge of exact numbers of labeled and unlabeled chromosomes per unit volume will determine the slope of the line, but is not necessary for interpretation of the results. The only way that a linear relationship may be obtained is if there is no exchange of label (i.e., chromosomal components) among chromosomes.

with unlabeled chromosomes both in the presence and absence of metrizamide, and a linear function is obtained as shown in Fig. 6, it demonstrates conclusively that the proteins which are associated with a particular chromosome remain associated with that chromosome.

WAYNE WRAY

D.   Identification of Chromosomal Proteins

Chromosomal proteins from HeLa were analyzed by SDS slab gel electrophoresis (Figs. 7 and 8) as described in the general methods (Section II). As compared to the control chromosomes sample (slot b), chromosomes after exposure to metrizamide (slot c) had identical polypeptide populations.

IV.   Chromosome Separation

A.   Banded Chromosomes of Different Densities

There are a number of ways by which the density of chromosomes may be specifically altered. This type of experiment could play a very important role in the fractionation of a genome; so that the following experiment was

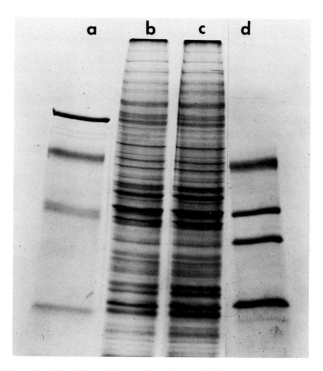

FIG. 7.   Polyacrylamide slab gel electropherogram of chromosomal proteins from HeLa. Slot a. Molecular weight standards; phosphorylase A, 94,000 MW; bovine serum albumin, 68,000 MW; ovalbumin, 45,000 MW; α-chymotrypsinogen, 25,700 MW. Slot b. Control isolated chromosomes. Slot c. Chromosomes exposed to metrizamide. Slot d. Molecular weight standards: bovine serum albumin, 68,000 MW; fumarase, 49,000 MW; liver alcohol dehydrogenase, 41,600 MW; and carbonic anhydrase, 29,000 MW.

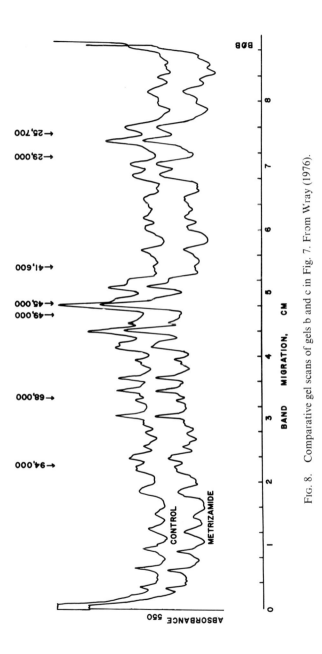

FIG. 8.   Comparative gel scans of gels b and c in Fig. 7. From Wray (1976).

done to demonstrate that chromosome populations with altered buoyant densities may be separated on metrizamide gradients. CHO chromosomes labeled with [^{14}C]-thymidine were treated with 0.25 N HCl for 1 hour at 4 °C to remove histones. They were then washed once and mixed with [^{3}H]-thymidine labeled CHO chromosomes in buffer containing 0.75 M metrizamide, 5 mM NDA, and isopycnically buoyed as described earlier. Since proteins band isopycnically in metrizamide at a greater density than DNA, removal of protein would yield a less dense chromosome. The results shown in Fig. 9 indicate that the chromosomes which had some histone removed have a lower density than the control and that it is possible to separate the two populations by their buoyant density in metrizamide. It should be noted that the densities of both chromosome populations are higher than 1.24 gm/ml as shown in Fig. 4. This difference is apparently due to the presence of NDA in the gradient.

V.   Discussion

The introduction of iodinated density gradient media for isopycnic centrifugation of metaphase chromosomes represents a significant advance

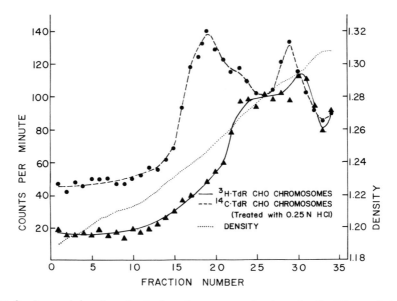

FIG. 9.   Isopycnic banding of metaphase chromosomes showing a density difference induced by removal of histone proteins from the chromosomes. Centrifuge conditions are described in the text. From Wray (1976).

over previous technology. We have shown that metrizamide is a useful and dependable medium for banding chromosomes according to their buoyant density. The morphology of the metaphase chromosome after exposure to metrizamide is excellent and the total protein complement and the DNA: protein ratio is not altered.

ACKNOWLEDGMENTS

This work was supported by Grants NSF BMS 75–05622, American Cancer Society VC-163, and CA 18455 from the National Cancer Institute. The technical assistance of Mr. M. Tomlinson and Ms. M. G. Fields is gratefully acknowledged, and the use of the electron microscope facilities of the Department of Pharmacology, Baylor College of Medicine, provided by Dr. Y. Daskal is sincerely appreciated. Technical assistance from Dr. V. Wray is gratefully acknowledged.

REFERENCES

Anderson, T. F. (1951). *Trans. N.Y. Acad. Sci.* [2] **13**, 130.
Birnie, G. D., Rickwood, D., and Hell, A. (1973). *Biochim. Biophys. Acta* **331**, 283.
Burton, K. (1956). *Biochem. J.* **62**, 315.
Groves, W. E., Davis, F. C., and Sells, B. H. (1968). *Anal. Biochem.* **22**, 195.
Hancock, R. (1970). *J. Mol. Biol.* **48**, 357.
Hossainy, E., Zweidler, A., and Bloch, D. P. (1973). *J. Mol. Biol* **74**, 283.
Laemmli, U. K. (1970). *Nature (London)* **227**, 680.
Lowry, O. H., Rosebrough, H. J., Farr, A. L., and Randall, R. (1951). *J. Biol. Chem.* **193**, 265.
MacGillivray, A. J., Cameron, A., Krauze, R. J., Rickwood, D., and Paul, J. (1972). *Biochim. Biophys. Acta* **277**, 384.
Rickwood, D., Hell, A., and Birnie, G. D. (1973). *FEBS Lett.* **33**, 221.
Stubblefield, E. (1968). *In* "Methods in Cell Physiology" (D. M. Prescott, ed.), Vol. 3, pp. 25–43. Academic Press, New York.
Stubblefield, E., and Wray, W. (1973). *Cold Spring Harbor Symp. Quant. Biol.* **38**, 835.
Wray, W. (1973). *In* "Methods in Cell Biology" (D. M. Prescott, ed.), Vol. 6, pp. 283–306. Academic Press, New York.
Wray, W. (1976). *FEBS Lett.* **62**, 202.
Wray, W., and Stubblefield, E. (1970). *Exp. Cell Res.* **59**, 469.

Chapter 8

Isolation of Interphase Chromatin Structures from Cultured Cells

R. HANCOCK, A. J. FABER, AND S. FAKAN

Swiss Institute for Experimental Cancer Research,
Lausanne, Switzerland

I. Introduction

A large number of procedures are currently in use for isolation of the chromatin of eukaryotic cells. In general, these procedures are based on either of two principles. As a result of the partial or complete aggregation of chromatin in solutions of ionic strength in the range of 30–160 mM NaCl (Itzhaki, 1967; Honda *et al.*, 1975), or in the presence of low concentrations of divalent cations, homogenization of whole cells or tissues in these ionic conditions liberates the chromatin in the form of an insoluble complex, which may be purified by repeated direct sedimentation (Zubay and Doty, 1959) or by sedimentation in sucrose gradients (Bonner *et al.*, 1968). On the other hand, as a result of the enclosure of the chromatin within the nuclear envelope, an extensive purification may be achieved by first isolating purified nuclei. These contain, in addition to the chromatin, soluble proteins and RNPs that may be removed by extraction in appropriate buffers (for example, Commerford et al., 1963; Dingmann and Sporn, 1964; Paul and Gilmour, 1968; Clark and Felsenfeld, 1971; Oudet et al., 1975; Bakayev et al., 1975). Finally the nuclei are lysed, usually by mechanical disruption, in a medium of low ionic strength in which the fragmented chromatin is soluble. Nuclear envelope fragments may be removed by centrifugation, although because of their intimate contact with the chromatin their complete removal may be difficult (Jackson et al., 1968; Tata et al., 1972). Alternatively, the chromatin may be fragmented within purified nuclei by nicking of the chromosomal DNA using an endogenous DNase (Hewish and Burgoyne, 1973) or micro-coccal DNase (Noll, 1974), and the chromatin fragments recovered from the nuclei.

It is probable that no single existing procedure for chromatin isolation yields a product suitable for every type of investigation. For example, single chromatin subunits (nu bodies: Olins and Olins, 1974; nucleosomes: Oudet et al., 1975; or monosomes: Bakayev et al., 1975) are probably not appropriate for studies of processes such as transcription or DNA replication which occur over spatially extensive regions. On the other hand, preparation of chromatin under conditions allowing interchain interactions may modify the linear molecular topology existing in vivo, by the formation of interstrand cross-links, foldbacks, or zigzags in the originally linear molecule. Mechanical fragmentation of chromatin cross-linked in this way may result in the formation of fragments containing regions of DNA that were not originally contiguous in vivo. Interstrand cross-links are predominantly due to ionic interaction between chromatin subunits when their net charge is sufficiently reduced by the presence of counterions to allow close contact. The importance of this factor was recognized early by Zubay and Doty (1959), who observed, by viscosity and light-scattering measurements, intermolecular association of chromatin fragments at concentrations of potassium phosphate above 1.1 mM. Reversible changes indicating aggregation of chromatin were seen by electron microscopy and circular dichroism spectra at ammonium sulfate concentrations of 5 mM and greater (Slayter et al., 1972), and aggregation of purified chromatin subunits was observed at concentrations of NaCl above 30 mM (Honda et al., 1975). Cross-links of this type may be impossible to dissociate completely again, even under conditions of extreme mechanical fragmentation, and only at the expense of interchain transfer of histone 1 (Hancock, 1974). Interchain cross-links may also be caused by divalent cations; Mg^{2+} and Ca^{2+} ions at concentrations above 1 mM cause aggregation of purified chromatin subunits (Honda et al., 1975).

Fragmented chromatin is heterogeneous in its response to monovalent cations (Marushige and Bonner, 1971) and to divalent cations (Gottesfeld *et al.,* 1974), the template-inactive fraction being selectively precipitated. S–S dimers between H3 molecules may represent a third possible source of cross-links. Garrard *et al.* (1975) found extensive formation of H3 dimers during homogenization of chromatin under conditions favoring aerial oxidation, but it is not yet clear what proportion of these dimers represent intersubunit rather than intrasubunit cross-links. As a consequence of the formation of these different types of cross-links during its purification, chromatin must often be exposed to vigorous mechanical fragmentation (shearing) in order to obtain fragments of a size suitable for further manipulations. It should be observed, however, that the *reversible* formation of cross-links may play important roles *in vivo* in the packing of chromatin and chromosomes.

To obtain chromatin fragments in which the linear sequence of regions existing *in vivo* is conserved, for studies of chromatin replication and of the organization of transcribed and nontranscribed regions, we explored procedures for isolation of chromatin from cultured cells that would eliminate as far as possible the formation of interchain links, using conditions of very low ionic strength, absence of divalent cations, and minimal exposure to aerial oxidation. In the procedure described here, the concentration of monovalent ions is maintained below 1 mM, and a nonionic detergent is used to lyse cells under conditions where both the cytoplasmic and nuclear membrane components, as well as the cytoplasmic and nonchromatin nuclear contents, are solubilized. The nuclear pore complex is detached from the peripheral chromatin, and the chromatin remains as a sedimentable macromolecular structure which may be further purified.

II.   Isolation and Purification of Chromatin Structures

A.   Preparation of Cells

The experimental procedures and the properties of the chromatin described here have been studied predominantly using mouse cells of line P815 (Schindler *et al.,* 1959); preliminary experiments have been carried out with some other cell lines (Section V). P815 cells do not attach to surfaces and are grown either as stationary suspension cultures, or preferably in spinner cultures for greater homogeneity and to allow precise pulse-labeling conditions. In our experience it is important for optimal results that the cells be in good exponential growth at the time of harvesting (Section II,C). In the case of other cell lines growing as monolayers, cells are detached using trypsin [Merck; 0.2 % in PBS (Dulbecco and Vogt, 1954)] without Ca^{2+} or

Mg^{2+}, with 1 % calf serum) at room temperature. As soon as visual examination shows detachment of the cell monolayer, a 5-fold excess (by weight) of trypsin inhibitor (Calbiochem; from egg-white) is added, followed by ice-cold growth medium (modified Eagle's medium; Dulbecco and Freeman, 1959) containing 10% calf serum. Proteolytic activity of trypsin is completely blocked under these conditions, as evidenced by the identity of both the absolute and relative content of histones in chromatin prepared by this procedure with those of chromatin prepared from cells harvested by mechanical detachment followed by the Zubay and Doty (1959) procedure. Cells are harvested by centrifugation (1500 g for 7 minutes at 4 °C) and washed once in growth medium without serum. Two standard protocols which we use routinely for purification of chromatin from either relatively small numbers of cells, or on a larger scale, are described in Table I.

B. Lysis of Cells

1. SOLUTIONS

Solution A1: 0.1 *M* sucrose (Mann; RNase free) in 0.2 m*M* phosphate buffer (prepared by titrating NaH_2PO_4 with KOH to pH 7.5, to give a final phosphate concentration of 0.2 m*M*)

Solution A2: Solution A1 adjusted to pH 8.5 and used for purification of the chromatin (Section II,C).

Solution B: 0.5% Nonidet P40 (NP40; Shell Chemical Co.) in 0.2 m*M* EDTA, pH 7.5. The NP40 is added dropwise to the EDTA solution with vigorous stirring.

Because of the low buffering capacity of these solutions, their pH should be verified before use. For certain experiments, for example, for the characterization of template-bound nascent RNA (Section IV,D), or of *in vitro*

TABLE I

STANDARD PROTOCOLS FOR CHROMATIN ISOLATION

Number of cells used	5×10^6–5×10^7	5×10^8
Centrifuge tubes used for lysis and subsequent centrifugations[a]	12 ml (15 mm × 100 mm)	40 ml (25 mm × 105 mm)
Volume of solution A1 for resuspension of washed cells	1 ml	10 ml
Volume of solution B added to lyse cells and to resuspend chromatin pellet	1 ml	10 ml
Volume of solution A2 on which lysate is layered	2 ml lysate on 10 ml	5 ml lysate on 35 ml (4 tubes)

[a]Centrifugations at 3500 rpm for 15–20 minutes at 4°C, swingout rotor, maximum radius 22 cm.

transcription (Section IV,E), it is preferable to sterilize these solutions by membrane filtration.

2. PROCEDURE

The pellet of cells is washed again by resuspension in solution A1 by gentle intermittent mixing on a rotary mixer (Vortex, or similar type) until the suspension is completely homogeneous. (The presence of nondispersed cells at this stage may result in incomplete lysis in the following step.) This medium is hypotonic, and microscopic examination (phase contrast) of the cell suspension at this stage shows some swollen cells with dispersed cytoplasm and decondensed chromatin. The cells are centrifuged (1500 g for 7 minutes at 4 °C). The cell pellet is resuspended in the appropriate volume (Table I) of solution A1, and again thoroughly dispersed by gentle rotary mixing. An equal volume of detergent solution (solution B) is then added dropwise, at a rate of approximately 1 ml/min, with intermittent gentle rotary mixing, just sufficient to disperse the added drops of detergent solution. The turbidity of the suspension decreases, but the suspension remains opalescent. The events occurring during lysis are described in Section III.

C. Purification of Chromatin

To isolate the chromatin structures from the cell lysate, this is layered over solution A2 and centrifuged, using the conditions described in Table I. The choice of a pH of 8.5 for solution A2 is based on the observation that the removal of nonchromatin nonhistone proteins is slightly improved at pHs above neutrality; a similar observation has been made in other procedures of chromatin purification (Dingmann and Sporn, 1964). The chromatin forms a translucent pellet which should appear clearly separated from the supernatant solution; the other material in the lysate remains at the surface. To remove this material as cleanly as possible (it tends to adhere to, and move down, the walls of the tube), we remove the supernatant in two steps. First, the surface layer and about half of the clear supernatant are removed by suction, using a pipette and bulb. The inside walls of the tube are then carefully wiped with absorbent paper, and the remainder of the supernatant is removed by suction. Care is necessary on approaching the surface of the chromatin pellet, and it is necessary to leave a few millimeters of supernatant to avoid disturbing the pellet.

The chromatin pellet at this stage contains some nonhistone proteins which, by the operational criterion of nonattachment to DNA after formaldehyde treatment, do not represent a component of the chromatin (Hancock, 1974). Some RNA may also be present which does not represent template-bound nascent RNA (Section IV,D). Depending on the nature of the experiment, it may be necessary to repeat the purification step to remove these components. Since fragments of chromatin are soluble in the ionic conditions used for purification, it is important to reduce the chances of fragmentation in further manipulations. The volume of detergent solution B indicated in Table I is added to the pellet, which is detached from the tube by gentle hand mixing. The pellet is then taken, as intact as possible, into a polystyrene pipette with a wide bore (approximately 2 mm; Falcon Plastics), layered again over solution A2, and again centrifuged. This step may be repeated once more if necessary; if the chromatin is to be finally fragmented by homogenization, the last centrifugation may conveniently be performed in a Dounce homogenizer. During these further purification steps, the chromatin structures are partially broken, but remain as large sedimentable fragments.

The final chromatin pellet can be fragmented and solubilized in an appropriate medium. We routinely perform this step by 10–20 gentle strokes in a Dounce homogenizer [Kontes Glass Co., New Jersey; 15-ml volume, tight (B) piston] in 0.2 mM EDTA, pH 7.2; the degree of homogenization determines the length of the resulting linear fragments (Section IV,B). After homogenization, the preparation may be centrifuged (Spinco rotor SW 65, 15 krpm, 15 minutes) to remove a very small pellet containing remnants of nucleoli and pore complex, but a negligible amount of chromatin. The fragmented chromatin is finally dialyzed against 0.2 mM EDTA, pH 7.2.

We have very rarely experienced failure to recover discrete pellets of chromatin on repeated centrifugation. This problem could be correlated with the presence of unhealthy cells in the starting population (due, for example, to harvesting at too high a cell density), and we have ascribed it (without evidence) to the liberation by these degenerating cells of nucleolytic and/or proteolytic enzymes that degrade the chromatin and render it nonsedimentable.

The rationale for including EDTA in solution B is to reduce the possibility of interchain interactions due to divalent cations. However, the procedures of lysis and purification of the chromatin may be carried out equally successfully in the absence of EDTA (in solution B) if desired. These conditions may be preferable for some experiments—for example, where enzymic properties of the chromatin are to be studied.

III. Events during Lysis

If lysis is carried out under the phase-contrast microscope, it is observed that upon contact with the detergent solution the cells swell and their cytoplasm disperses instantaneously. The nuclei also swell and lose virtually all optical contrast, but are still visible as faint spherical structures. These structures may be visualized in the electron microscope after separating them from the lysate by centrifugation through solution A2 (Section II,C).

The sedimented chromatin structures have a diameter greater than that of nuclei (Figs. 1 and 2). They consist of a network of chromatin fibrils, with apparently no other detectable electron-dense material. Regions containing a network of fibrils relatively homogeneous in size and distribution, but sometimes arranged more densely, are observed; these regions do not in general appear to be distributed in the same manner as condensed and dispersed chromatin regions in the cell nucleus. Material that could represent remnants of nuclear envelope components can be detected only extremely rarely at the periphery of chromatin structures (Fig. 3), and less than 10% of the nuclear content of choline-containing phospholipids (the predominant category of lipids in the nuclear envelope; Kleinig, 1970) remain associated with them (Table II). Further, close examination of electron micrographs does not show the presence of the nuclear lamina-pore network (Barton et al., 1971; Aaronson and Blöbel, 1975) or of the dense lamina (Fawcett, 1966; Kalifat et al., 1967; Stelly et al., 1970), which lie between the envelope and the peripheral chromatin. Detached material representing the pore network, containing pores and connecting fibrils, is recovered when the cell lysate is centrifuged directly onto the electron microscope grid (Fig. 4). It seems probable that the decondensation and expansion of the chromatin at lysis detaches this layer from the peripheral chromatin after solubilization of the nuclear envelope. In contrast, when the nuclear envelope is removed by detergents in the presence of Mg^{2+} ions, the pore network remains attached to the chromatin (Aaronson and Blöbel, 1975; Riley et al., 1975).

The expansion in volume of the chromatin upon lysis is probably due to electrostatic repulsion between fibers as the counterion concentration is lowered. An analogous decondensation of intranuclear chromatin occurs in solutions of low monovalent cation concentrations (Olins and Olins, 1972). The precise nature of the peripheral region of these structures is not yet clear. Although close examination of electron micrographs (Fig. 3) shows apparently only closely packed chromatin fibrils in, and sometimes extending outside, this region, it cannot be excluded that protein components of the nuclear envelope, remaining after extraction of the envelope lipids, are attached to the peripheral layer of chromatin. However, in view of other evidence that a more compactly structured layer of chromatin exists at the periphery of intranuclear chromatin (Everid et al., 1970), we would favor the view (Franke and Schinko, 1969) that form determination is an integral property of the peripheral chromatin. We are at present studying the nature of the dense peripheral layer of chromatin structures, using techniques of ultrastructural cytochemistry.

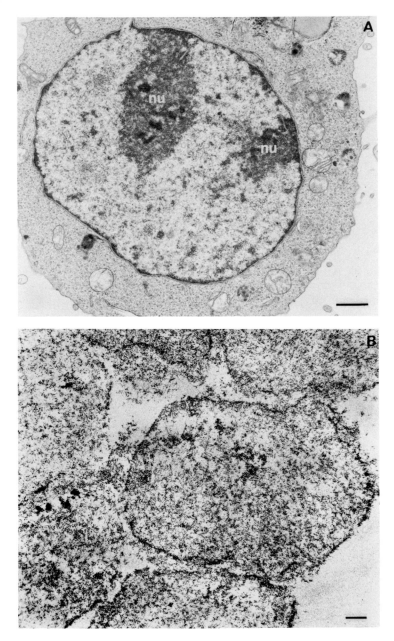

Fig. 1. Thin sections through the nuclear region (A) and through chromatin structures (B) of P815 cells. Chromatin structures or cells were fixed in glutaraldehyde, washed several times with phosphate buffer, postfixed with osmium tetroxide, dehydrated in acetone, and embedded in Epon. Ultrathin sections were stained with uranyl acetate and lead citrate. n, nucleolus. Reference bar, 1.0 μm. (A) × 8300; (B) × 5400.

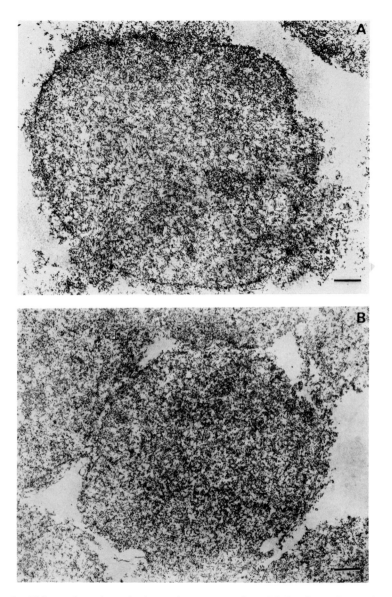

FIG. 2.   Thin sections through chromatin structures from P815 cells. Reference bar, 1.0 μm. (A) × 7000; (B) × 6800.

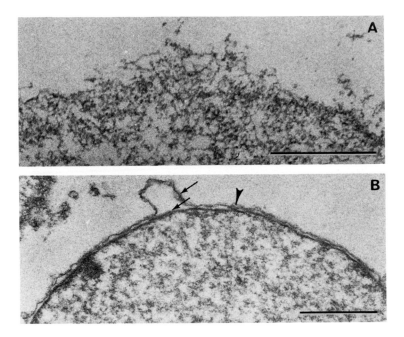

Fig. 3.   Thin sections through the peripheral region of a chromatin structure (A) and an isolated nucleus (B) from P815 cells. The two layers of the nuclear envelope (arrows), and the nuclear pore complex (arrowhead) are not seen in chromatin structures (nuclear preparation by Dr. J. Humbert). Reference bar, 1.0 μm. (A) × 28,000; (B) × 19,500.

IV.   Properties of Chromatin

A.   Composition

The relative contents of histone, nonhistone protein, RNA, and DNA in soluble chromatin prepared from chromatin structures purified by 3 cycles of washing correspond well with those for chromatin prepared from purified nuclei by a conventional technique (Table II). After only a single cycle of washing, the chromatin structures contain additional nonhistone proteins that do not attach to DNA on fixation by formaldehyde (Hancock, 1974). The RNA/DNA ratio is also higher at this first step. The individual constituents of the chromatin are considered in more detail below.

As a marker for nuclear envelope phospholipids, we have used radioactive choline, because choline-containing lipids represent the major class in the nuclear envelope (Kleinig, 1970), and choline is a specific precursor of

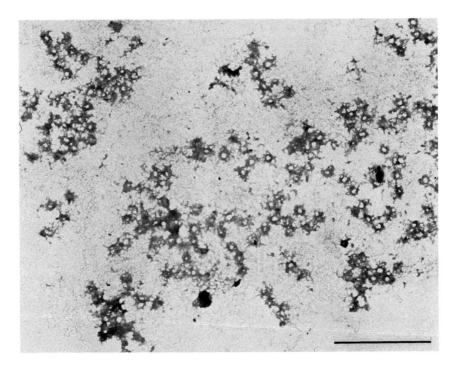

FIG. 4. Fragments of the nuclear pore complex detached from the peripheral chromatin after dispersion of the nuclear envelope. An aliquot of a cell lysate was centrifuged directly onto the electron microscope grid by the technique of Miller and Bakken (1972). Reference bar, 1.0 μm × 25,700.

phospholipids in P815 cells (R. Hancock, unpublished) and in others (Plagemann, 1968). By this criterion, the chromatin structures as first sedimented from a cell lysate have, by comparison with purified nuclei, already lost over 90% of the nuclear envelope lipids, in concordance with the absence of identifiable envelope components in electron micrographs (Figs. 1–3).

B. Structure

Chromatin prepared by shearing chromatin structures shows the characteristic repeating subunit structure described by Olins and Olins (1972) and by Oudet et al. (1975) (Fig. 5). The size of the fragments of chromatin depends on the conditions of shearing. After 8 strokes in the Dounce homogenizer (Section II,C), 85% of the chromatin DNA has molecular weights in the

TABLE II

COMPOSITION OF CHROMATIN FROM P815 CELLS

	Method of chromatin preparation	
	Lysis using NP40	From purified nuclei[a]
Mass ratio		
Total protein/DNA	$1.55 \pm 0.21\,(6)$[b]	1.50
Histone/DNA	$1.05 \pm 0.10\,(6)$	1.12
RNA/DNA	$0.075 \pm 0.02\,(6)$	0.06
Buoyant density in CsCl (gm/ml) after fixing with formaldahyde	$1.385 \pm 0.003\,(6)$	1.384
Recovery of cellular DNA in chromatin (%)	$81 \pm 7\,(6)$	87
Ratio of choline-containing phospholipids/DNA as % of ratio in purified nuclei[c]	9.5 (2)	(100)

[a] Chromatin prepared from nuclei after extraction with 80 mM NaCl–20 mM EDTA (Hancock, 1969).

[b] Values in parentheses represent the number of determinations. The experimental conditions and results are described in Hancock (1974).

[c] Chromatin structures sedimented once only.

range 3.7×10^5 to 5.5×10^6 (peak at 1.9×10^6), measured by sedimentation in isokinetic alkaline sucrose gradients (Noll, 1967) by reference to a 16 S marker of polyoma virus DNA form II. After more extensive shearing (16 Dounce strokes) the corresponding molecular weight range is 2.8×10^5 to 4.5×10^6 (peak at 1.6×10^6).

Chromatin structures sediment from the cell lysate in 15–20% sucrose gradients containing 1 M NaCl (Spinco rotor SW 65, 5000 rpm, 30 minutes) as a discrete band. The sedimentation velocity is changed in the presence of ethidium bromide in a manner compatible with the titration of superhelical DNA; chromatin structures may thus contain topologically-closed DNA regions which retain their superhelicity (Germond *et al.,* 1975) even after dissociation of histones.

FIG. 5.   Electron micrographs of chromatin prepared by gentle shearing of chromatin structures. The chromatin samples were prepared for electron microscopy as described by Oudet *et al.* (1975). In freshly prepared (undialyzed) chromatin (A) the nucleosomes appear more closely spaced than after a period of dialysis against 0.2 mM EDTA (B). Micrographs by kind courtesy of Drs. P. Oudet and P. Chambon. Reference bar 0.37 μm. × 67,500.

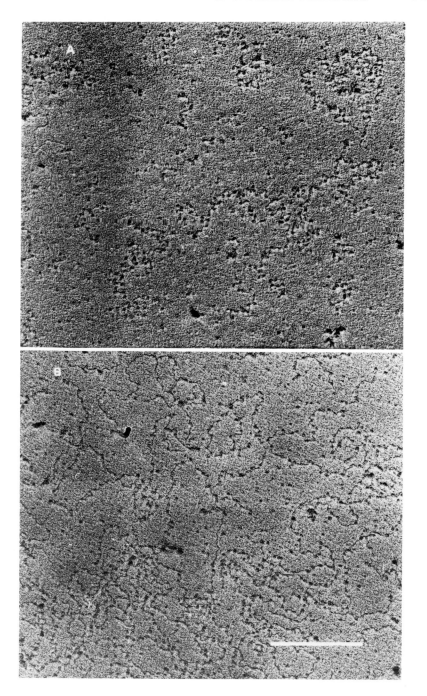

C.   Chromatin Nonhistone Proteins

The spectrum of nonhistone proteins, separated by electrophoresis on sodium dodecyl sulfate-polyacrylamide gels, is closely similar to that of chromatin prepared from purified nuclei (Fig. 6); some minor differences are detectable. The available evidence suggests that proteins associated with free or nascent RNA do not contribute significantly to the nonhistone protein population. Chromatin was prepared containing nonhistone proteins labeled during several generations with [^{3}H]-tryptophan, and treated with RNase (Worthington; 3 × cryst., 75 μg/ml) until no acid-precipitable RNA remained (as monitored in a parallel chromatin preparation from cells prelabeled with [^{3}H]-uridine). It was then layered over a 60% sucrose cushion and centrifuged (Spinco SW 65 rotor, 45 krpm, 27 hours, 4°C) so that any free nonhistone proteins liberated from RNase-digested RNPs should remain on top of the cushion. Only 3% of the total nonhistone protein was found in this position.

The origin of the histones and nonhistones proteins of chromatin prepared by this procedure has been studied. Mixing experiments in which two populations of cells, one containing density-labeled DNA and the other radioactively labeled proteins, were lysed together, showed (within the limits of interpretation of reconstruction experiments of this type) that no detectable fraction of either the histones or of the nonhistone proteins in the purified chromatin are acquired from exogenous sources (i.e., either by transfer from other chromatin molecules, or from other proteins present in the cell lysate)

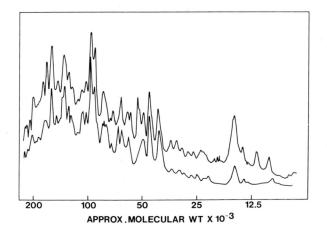

FIG. 6.   Separation of chromatin nonhistone proteins on 15% sodium dodecyl sulfate–polyacrylamide gels (Laemmli and Favre, 1973) (migration from left to right). Soluble chromatin was prepared from 3 times-washed chromatin structures (upper trace) or from purified nuclei (lower trace). Experimental details are given in Hancock (1974).

(Fig. 7). These experiments also demonstrate that no intermolecular cross-links are formed during chromatin purification. The very low viscosity of this chromatin at high concentration (R. Hancock, unpublished), compared with that of chromatin of even shorter DNA fragment length prepared by the Zubay and Doty (1959) procedure, is compatible with this conclusion.

The ionic conditions used during purification of chromatin may influence the retention of proteins whose binding to DNA or chromatin is ionic strength-dependent. An example of this phenomenon is the enzyme DNA untwistase (Champoux and Dulbecco, 1972), which is believed to operate during DNA replication. This enzyme is retained in chromatin purified by the present procedure, but may be liberated on subsequent exposure of the chromatin to 150 mM or 300 mM phosphate (Germond et al., 1975, and personal communication).

D.   Chromatin RNA

After pulse-labeling growing cells with [^{3}H]-uridine for 3 minutes the major part of the newly synthesized RNA is recovered in the chromatin. Our present evidence, although not rigorous, is compatible with the model that the majority of the chromatin RNA represents growing RNA molecules still attached through RNA polymerase to the chromatin template.

When chromatin from cells pulse-labeled for 3 minutes with [^{3}H]-uridine was analyzed by equilibrium density centrifugation in a Cs_2SO_4 gradient, 60% of the labeled RNA banded in a broad peak associated with the DNA at a density indicating that a significant amount of protein is also present (Fig. 8A). The remaining 40% of the labeled RNA may be derived from RNP particles. These particles (prepared from purified nuclei by the method of Samarina et al., 1968) dissociate completely into RNA and protein in Cs_2SO_4 gradients. When this chromatin was treated with proteinase K (a broad-spectrum protease in which neither RNase nor DNase activities can be detected: Wiegers and Hilz, 1971; Gross-Bellard et al., 1973), more than 97% of the [^{3}H]-RNA originally associated with the DNA was detached and sedimented as free RNA, and the DNA banded at its normal density (Fig. 8B).

When chromatin structures are sedimented from the cell lysate onto electron microscope grids by the procedures of Miller and Bakken (1972), transcription of both presumptive ribosomal and non-ribosomal RNAs may be visualized (Fig. 9) (Villard and Fakan, 1976).

E.   RNA Synthesis

Both once-purified chromatin structures and chromatin fragmented after 3 cycles of washing can support RNA synthesis using exogenous E. coli RNA polymerase (Table III). The level of RNA synthesis without added polymer-

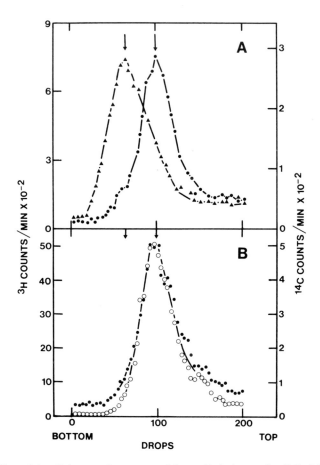

Fig. 7. The origin of the proteins recovered in purified chromatin. Cells labeled during several generations with [³H]-arginine (incorporated into both histones and nonhistone proteins) were mixed with a 10-fold excess of unlabeled cells whose DNA (and thus chromatin) had been density-labeled with 5-iododeoxyuridine during one replication time. If any exchange or non-specific binding of the proteins of chromatin occurs during its purification, the density-labeled chromatin would acquire ³H-labeled proteins. However, when the chromatin is fixed with formaldehyde and banded to buoyant density equilibrium in a CsCl gradient, together with a chromatin marker of normal density (panel B), no ³H radioactivity can be detected in the region of the gradient where dense chromatin is located (indicated by the arrow derived from the position of the dense marker in panel A). (A) ▲ and ●, ³H and ¹⁴C in dense and normal chromatin markers, respectively. (B) ○ and ●, ³H and ¹⁴C in experimental sample and in normal chromatin marker, respectively. Reproduced from Hancock (1974).

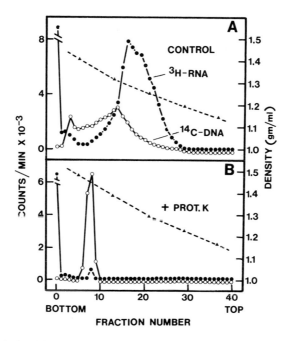

FIG. 8.   Analysis of chromatin containing pulse-labeled RNA by equilibrium density centrifugation in Cs_2SO_4 gradients. Sheared chromatin was prepared from cells prelabeled for 2 generations with [^{14}C]-thymidine and then pulse-labeled for 3 minutes with [^{3}H]-uridine. One part of the chromatin was treated with proteinase K (Merck; electrophoretically pure, 100 μg/ml in 5 mM Tris pH 8.0, 5 mM EDTA, 5 mM NaCl, at 37°C for 1 hour). Samples of control and proteinase-K treated chromatin (containing 50 μg of DNA) were mixed with Cs_2SO_4 (in 10 mM Tris pH 7.4, density 1.310 gm/ml in 5 ml) and centrifuged (Spinco SW 65 rotor, 40 krpm, 4 days, 20°). Fractions were collected from the bottom of the gradients and trichloroacetic acid-insoluble radioactivity was determined. (A) Original chromatin. (B) Chromatin treated with proteinase K. ●, ^{3}H; ○, ^{14}C.

ase is relatively higher in chromatin structures, suggesting that these may retain endogenous RNA polymerase, or other stimulatory factors, which are lost during further purification.

V.   Applications

This procedure does not represent an exception to the reflection made earlier (Section I) that all methods for chromatin isolation have certain limitations. For example, although we have found the procedure suitable for

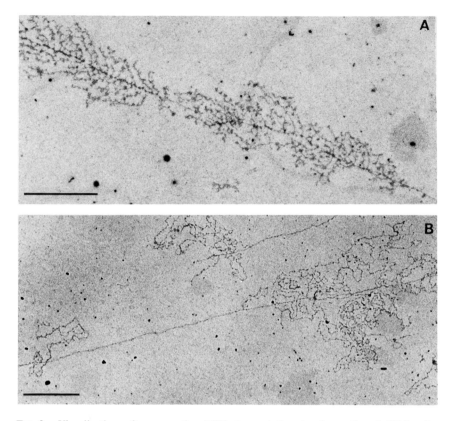

Fig. 9. Visualization of presumptive RNA transcription in chromatin of P815 cells. Chromatin structures were sedimented from a cell lysate onto electron microscope grids, processed, and stained with phosphotungstic acid, according to Miller and Bakken (1972). A and B represent probable regions of ribosomal and nonribosomal transcription, respectively (preparations kindly contributed by Danielle Villard). Reference bars 0.5 μ. × 40,000 (A), × 30,000 (B).

several types of experiment where relatively small numbers of cells are used, we have not yet resolved some problems encountered in scaling up to very large numbers of cells (> 5 × 10^9). Although we have some evidence that it may be applied to tissues that can be converted to a homogeneous cell suspension by mechanical or enzymic dissociation (for example thymus or liver), we have not further explored these applications. We have used this procedure routinely for purification of chromatin from CHO cells, both growing as monolayers and as suspensions of selectively detached mitotic cells. Electron microscopy and analysis of purified chromatin show that the

TABLE III

RNA Synthesis by Chromatin Structures and Soluble Chromatin[a]

Chromatin preparation	[^{3}H]-UTP incorporated ($\mu\mu$moles/10 minutes/0.25 ml)[b]	
	E. coli RNA polymerase (35 μg)	
	−	+
Chromatin structures (first pellet)	300	1126
Sheared chromatin (3 × washed)	30	1300
+ Actinomycin D (25 μg/ml)	—	52
+ RNase (50 μg/ml)	—	82
Sheared chomatin prepared from nuclei extracted with 80 mM NaCl–20 mM EDTA	12	1850

[a] The experimental conditions and results are presented in detail in Hancock (1974).
[b] Using chromatin containing 30 μg DNA. Incorporation is approximately linear during this period.

procedure can also be applied to cells of lines BSC, 3T3, L, and HeLa, but the optimum conditions and the properties of the chromatin have not been characterized in detail for these cell lines.

The procedure was originally developed for kinetic studies of chromatin assembly using density-labeling techniques (Hancock, 1970) and has proved suitable for this type of experiment (R. Hancock, unpublished; Jackson et al., 1975). Its applicability to isolation of chromatin of mitotic cells has facilitated comparisons of the nonhistone proteins (Nobis et al., 1973) and of histone 3 dimerization (Garrard et al., 1975) in interphase and mitotic chromatin. The linear chromatin fragments produced are appropriate for the separation of fragments which bear nascent RNA from the bulk of the chromatin (Turner and Hancock, 1974; A. J. Faber, unpublished). Although we are aware of only one report of an enzyme which is retained in chromatin prepared at the low ionic strength of this procedure, but dissociated at higher ionic strengths (Germond et al., 1975), other similar situations may be discovered.

Acknowledgments

We are very grateful to Drs. Pierre Oudet and Pierre Chambon for collaboration with electron microscopy and, together with Drs. Lee Compton and Werner Franke, for valuable discussions and comments.

This work was supported by the Swiss National Science Foundation.

REFERENCES

Aaronson, R. P., and Blöbel, G. (1975). *Proc. Natl. Acad. Sci. U.S.A.* **72**, 1007–1011.

Bakayev, V. V., Melnickov, A. A., Osicka, V. D., and Varshavsky, A. J. (1975). *Nucleic Acids Res.* **2**, 1401–1419

Barton, A. D., Kisielski, W. E., Wassermann, F., and Mackevicius, F. (1971). *Z. Zellforsch. Mikrosk. Anat.* **115**, 299–306.

Bonner, J., Chalkley, G. R., Dahmus, M. E., Fambrough, D., Fujimara, F., Huang, R. C., Huberman, J., Jensen, R., Marushige, K., Ohlenbusch, H., Olivera, B., and Wildholm, J. (1968). *In* "Methods in Enzymology" (L. Grossman and K. Moldave, eds.), Vol. 12, Part B, pp. 3–65. Academic Press, New York.

Champoux, J. J., and Dulbecco, R. (1972). *Proc. Natl. Acad. Sci. U.S.A.* **69**, 143–146.

Clark, R. J., and Felsenfeld, G. (1971) *Nature (London)* **229**, 101–105.

Commerford, S. L., Hunter, M. J., and Oncley, J. L. (1963). *J. Biol. Chem.* **238**, 2123–2133.

Dingmann, C. W., and Sporn, M. B. (1964). *J. Biol. Chem.* **239**, 3483–3492.

Dulbecco, R., and Freeman, G. (1959). *Virology* **8**, 396–397.

Dulbecco, R., and Vogt, M. (1954). *J. Exp. Med.* **99**, 167–182.

Everid, A. C., Small, J. V., and Davies, H. G. (1970). *J. Cell. Sci.* **7**, 35–48.

Fawcett, D. W. (1966). *Am. J. Anat.* **119**, 129–146.

Franke, W. W., and Schinko, W. (1969). *J. Cell Biol.* **42**, 326–331.

Garrard, W. T., Nobis, P., and Hancock, R. (1975). *Fed. Proc., Fed. Am. Soc. Exp. Biol.* **34**, 611.

Germond, J. E., Hirt, B., Oudet, P., Gross-Bellard, M., and Chambon, P. (1975). *Proc. Natl. Acad. Sci. U.S.A.* **72**, 1843–1847.

Gottesfeld, J., Garrard, W., Bagi, G., and Bonner, J. (1974). *Proc. Natl. Acad. Sci. U.S.A.* **71**, 2193–2197.

Gross-Bellard, M., Oudet, P., and Chambon, P. (1973). *Eur. J. Biochem.* **36**, 32–38.

Hancock, R. (1969). *J. Mol. Biol.* **40**, 457–466.

Hancock, R. (1970). *J. Mol. Biol.* **48**, 357–360.

Hancock, R. (1974). *J. Mol. Biol.* **86**, 649–663.

Hewish, D. R., and Burgoyne, L. A. (1973). *Biochem. Biophys. Res. Commun.* **52**, 504–510.

Honda, B. M., Baillie, D. L., and Candido, E. P. M. (1975). *J. Biol. Chem.* **250**, 4643–4647.

Itzhaki, R. F. (1967). *Biochem. J.* **105**, 741–748.

Jackson, V., Earnhardt, J., and Chalkley, R. (1968). *Biochem. Biophys. Res. Commun.* **33**, 253–259.

Jackson, V., Granner, D. K., and Chalkley, R. (1975). *Proc. Natl. Acad. Sci. U.S.A.* **72**, 4440–4444.

Kalifat, S. R., Bouteille, M., and Delarue, J. (1967). *J. Microsc. (Paris)* **6**, 1019.

Kleinig, H. (1970). *J. Cell Biol.* **46**, 396–402.

Laemmli, U. K., and Favre, M. (1973). *J. Mol. Biol.* **80**, 575–599.

Marushige, K., and Bonner, J. (1971). *Proc. Natl. Acad. Sci. U.S.A.* **68**, 2941–2944.

Miller, O. L., and Bakken, A. H. (1972). *Acta Endocrinol. (Copenhagen), Suppl.* **168**, 155–173.

Nobis, P., Hancock, R., and Dewey, W. C. (1973). *J. Cell Biol.* **59**, 248a.

Noll, H. (1967). *Tech. Protein Biosynth.* **2**, 101–178.

Noll, M. (1974). *Nature (London)* **251**, 249–251.

Olins, D. E., and Olins, A. L. (1972). *J. Cell Biol.* **53**, 715–736.

Olins, D. E., and Olins, A. L. (1974). *Science* **183**, 330–332.

Oudet, P., Gross-Bellard, M., and Chambon, P. (1975). *Cell* **4**, 281–300.

Paul, J., and Gilmour, R. S. (1968). *J. Mol. Biol.* **34**, 305–316.

Plagemann, P. G. W. (1968). *Arch. Biochem. Biophys.* **128**, 70–87.

Riley, D. E., Keller, J. M., and Byers, B. (1975). *Biochemistry* **14**, 3005–3013.

Samarina, O. P., Lukanidin, E. M., Molnar, J., and Georgiev, G. P. (1968). *J. Mol. Biol.* **33**, 251–263.

Schindler, R., Day, M., and Fischer, G. A. (1959). *Cancer Res.* **19**, 47–51.

Slayter, H. S., Shih, T. Y., Adler, A. J., and Fasman, G. D. (1972). *Biochemistry* **11**, 3044–3054.

Stelly, N., Stevens, B. J., and Andrée, J. (1970). *J. Microsc. (Paris)* **9**, 1015–1028.

Tata, J. R., Hamilton, M. J., and Cole, R. D. (1972). *J. Mol. Biol.* **67**, 231–246.

Tsanev, R. G., and Petrov, P. (1976). *J. Microsc. Biol. Cell* **27** (in press).

Turner, G., and Hancock, R. (1974). *Biochem. Biophys. Res. Commun.* **58**, 437–445.

Villard, D., and Fakan, S. (1976). In preparation.

Wiegers, U. and Hilz, H. (1971). *Biochem. Biophys. Res. Commun.* **44**, 513–519.

Yaneva, M., and Dessev, G. (1976). *Eur. J. Biochem.* **66**, 535–542.

Zubay, G., and Doty, P. (1959). *J. Mol. Biol.* **1**, 1–20.

NOTE ADDED IN PROOF

Chromatin may be isolated from rat ascites tumor cells and from rat liver using modifications of this procedure (Yaneva and Dessev, 1976; Tsanev and Petrov, 1976). Such chromatin differs from chromatin prepared by a salt precipitation method in showing a less heterogeneous thermal melting profile and a discrete pattern of nucleosomal DNA fragments upon digestion with micrococcal nuclease.

Chapter 9

Quantitative Determination of Nonhistone and Histone Proteins in Chromatin

JOHANN SONNENBICHLER, F. MACHICAO, AND I. ZETL

Max-Planck-Institut für Biochemie,
Martinsried, West Germany

I.   Introduction

During the last 10 years nuclear and chromosomal nonhistones have assumed great importance in discussions on chromosome structure and gene regulation. Thus, many laboratories are engaged in elucidating the multiplicity and functions of these proteins. In this connection also, the importance of quantitative analytical data concerning the nonhistones has increased (cf. Bonner *et al.*, 1968a). Earlier data on the nonhistone proportions are often called into question. This may mainly be due to problems connected with isolation of the nucleoprotein material. But some of the contradictory results also arise from different techniques used for analysis.

This chapter will give a short survey of the most common, existing analytical methods for quantitative nonhistone determinations, and thereafter a procedure will be introduced which seems to be more advantageous in the analysis of the nonhistone moiety of nucleoprotein materials, especially in the presence of histones.

II.   Existing Methods for Quantitative Determination of the Nonhistone Moiety in Nucleoprotein Material

A.   Selective Solubilization of Histones and Precipitation of DNA and Nonhistones with Dilute Mineral Acids

Since the early 1950s, acid treatment of nucleoprotein material has been used to dissociate and solubilze the histone proteins and to precipitate the nonhistones together with the DNA (Bonner *et al.*, 1968b). Subsequent protein analyses according to Lowry *et al.* (1951) or other methods were used to determine the protein portion of this residual nucleoprotein complex, which was equated with the nonhistone moiety. Until now, most of the quantitative statements still are made, using such a method. We know, however, that on the one hand part of the nonhistone is soluble in dilute mineral acids (Wilson and Spelsberg, 1973) and on the other hand that, because of denaturation and aggregation phenomena, a certain amount of histone tends to bind to the nonhistone moiety and precipitates together with the DNA, thus falsifying nonhistone quantities (Sonnenbichler and Nobis, 1970). This is true especially for reconstituted nucleoprotein material (Fujitani and Holoubek, 1974). Furthermore, the acid-precipitated material often resists complete solubilization, which is necessary for reliable protein analysis.

Therefore in our opinion the use of acid precipitation of nucleoprotein material for the quantitative analysis of nonhistone should be avoided.

Besides the simple treatment of nucleoprotein with dilute mineral acids, there also exist modifications that allow specific recovery of the nonhistone moiety after precipitation fron the DNA. Suria and Liew (1974) compared some of these techniques, which cannot be cited because of lack of space. In general their conclusion was that the yields depend very much on the methods used.

B.   Quantitative Determination of the Nonhistones with Polyacrylamide Gel Electrophoresis

The quantitative determination of the nonhistone moiety by acrylamide gel electrophoreses performed with subsequent staining and optical scanning is also of doubtful accuracy. In this case, the main problem is that part of the nonhistone protein remains on the gel surface or even enters the upper buffer during electrophoresis. Furthermore, a reproducible staining of the protein bands is difficult because of diffusion. Problems arise also with destaining, during which some colored protein bands start to fade with time. These problems also exist when gel slices are extracted with dimethyl sulfoxide according to an old method of Johns (1967) or with other solvents. Especially impracticable for quantitative nonhistone determination is the scanning of SDS gels where histones and nonhistones can hardly be differentiated when they have similar molecular weights.

C.   Quantitative Determination after Preparative Fractionation of Chromatin Proteins

There exists a series of useful preparative separation procedures for nonhistone proteins from histones by chromatography on Bio-Rex 70 (Richter and Sekeris, 1972; Vandenbroek et al., 1973; Levy et al., 1972), on Bio-Gel (Ishitani and Listowsky, 1975), hydroxyapatite (Paul, 1972), QAE-Sephadex (Richter and Sekeris, 1972), SE-Sephadex (Elgin and Bonner, 1972), CM-cellulose (Yoshida and Shimura, 1972), or ECTHAM-Cellulose (Reeck et al., 1974). Because of incomplete elution caused by some irreversible adsorption, however, quantitative assertions based on these preparative procedures are not reliable and should not be taken as a basis for analytical calculations.

III.   Cellogel Electrophoresis for Quantitative Nonhistone Determinations

In 1971 we described a simple procedure that avoids most of the above-mentioned problems and allows a quantitative determination of chromatin nonhistones in the presence of histones. This method is based mainly on cellogel electrophoresis originally developed for histone analysis (Machicao and Sonnenbichler, 1971; Sonnenbichler *et al.*, 1975). In this method a prerequisite for the analyses is the fractionation of the nucleoprotein complex into the nucleic acids and the protein moiety.

A. Methods for Isolation of the Protein Moiety from Nucleoprotein Materials

The most useful method for separation of all the proteins from the nucleic acids seems to be ultracentrifugation of the nucleoprotein material in high salt and urea concentrations.

1. Centrifugation in $2 \ M$ NaCl $+ 5 \ M$ Urea

For large amounts of nucleoprotein material (corresponding to up to 100 mg of DNA) as in the case of calf thymus chromatin, ultracentrifugation in $2 \ M$ NaCl $+ 5 \ M$ urea $+ 0.01 \ M$ NaHSO$_3$ is recommended. After 35 hours at 100,000 g (angle rotor) with DNA concentrations of 200–500 μg/ml, the DNA has pelleted while the proteins are distributed in the whole supernatant. Proteolytic degradation is avoided by the addition of 0.01 M NaHSO$_3$ or 10^{-4} M phenylmethansulfonyl fluoride. In order to separate all the proteins that might still be partially intermingled with the DNA pellet, the sedimented nucleic acids must be redissolved and recentrifuged in the same way (Bornkamm *et al.*, 1972; Chaudhur, 1973).

2. Centrifugation in $2 \ M$ CsCl $+ 5 \ M$ Urea

With smaller amounts of nucleoprotein material, the complete separation of all proteins from the DNA can be done better by a density gradient centrifugation. After 70 hours (DNA concentrations of 150–350 μg/ml) at 100,000 g, the DNA forms a pellet free of protein, and the protein moiety bands in the upper region of the gradient. In this case a recentrifugation of the DNA pellet is not necessary.

With both centrifugation procedures the protein moiety can be isolated quantitatively. A short dialysis (up to 3 hours) reduces the salt concentrations $< 0.6 \ M$, and then the protein can be precipitated with 6 volumes of acetone.

After washing with pure acetone and drying *in vacuo* the proteins are ready for analysis. As shown with histone mixtures, the different proteins can be precipitated completely by acetone even in the presence of high concentrations of urea (Table I).

B. Quantitative Electrophoretic Analysis of the Protein Moiety

Kohn (1957) first applied cellulose acetate sheets for electrophoresis of proteins. In 1969 Kleinsmith and Allfrey applied cellogel electrophoresis

TABLE I

COMPOSITION OF HISTONE MIXTURES BEFORE AND AFTER PRECIPITATION WITH ACETONE[a]

Sample	H 1	H 2A	H 2B	H 3	H 4
Original mixture	15.6	11.9	26.0	32.3	14.0
After precipitation from aqueous solutions	15.5	12.0	25.5	33.7	13.4
After precipitation from solutions containing $5M$ urea	16.9	13.9	25.7	30.1	13.3

[a]Extinction values [%] of corresponding stained bands after cellogel electrophoresis. Average values for six determinations.

at neutral pH to separate phosphoproteins and histones. A few years later, Machicao and Sonnenbichler (1971) developed an electrophoretic separation system on cellogel strips at pH 10 which has proved to be a convenient method to analyze the different histone types (Fig. 1a). This system also proved to be very useful for quantitative differentiation of the nonhistone moiety in the presence of histones (Sonnenbichler et al., 1975). While for the characterization of histones a high resolution of the electrophoretic bands is desired, with the objective of discriminating between the bulk of histone and nonhistone proteins, the electrophoretic analysis is

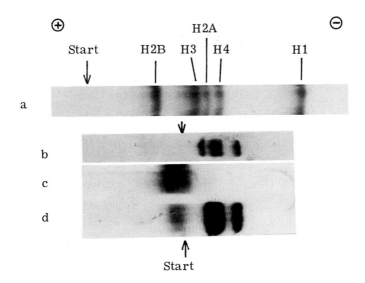

FIG. 1.   Electrophoresis on cellulose acetate sheets at pH 10.0 of (a) histones at high-resolution conditions and (b) histones, (c) isolated nonhistones, (d) total proteins from rat liver chromatin at low-resolution conditions.

better done at low resolution conditions. As can be seen from Fig. 1b, c, and d, a sufficient separation of histones and nonhistones on the cellogel strips can be achieved very easily. Moreover, in this case quantitative evaluations seem to be more reliable than if the bands are more scattered.

1. PERFORMANCE OF THE ELECTROPHORESIS

For analysis the proteins must be dissolved completely in 1% acetic acid, % β-mercaptoethanol + 8 M urea to concentrations of about 1 mg/100 μl. The electrophoreses are performed on a refrigerated Teflon surface in a moist chamber first described by Heil and Zillig (1970). The apparatus is shown in Fig. 2.

The electrophoreses are performed in 0.6 M ammonium borate buffer with 6 M urea, 0.01 M EDTA, and 0.01 M mercaptoethanol, pH 10. The buffer is adjusted with 25% ammonia. The cellogel strips (4 cm × 17 cm from Chemetron, Milan, Italy) are thoroughly equilibrated with buffer and put in the chamber, where they are carefully blotted with filter paper in order to avoid air bubbles between sheet and surface. About 1 nmole of protein in 1–2 μl should be applied. The actual pH on the cellogel during electrophoresis was found to be 9.3, contrary to our preliminary statement derived from the buffer pH, apparently owing to interactions with CO_2 from the atmos-

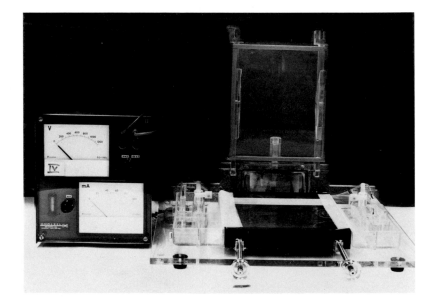

FIG. 2.   Apparatus used for cellogel electrophoretic analysis according to Heil and Zillig (1970), with a refrigerated Teflon surface.

phere. While for histone analysis an electric field of about 60 V/cm is applied for 1 hour, for low resolution electrophoresis the time is shortened to about 15 minutes.

As can be seen from Fig. 1, the histones migrate to the cathode during electrophoresis, relatively fast under low resolution conditions. Most of the nonhistones migrate to the anode or remain near the starting line; some nonhistones apparently with $pK > 9.3$ also migrate to the negative side, but much more slowly than do the histones, and can therefore be easily differentiated. The presence of high molar urea avoids histone–nonhistone aggregations (Naito and Sonnenbichler, 1972), which can additionally be proved by analyses of artificial mixtures.

2. STAINING PROCEDURES AND QUANTITATIVE EVALUATIONS

After electrophoresis the cellogel strips are stained for 10 minutes with 0.5 % (w/v) amido black in a mixture of 45 % methanol, 45 % water, and 10 % glacial acetic acid. Excess dye is removed by successive washing with the same solvent for 15 minutes.

For quantitative evaluation the colored areas are cut out, and the cellogel pieces and corresponding blanks for references are dissolved in glacial acetic acid. The intensities measured at 630 nm parallel the protein content. However, different binding of amido black to the histones and nonhistones must be taken into account. The reproducibility of the staining evaluations is very satisfactory as shown in Table II.

The staining ability of amido black for the nonhistones and the bulk of histones, or for the purified individual histones, respectively, was calculated by staining known amounts of the corresponding proteins spotted on cellogel strips with and without electrophoresis.

The isolation of nonhistones for this purpose was performed in two ways: (a) calf thymus chromatin was stirred intensively with 0.30 M NaCl,

TABLE II

E_{630} DATA OF CORRESPONDING STAINED AREAS[a] AS EXAMPLES FOR THE REPRODUCIBILITY OF THE DESCRIBED METHOD[b]

E_{630} of	1	2	3	4	5	6
Histone areas	0.580	0.395	0.380	0.332	0.265	0.280
Nonhistone areas	0.107	0.076	0.074	0.062	0.047	0.055
% Nonhistone	15.5	16.0	16.0	16.0	15.0	16.5

[a] Dissolved in glacial acetic acid after electrophoreses of the proteins from calf thymus chromatin (washed twice with 0.15 M NaCl).

[b] Usually six electrophoreses are performed with each protein sample. Average value 15.8 % ± 0.51 (without correction of the different staining abilities of histones and nonhistones).

and after dialysis the extracted nonhistones were precipitated with 6 volumes of acetone; (b) after separation of all proteins from calf thymus chromatin (see above) the nonhistone proteins were separated by chromatography on Bio-Rex 70 (Richter and Sekeris, 1972). In both procedures the purity and the absence of histones was ascertained by acrylamide gel electrophoreses according to Panyim and Chalkley (1968). Histones were fractionated and purified according to Oliver *et al.* (1972).

For the histones an average value is used for some calculations because one can assume that, in general, the quantitative composition of these proteins is very similar in different chromatins. As shown in Fig. 3, the staining capacity of the nonhistone moiety is quite different from those of serum albumin and histones.

Since the nonhistones are far more heterogeneous than the histones, different color values for each polypeptide cannot be determined. Possibly, certain species of nonhistone proteins may differ widely in their staining ability. However, the described preparations of the nonhistone moieties, including those from different tissues, always showed identical color values, presumably as a consequence of their great heterogeneity and statistical averaging.

In order to get correct protein quantities, the following calculated multi-

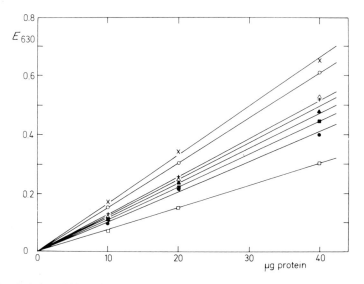

FIG. 3. Staining ability of the different proteins with amido black under the described conditions. ×, H 1; ○, H 2B; △, H 3; +, histone mixture; ▲, H 2A; ■, bovine serum albumin; ●, H 4; □, nonhistones.

plication factors are used: (bovine serum albumin is included for comparison):

Histones $= 1.0$

H 1	$= 0.91$	H 4	$= 1.26$
H 2A	$= 1.12$	Nonhistones	$= 1.69$
H 2B	$= 0.90$		
H 3	$= 1.02$	BSA	$= 1.17$

In contrast to our results on cellulose acetate sheets, Fambrough *et al.* (1968) could not find differences in histone staining in acrylamide gels.

If the staining and destaining procedures are held constant, the described correction factors lead to reliable data for quantitative calculations of the protein composition. Thus the determination of histones and nonhistones in mixtures can be done very easily. Further, the electrophoretic separation can be performed in a rather short time, and different samples may be analyzed in the same electrophoretic run.

IV.   Examples of the Quantitative Determination of Histones and Nonhistones by Cellogel Electrophoresis

A.   Determination of the Protein Composition in Chromatin

Nuclei from different tissues are prepared by homogenization in 0.32 M sucrose + 1 mM MgCl$_2$. After centrifugation and washing in the same solvent, the nuclei are purified by ultracentrifugation in 2.2 M sucrose at 60,000 g for 90–120 minutes. All procedures are carried out at 0°C. Chromatin may be isolated according to Zubay and Doty (1959) with slight modifications, such as washing the crude material only once with 0.024 M EDTA + 0.075 M NaCl, pH 7.0, and then twice in 0.15 M NaCl. The chromatin fibers are swollen in water and brought to 2 M CsCl + 5 M urea + 0.01 M NaHSO$_3$. Intensive stirring for 30 minutes or mixing with a blender is used to dissolve the material *completely*. Subsequently, the proteins are separated from the nucleic acids by ultracentrifugation as described before. After separation the isolated proteins can be applied to cellogel electrophoresis. The histone and nonhistone values for some chromatins, determined in the described way, are listed in Table III.

In spite of the simple procedures that we used for chromatin purification, which are commented upon in the following section, the values for nonhistone portions in Table III, are, in general, somewhat lower than those normally cited (Bonner *et al.*, 1968a). This may be related to the problems of chromatin isolation.

TABLE III

HISTONE AND NONHISTONE PORTIONS OF SOME CHROMATINS[a]

Chromatin	Parts protein/ 100 parts DNA	% Nonhistone	% Histone
Calf thymus	189	24	76
Rat liver	205	46	54
Pig brain	205	34	66

[a]Values after a primary rinsing of the material in 0.075 M NaCl + 0.024 M EDTA and subsequently twice in 0.15 M NaCl determined by electrophoretic analyses of the total chromatin proteins.

B.   Dependence of the Nonhistone Ratios from the Washing Procedure for Chromatin

Often contamination by absorption of nonchromosomal material to chromatin during isolation, or, conversely, a loss of loosely bound chromosomal proteins during purification may falsify the analytical data. We know that under salt conditions higher than 0.2 M NaCl or in the presence of detergents, part of the nonhistones are released (Goodwin and Johns, 1972). With urea or with salt concentrations higher than O.4 M NaCl histones dissociate (especially H l) and may be lost.

In order to determine the influence of washing the crude chromatin with NaCl solutions > 0.15 M NaCl, we analyzed the histone and nonhistone contents of materials treated for 15 minutes with cold 0.30 M and 0.35 M NaCl solutions, salt conditions which are believed to prevail for the nuclear interior (Langendorf *et al.*, 1961). As can be seen from Table IV, with repeated washing steps, more nonhistones are released and their proportion decreases.

Apart from these analytical data, the question remains, as to what part of the nonhistones can be seen as a true constituent of the nucleoprotein complex.

C.   Analysis of the Residual Complex from Acid-Treated Chromatin

In 1970 we published a paper concerning the "residual complex" as a product of acid-treated chromatin (Sonnenbichler and Nobis, 1970). As a consequence of denaturation and aggregation phenomena histones coprecipitate with nonhistones and together with DNA, thus adding to the nonhistone fration. For the following analysis the "residual complex"

TABLE IV

DEPENDENCE OF THE NONHISTONE RATIOS ON THE CHROMATIN WASHING PROCEDURE[a]

Calf thymus chromatin	Successive washing steps in 1 M NaCl	Parts protein/ 100 parts DNA	% Nonhistone	% Histone
a	1 × 0.15	198	29	71
	2 × 0.15	189	24	76
b	1 × 0.15			
	1 × 0.30	184	20	80
c	2 × 0.15			
	1 × 0.30	172	18	82
d	2 × 0.15			
	2 × 0.30	170	16	84
e	1 × 0.15			
	2 × 0.30		16	84
f	2 × 0 15			
	2 × 0.30	166	12	88
	1 × 0.35			

[a] Values after a primary rinsing of the material in 0.075 M NaCl + 0.024 M EDTA.

of calf thymus chromatin was used after treatment with 0.1 M HCl at pH 0.8 for 30 minutes. The acid-insoluble nucleoprotein complex can be separated at pH 10.5 by density gradient centrifugation in 2 M CsCl + 5 M urea (Sonnenbichler and Nobis, 1970). By cellogel electrophoretic analysis of the protein moiety, it can be demonstrated that besides 88 % nonhistones, 12 % histones are present in this acid-insoluble material. This result contradicts the statement of Fujitani and Holoubek (1975), who maintain that only in reconstituted nucleohistones do some histones remain bound to the DNA during acid treatment.

D.   Quantitative Analysis of the Protein Composition of Calf Thymus Chromatin

If high resolution conditions (see Fig. 1a) are used for the electrophoretic analyses, the portions of the different histone types can be determined in addition to the nonhistone moiety (Sonnenbichler and Zetl, 1975). With the proteins of calf thymus isolated as described above, electrophoretic analyses were performed under high resolution conditions. The chromatin of calf thymus used in these experiments had a protein: DNA ratio of $185 \pm 8 : 100$. Table V shows the relative extinction values of the separated protein bands. The relative protein weight ratios were calculated by using the different staining factors of the proteins.

TABLE V

EXTINCTION VALUES FROM HIGH RESOLUTION ELECTROPHORESES,
AND CALCULATION OF PROTEIN WEIGHT RATIOS OF CALF THYMUS
CHROMATIN WITH A PROTEIN/DNA RATIO OF $185 \pm 8 : 100$

	Extinction values (%) of proteins separated on cellogel strips	Staining correction factors	Weight ratio (%) of chromatin proteins
H 1	16.5 ± 0.5^{a}	0.76	11.4
H 2A	15.1 ± 2.0	1.23	17.0
H 2B	21.4 ± 0.9	1.06	20.8
H 3	22.6 ± 1.6	1.02	21.1
H 4	10.2 ± 1.1	0.82	7.7
Nonhistones	14.2	1.69	22

^{a} Standard deviation.

$$s = \left\{ \frac{\Sigma x^2 - [(\Sigma x)^2/n]}{n-1} \right\}^{1/2}$$

V.   Discussion

The analysis of chromosomal proteins by cellogel electrophoresis after a complete, prior separation of the protein moiety from the DNA by ultracentrifugation in high salt and urea concentrations represents a simple and very reliable way to determine histone and nonhistone quantities. Because the described electrophoretic system works at alkaline pH and because of the mobility of the histones, which have pK values > 10, a clear distinction of these very basic proteins from all nonhistones is possible even if a part of the latter has a basic character. Low resolution conditions proved to be especially successful for quantitative analyses of the histone and nonhistone moieties. The electrophoreses can be carried out in a rather short time, and it is possible to compare different samples in the same electrophoretic run. In contrast to a method involving preceding fractionation of the nucleoprotein by acid treatment or column chromatography, no material is lost with the recommended procedures, and therefore the analyses must reflect the true protein compositions.

The question remains as to what part of the proteins can be seen as true constituents of the chromosomal material under investigation. This is a problem that centers on the isolation procedures for nucleoproteins. Certainly everyone working in the field of chromatin analysis will agree that we do not yet have really satisfactory methods for separating chromo-somal material, for subfractionating eu- and heterochromatin or even single chromosomes. The more biochemical techniques available which enable us to fractionate nuclear material properly, the more understanding we will gain about functional aspects of the eukaryotic gene.

Analyses: Protein was determined according to Lowry *et al.* (1951). In this case, calf thymus histones, nonhistones, and bovine serum albumin were used as standards. DNA was determined according to Dische (1930). Since high salt concentrations produce a considerable quenching of the diphenylamine reaction, all analyses for DNA were performed in water or in low-salt conditions.

REFERENCES

Bonner, J., Dahmus, M. E., Fambrough, D., Huang, R. C., Marushige, K., and Tuan, D. Y. H. (1968a). *Science* **159**, 47–56.

Bonner, J., Chalkley, G. R., Dahmus, M., Fambrough, D. Fujimura, F., Huang, R. C., Huberman, J., Jensen, R., Marushige, K., Ohlenbusch, H., Olivera, B., and Widholm, J. (1968b). *In* "Methods in Enzymology" (L. Grossman and K. Moldave, eds.), Vol. 12 part B. pp. 3–65. Academic Press, New York.

Bornkamm, G. W., Nobis, P., and Sonnenbichler, J. (1972). *Biochim. Biophys. Acta* **278**, 258–265.

Chaudhur, S. (1973). *Biochim. Biophys. Acta* **322**, 155–165.

Dische, Z. (1930). *Mikrochemie* **8**, 4–12.

Elgin, S. C. R., and Bonner, J. (1972). *Biochemistry* **11**, 772–781.

Fambrough, D., Fujimura, F., and Bonner, J. (1968). *Biochemistry* **7**, 575–585.

Fujitani, H., and Holoubek, V. (1974). *Tex. Rep. Biol. Med.* **32**, 461–477.

Goodwin, G. H., and Johns, E. W. (1972). *FEBS Lett.* **21**, 103–104.

Heil, A., and Zillig, W., (1970). *FEBS Lett.* **11**, 165–168.

Ishitani, K., and Listowsky, I. (1975). *Fed. Proc. Fed. Am. Soc. Exp. Biol.* **34**, 610.

Johns, E. W. (1967). *Biochem. J.* **104**, 78–82.

Kleinsmith, L. J., and Allfrey, V. G. (1969). *Biochim. Biophys. Acta* **175**, 123–135.

Kohn, J. (1957). *Clin. Chim. Acta* **2**, 297–305.

Langendorf, H., Siebert, G., Lorenz, J., Hannover, R., and Beyer, R. (1961). *Biochem. Z.* **335**, 273–284.

Levy, S., Simon, R. T., and Sober, H. A. (1972). *Biochemistry* **11**, 1547–1554.

Lowry, H. O., Rousebrough, N. J., Farr, A. L., and Randall, R. J. (1951). *J. Biol. Chem.* **193**, 265–275.

Machicao, F., and Sonnenbichler, J. (1971). *Biochim. Biophys. Acta* **236**, 360–361.

Naito, J., and Sonnenbichler, J. (1972). *Hoppe-Seyler's Z. Physiol. Chem.* **353**, 1228–1234.

Oliver, D., Sommer, K. R., Panyim, S., Spikert, S., and Chalkley, R. (1972). *Biochem. J.* **129**, 349–353.

Panyim, S., and Chalkley, R. (1968). *Arch. Biochem. Biophys.* **130**, 337–346.

Paul, J. (1972). *Biochim. Biophys. Acta* **277**, 384–403.

Reeck, G. R., Simpson, R. T., and Sobert, H. A. (1974). *Eur. J. Biochem.* **49**, 407–414.

Richter, K. H., and Sekeris, C. E. (1972). *Arch. Biochem. Biophys.* **148**, 44–53.

Sonnenbichler, J., and Nobis, P. (1970). *Eur. J. Biochem.* **16**, 60–69.

Sonnenbichler, J., and Zetl, I. (1975). *Hoppe-Seyler's Z. Physiol. Chem.* **356**, 599–603.

Sonnenbichler, J., Zetl, I., and Machicao, F. (1975). *Anal. Biochem.* **64**, 74–79.
Suria, D., and Liew, C. C. (1974). *Can. J. Biochem.* **52**, 1143–1151.
Vandenbroek, H. W., Nooden, L. D., Bonner, J., and Swall, S. (1973). *Biochemistry* **12**, 229–236.
Wilson, E. H., and Spelsberg, T. C. (1973). *Biochim. Biophys. Acta* **322**, 145–154.
Yoshida, M., and Shimura, K. (1972). *Biochim. Biophys. Acta* **263**, 690–695.
Zubay, G., and Doty, P. (1959). *J. Mol. Biol.* **1**, 1–20.

Chapter 10

Solubilization of Chromatin with Heparin and the Isolation of Nuclear Membranes

MICHEL BORNENS

Department of Molecular Biology,
Pasteur Institute,
Paris, France

I. Introduction

The interphase nucleus of a mammalian cell is a very complex organelle in which a very large amount of DNA is packed in a controlled manner through association with proteins. Many aspects of the whole nuclear organization, such as chromatin structure, the relationship between chromatin and the inner nuclear membrane, or chromatin and pore complexes, are still poorly understood.

To unravel the whole nuclear organization, one needs tools that would specifically unlock fundamental nucleoprotein associations. The importance of histones in the chromatin structure makes polyanions candidates for such an action. Indeed, a great amount of work has been done on interactions of natural and synthetic polyanions with nuclei and chromatin, at both the structural and functional levels (Chambon *et al.,* 1968; Kraemer and Coffey, 1970a; Barry and Merriam, 1972; Miller *et al.,* 1972; Skalka and Cejkova, 1973). Theories have been proposed for genetic derepression by polyanions (Frenster, 1965; Coffey *et al.,* 1973). An extensive morphological study by Arnold *et al.* (1972) has shown that the ultrastructure of chromatin is greatly modified after interaction of nuclei with polyanions. However, the presence of phosphate dramatically modifies the effect of heparin on isolated nuclei, allowing complete solubilization of chromatin and the isolation of pure nuclear envelopes (Bornens, 1973).

II.   Isolation of Nuclei

Nuclei from rat liver were prepared according to Blobel and Potter (1966) with slight modifications to adapt the procedure to larger amounts of tissue.

III.   Solubilization of Chromatin

A.   Nuclear Suspensions

Low ionic strength media and an absence of divalent cations are needed to obtain an optimal effect of heparin on nuclei (Kraemer and Coffey, 1970b) as well as chromatin solubilization; 0.010 M Tris-HCl, pH 7.5 or 8.0, were used. The effect of phosphate on heparin action has been demonstrated also with nuclear suspensions in unbuffered 0.25 M sucrose. Nuclei were resuspended in the appropriate medium by gentle rehomogenization with a small glass–Teflon homogenizer (three strokes). At this step, nuclei were already swollen and more or less disrupted. Heparin was added to such nuclei maintained at 4°C.

B.   Monitoring Method for Chromatin Solubilization

The basic features of heparin action on nuclei, beside functional modifications, are swelling and release of nucleoproteins in a soluble form. Such a release probably involves some kind of unpacking of chromatin through complexing of histones by heparin.

Solubilization of chromatin was measured by viscometry. This was carried out, after heparin action, on the supernatant of the nuclear suspension after sedimentation of the insoluble material at 45,000 g for 15 minutes. Such a supernatant is a complex mixture of nucleoproteins, proteins, heparin, and possibly free nucleic acids. Viscosity of such a solution is obviously a complex function of the various macromolecule concentrations, their conformations, the shear rate for DNA-like molecules, and the ionic environment. It is easy to show, however, that DNA concentration is by far the main factor contributing to the viscosity of such a solution. Therefore, viscometry is a very convenient way to monitor the effect of increasing concentration of heparin on the solubilization of nuclear DNA (or DNP). Viscosity was monitored by recording the flow time of the solutions in a Cannon-Ubbelhode semi-micro dilution viscometer (size 150) maintained at 10°C.

C.   Effect of Phosphate

A typical experiment is shown in Fig. 1. The effect of increasing amounts of heparin in the absence of phosphate is a slight and linear increase in the flow times. When Na_2HPO_4 was present (0.002 M), flow times increased much more, following a sigmoidal curve. A plateau was reached for a ratio of heparin to DNA close to 1, on a weight basis. One can observe also in Fig. 1 that the viscosity due to heparin itself was negligible; conversely, supernatants of nuclear suspensions in which no heparin had been added showed increased viscosity, due to extraction of nuclear material when nuclei were suspended in a medium of low ionic strength. The same values were obtained in the presence or in the absence of phosphate.

Such a difference in the shape of the two curves could be due to a differential release of DNP or to a phosphate effect on the conformation of released DNP. The DNA content of each supernatant is shown in Fig. 2. Clearly, viscosity parallels DNA release in both cases. Recovery of the DNA in the supernatant was near 100 % at the plateau, when phosphate was present. The lowest heparin concentration which, in the presence of phosphate, enabled complete release of DNA, in the absence of phosphate released only 10–15 % of the DNA. We have shown (Bornens, 1973) that RNA release is different; early release occurs either with or without phosphate. The effect of RNA release on the viscosity is probably very small since Figs. 1 and 2 are almost competely superimposable. The correspondence between viscosity and DNA concentration is roughly the same for all the determinations, in the presence or in the absence of phosphate, except for values at the plateau for

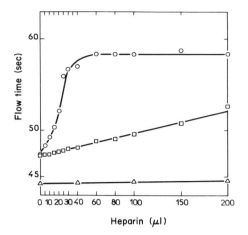

FIG. 1.   Viscometric titrations. Nuclear suspension in 0.25 M sucrose contained 295 μg of DNA per milliliter. Aliquots of 1.5 ml were prepared and divided into two series: one series received 0.17 ml of 0.020 M Na$_2$HPO$_4$ in each tube to give a final concentration of 0.002 M PO$_4$. Increasing amounts of heparin (Na$^+$ salt, grade I, from Sigma: 10 mg/ml) were added in parallel to the two series. After mixing by vortex, 5 ml of water were added to each tube and mixed. All the preceding operations were carried out at 4°C. The tubes were then spun for 15 minutes at 45,000 g (21,000 rpm, rotor JA 21, Beckman-Spinco J21). Supernatants were kept in an ice bath, and their viscosity was checked by recording their flow time in a Cannon-Ubbelhode semimicrodilution viscometer (size 150) maintained at 10°C. □———□, No phosphate; ○———○, 0.002 M Na$_2$HPO$_4$; △———△, heparin alone added to 6.5 ml of 0.002 M Na$_2$HPO$_4$.

which the viscosity is higher by 10 to 15%, meaning that conformation or shear rate is different.

Therefore, viscosity of the nuclear supernatants is, if not a measure of DNA release, at least a very close reflection of the pattern of DNA release when phosphate is present or absent. Viscometry can be used to compare such patterns of DNA release under different conditions.

D.   Role of the DNA Concentration

When increasing concentrations of nuclei were incubated in the presence of phosphate with a constant amount of heparin, the results shown in Fig. 3 were obtained. These results show that the release of DNA was not proportional to nuclei (or DNA) concentration; a strong dependency on the concentration of nuclei was observed for the release of DNA which followed a sigmoidal pattern. Maximum release of soluble DNA was obtained only

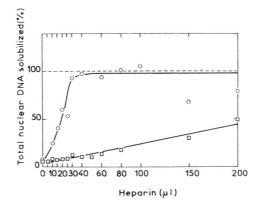

FIG. 2.   Solubilization of nuclear DNA in the experiment described in Fig. 1. Aliquots of the supernatants were used to determine DNA. After precipitation by 0.2 N perchloric acid, samples were extracted twice with chloroform–methanol (2:1) and then hydrolyzed with 0.3 N KOH at 37°C for 1 hour; after acidification, the insoluble products were rinsed twice with 0.1 N perchloric acid. DNA was hydrolyzed with 0.5 N per chloric acid at 70°C for 20 minutes. The hydrolysis was repeated once. DNA was measured by phosphorus determination and by the diphenylamine reaction. Results were in agreement. □ ———□, no phosphate; ○ ———○, 0.002 M Na$_2$HPO$_4$.

with nuclei concentrations corresponding to more than 200 μg of DNA per milliliter for a constant amount of heparin. For low concentrations of nuclei, corresponding to 50 μg of DNA per milliliter, release of DNA in the presence of phosphate was not significantly higher than release in the absence of phosphate.

Since phosphate promoted the cooperative release of DNA, one could expect increasing concentrations of phosphate to activate in a cooperative way the release of DNA for a given and low concentration of nuclei and for a fixed concentration of heparin. This is what was observed; nuclear suspension, corresponding to 50 μg of DNA per milliliter was added to fixed concentration of heparin in the presence of 2.5 and 10 mM Na$_2$HPO$_4$; 10 mM phosphate allowed near maximum solubilization although the concentration of nuclei was very low.

E.   Biochemical Modifications

To elucidate the actual molecular events responsible for DNA solubilization, careful analysis of the state of the different macromolecules, particularly proteins, in the "soluble chromatin" is needed; work in progress indicates that histones and nonhistone proteins are progressively removed by heparin (J. C. Courvalin and M. Bornens, unpublished data).

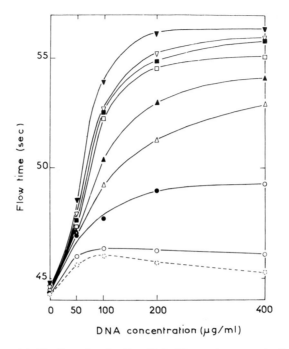

FIG. 3.   Viscometric titrations showing the effect of increasing concentrations of nuclei on chromatin solubilization in presence of PO_4^{2-}. Nuclei were resuspended in 0.002 M Na_2HPO_4 at concentrations corresponding to 50, 100, 200, 300, and 400 μg of DNA per milliliter and maintained at 4°C. Nuclear suspensions were added with increasing concentrations of heparin and then diluted with 0.002 M Na_2HPO_4 to a final concentration of 50 μg DNA per milliliter. After sedimentation of the insoluble material (45,000 g for 15 minutes), supernatants were kept in an ice-bath and their viscosity checked by recording their flow time in a Cannon-Ubbelhode semimicrodilution viscometer (size 150) maintained at 10°C. Each curve represents the flow times recorded when a constant volume of heparin solution (10 mg/ml) was added. --------- , No heparin; with heparin: $\bigcirc$———$\bigcirc$, 10 μl; $\bullet$———$\bullet$, 20 μl; $\triangle$———$\triangle$, 30 μl; $\blacktriangle$———$\blacktriangle$, 40 μl; $\square$———$\square$, 60 μl; $\blacksquare$———$\blacksquare$, 80 μl; $\triangledown$———$\triangledown$, 100 μl; $\blacktriangledown$———$\blacktriangledown$, 150 μl.

FIG. 4.   (a) Electron microscopy of the material solubilized when heparin action is complete. After sedimentation of the nuclear membranes, 5 μl of supernatant was added with 50 μl of 0.5 M ammonium acetate and spread on 0.2 M ammonium acetate. After 1 minute, the material was taken up on carbon-coated grids which were then air-dried and shadowed with palladium–platinum. $\times$ 92,400. Bar represents 0.1 μm. (b) Electron microscopy of the material solubilized at the beginning of heparin action (heparin: DNA, 0.30). After sedimentation of the insoluble material, 5 μl of supernatant were added with 50 μl of 0.5 M ammonium acetate and spread on 0.2 M ammonium acetate. The material was then processed as described in the legend of Fig. 4a. $\times$ 92,400. Bar represents 0.1 μm.

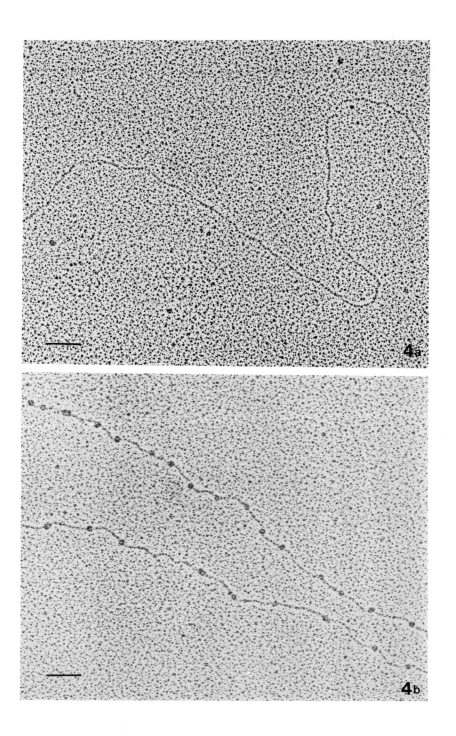

4a

4b

F.   Electron Microscopy of the Soluble Chromatin

Electron microscopic observations of soluble material after heparin action have shown that heparin indeed caused a very dramatic modification of chromatin organization; when solubilization was complete, the solubilized material looked like DNA (Fig. 4a); at the beginning of heparin action (ratio heparin:DNA, 0.30), discrete structures, mainly strings of regularly spaced particles, could be observed in the solubilized material (Fig. 4b). That the fundamental structure of chromatin is a flexible chain composed of globular particles (γ bodies or nucleosomes) connected by DNA filaments is now well documented by electron microscopy (Olins and Olins, 1974; Oudet et al., 1975). This confirms other data obtained from nuclease digestion which suggested the existence of repeating units (Hewisch and Burgoyne, 1973; Sahasrabuddhe and Van Holde, 1974). One can therefore expect chromatin solubilization by heparin, which avoids shearing and nuclease action and which is progressive, to allow some insights into the nucleoprotein association and the structure of native chromatin.

IV.   Isolation of Nuclear Envelopes

The critical importance of phosphate for chromatin solubilization was shown by the appearance of the 45,000 g pellets obtained after heparin action. When phosphate was absent, the pellets were big and showed a gel-like appearance. Electron microscopic observation of such pellets showed characteristic alterations of nuclear ultrastructure, as described by Arnold et al. (1972) and Cook and Aikawa (1973). When phosphate was present and suspensions of nuclei concentrated enough to correspond to more than 200 μg DNA per milliliter, the addition of optimal concentrations of herapin promoted the instantaneous solubilization of chromatin. The 45,000 g pellets obtained had, upon electron microscopic observation, the appearance of pure nuclear membranes.

A.   Isolation Procedure

All operations are performed at 4°C. Nuclei are resuspended in 0.010 M Tris-HCl, pH 8.0, containing 10 mM Na_2HPO_4 at a concentration corresponding to more than 200 μg DNA per milliliter. Resuspension is effected by gentle rehomogenization in a small glass–Teflon homogenizer. Heparin is then added in excess, corresponding to 1.5 times the amount of DNA present in the nuclear suspension. Dramatic increase in the viscosity is observed and the nuclear suspension becomes transparent. After brief vortexing the nuclear suspension is diluted with buffer to the desired volume and slowly stirred for 40 minutes; nuclear membranes are then sedimented by centrifugation for 40 minutes at 20,000 rpm (45,000 g).

Resuspension of the nuclei is the critical step; nuclear clumps occur easily at low ionic concentration in the absence of Mg^{2+}. This can impair penetration of heparin and should be avoided; a supplementary precaution of filtering the nuclear suspension after heparin action through nylon mesh can be taken; usually nothing is retained, this filtration, however, can avoid unexpected spoiling of the nuclear membrane pellet.

B. Incubation Time with Heparin

Completion of the action of heparin was shown to be a matter of minutes although most of the effect action was produced immediately. This was shown by analysis on sucrose density gradients (in 0.050 M Tris-HCl, pH 7.5, 25 mM KCl, 5 mM MgCl$_2$) of the nuclear membrane pellets obtained by centrifugation either immediately after addition of heparin or 40 minutes later, as described in the isolation procedure. A homogeneous fraction with an apparent density of 1.18 was obtained when heparin incubation was carried out for 40 minutes. No other material could be detected on the gradients or at the bottom of the tubes. The nuclear pellets obtained by centrifugation immediately after addition of heparin showed most of the material as a band with an apparent density of 1.18. However, a small amount of heavier components could be seen, as well as some heavy material at the bottom of the tube.

C. Binding of Heparin to Nuclear Membranes

A ratio of heparin to DNA of 1 is necessary to obtain a complete solubilization of chromatin. As described in the isolation procedure, a ratio of 1.5 was used to assure complete chromatin solubilization; heparin association with nuclear membranes was examined. Solubilization of nuclei was realized with [^{35}S]-heparin (purchased from the Radiochemical Centre, Amersham). After incubation and pelleting of the nuclear membranes, 100% of the radioactivity was found in the supernatant. When the nuclear membrane pellet was analyzed on a sucrose gradient, no trace of radioactivity was found associated with the nuclear membrane.

D. Electron Microscopic Observation of Nuclear Membranes

The 40,000 g pellets appeared to consist of very clean nuclear membrane preparations although no gradient purification had been used (Figs. 5a–c).

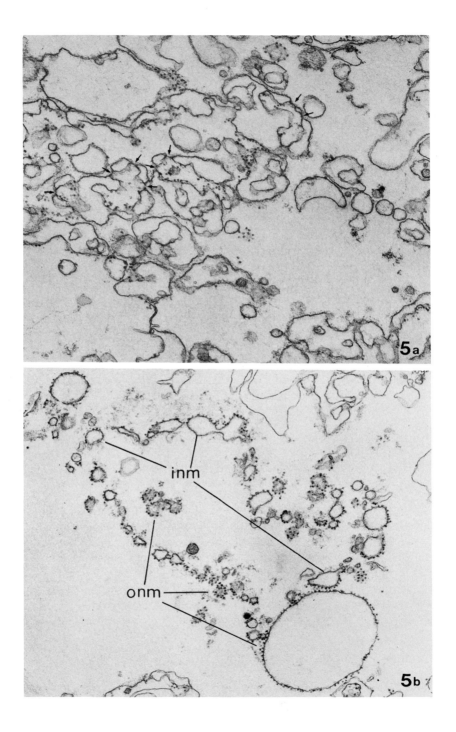

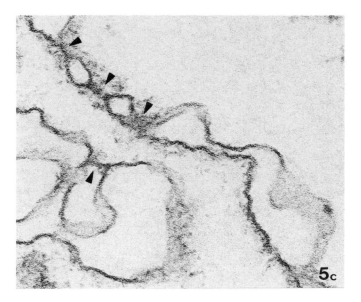

FIG. 5. (a) General appearance of the 45,000 g pellet. The fraction consists of nuclear envelopes in which ribosome-bearing outer membrane and nuclear pores (arrows) can be seen. ×27,720. (b) Details of the 45,000 g pellet showing one envelope with the outer nuclear membrane (onm) covered with ribosomes. Inm = inner nuclear membrane. ×23,100. (c) Details of the 45,000 g pellet obtained after heparin action, showing several nuclear pores (arrows) and the membrane unit structure of both leaflets of the nuclear envelope. ×97,000.

It was easy to identify whole nuclear envelopes with their perinuclear space preserved. The outer leaflet of the nuclear envelopes were often covered with ribosomes (see Fig. 5b). No material was consistently associated with the inner surface of the inner leaflet although sometimes fluffy material was seen (Fig. 5c).

Nuclear membranes pores were present and easily identifiable when cut in transverse section (see Figs. 5a and 5c); they showed dense material in the lumen of the pores. Holes in this material, however, could be observed frequently; in tangential sections, the lumen of the pores seemed often more or less empty. The nuclear pore complex was missing, but the membrane margin of the pore was visible; the average diameter of the membrane annulus was 700 Å.

E.   Yield

Virtually 100% of the nuclear DNA was solubilized in the isolation procedure described above (with nuclei from rat liver and rat thymus). As judged by gradient behavior and electron microscopy, nuclear membranes were the only nuclear structures which remained insoluble. It is therefore possible to obtain nuclear membranes with a virtually 100% yield by a simple sedimentation step.

F. Comments

This method avoids nuclear damage such as is produced by sonication or by high ionic concentration; this is probably why ribosome preservation on the outer leaflet is good. Moreover, it allows a very good morphological preservation of whole nuclear envelopes; these are not damaged by shearing such as that caused by procedures using sucrose gradients, which represent a drastic treatment for biological membranes. We have noticed that the appearance of isolated nuclear membranes was dramatically modified after gradient purification, a very significant number of vesicles being created. Avoiding such shearing can be important in facilitating observation; for example, centrioles have been observed to be associated possibly with nuclear envelopes (M. Bornens, unpublished data). Such an observation would probably no longer be possible after sucrose gradient purification. The method described is a simple one, involving only the incubation of nuclei with heparin for 40 minutes and the necessary time for pelleting nuclear membranes; it should be very useful when only small quantities of starting material are available.

V. Summary

Incubation of nuclei at 4°C with heparin in the presence of phosphate anions leads to immediate solubilization of virtually 100% of the nuclear DNA. At low concentrations of phosphate (2 mM) chromatin solubilization is strongly dependent upon nuclear concentration; it is maximal when the nuclear concentration corresponds to more than 200 μg/ml. On electron microscopic observation, heparin-solubilized chromatin appears as long particle-bearing fibers at low heparin to DNA ratios, and as long DNA-like fibers for heparin to DNA ratios close to 1.

Pure nuclear envelopes with a virtually 100% yield can be obtained by a simple sedimentation step following the incubation at 4°C for 40 minutes of nuclei with heparin. Such envelopes behave homogeneously on sucrose gradients and display an apparent buoyant density of 1.18. Envelopes do not bind heparin.

REFERENCES

Arnold, E. A., Yawn, D. H., Grown, D. G., Wyllie, R. C., and Coffey, D. S. (1972). *J. Cell Biol.* **53**, 737–757.

Barry, J. M., and Merriam, R. W. (1972). *Exp. Cell Res.* **55**, 90–96.
Blobel, G., and Potter, V. R. (1966). *Science* **154**, 1662–1665.
Bornens, M. (1973). *Nature (London)* **244**, 28–30.
Chambon, P., Ramuz, M., Mandel, P., and Doly, J. (1968). *Biochim. Biophys. Acta* **157**, 504–519.
Coffey, D. S., Barrack, E. R., and Heston, W. D. W. (1973). "Brady Research," Monogr. II, pp. 1–59. Johns Hopkins University School of Medicine, Baltimore, Maryland.
Cook, R. T., and Aikawa, M. (1973). *Exp. Cell. Res.* **78**, 257–270.
Frenster, J. H. (1965). *Nature (London)* **206**, 1269–1270.
Hewish, D. R., and Burgoyne, L. A. (1973). *Biochem. Biophys. Res. Commun.* **52**, 504–510.
Kraemer, R. J., and Coffey, D. S. (1970a). *Biochim. Biophys. Acta* **224**, 553–567.
Kraemer, R. J., and Coffey, D. S. (1970b). *Biochim. Biophys. Acta* **224**, 568–578.
Miller, G. J., Berlowitz, L., and Regelson, W. (1972). *Exp. Cell Res.* **71**, 409–421.
Olins, A. L., and Olins, D. E. (1974). *Science* **183**, 330–332.
Oudet, P., Gross-Belard, M., and Chambon, P. (1975). *Cell* **4**, 281–300.
Sahasrabuddhe, C. G., and Van Holde, K. E. (1974). *J. Biol. Chem.* **249**, 152–156.
Skalka, M., and Cejkowa, M. (1973). *Folia Biol. (Prague)* **19**, 68–77.

Chapter 11

Fractionation of Cell Organelles in Silica Sol Gradients

JÜRGEN M. SCHMITT AND REINHOLD G. HERRMANN

Institut für Botanik der
Universität Düsseldorf,
Düsseldorf, West Germany

I. Introduction

Centrifugation is presently the most important method in the fractionation of subcellular particles. The simplest way of fractionating homogenates is to use differential centrifugation. However, the resolving power of this procedure is generally unsatisfactory, and whenever the purity of particulate fractions obtained by this technique is adequately controlled, considerable contaminations are found. This is actually to be expected. It may be deduced from Stokes' equation for the sedimentation of spherical particles in a centrifugal field. This simplifying equation predicts that compounds of very different size and density, such as small heavy particles and larger lighter ones, may travel at a similar rate and reach similar positions.

The introduction of isopycnic centrifugation, which fractionates independently of the size parameter, was a great advance in this respect. Isopycnic centrifugation of subcellular particles is most frequently performed in sucrose gradients. Although this procedure may yield preparations of improved purity, osmotically responding organelles are inevitably exposed to a hypertonic environment that may seriously impair or even abolish both their structural integrity and biological activity.

This brief outline illustrates the dilemma encountered in the usual combination of isotonic differential and hypertonic isopycnic centrifugations: whereas the former procedure leads to unbroken and functional organelles of moderate purity, isopycnically prepared organelles meet a much higher purity standard, but their structural and functional details are poorly preserved. Hence, it is desirable to purify particles under isotonic conditions not only during the differential, but also during the isopycnic centrifugation steps.

Gradient materials that exert little osmotic stress are high-molecular-weight compounds, such as Ficoll, dextran, polyvinylpyrrolidone, or, relevant in the context of this chapter, silica sols. The application of polymers in cell fractionation is, however, still in its developmental stage and is complicated by several factors. Such compounds must have different properties besides osmotic inertness. For example, they should provide high-density solutions without troublesome increase in viscosity. They should be sufficiently soluble in aqueous solutions, should not interact with the biological material, and should not interfere with the analytical procedures used. Silica sols fulfill many of these requirements, and, in addition, are cheap. They were introduced by Mateyko and Kopac (1963) for the fractionation of viable cells and subsequently adapted by Pertoft (1966, 1967) and Pertoft and Laurent (1969) for the separation of cells and viruses.

The purpose of this contribution is to discuss the application of this promising gradient material and its advantages and present disadvantages in subcellular fractionation. Typical procedures for the preparation and subfractionation of organelles will be described in some detail. Emphasis will be put on plant organelles. Since much biochemical information can be gained by an analysis of organelles from mutants, we also consider fractionation of mutant tissue. This requires special discussion since the behavior of mutant material in gradients is often different from that of wild-type material. Since mutants of higher plants usually are difficult to grow, information is also given on recently developed techniques concerning cultivation and handling of mutant material.

II.   Use of Silica Sols

A.   General Properties of Silica Sols

Ludox (trademark of DuPont) is an aqueous sol of surface hydroxylated silica, manufactured for technical use. The silica particles are spherical without an inner surface. Acid protons on the surface are replaceable by sodium or ammonium. The negatively charged particles are stabilized in the sol form by the repulsive action of coulombic forces. The repulsion forces of these charges stabilize the sol. Ludox AM is surface modified by aluminate ions and has a permanent "built-in" charge.

Different types of Ludox are available. They vary in particle diameter (7–25 nm), density (1.21–1.38 gm · cm^{-3}), viscosity (4.5–50 cp), silica content and surface modification. For further details, the reader is referred to the DuPont product information. Silica sols from other commercial sources have been tried for gradient centrifugation, but so far with little success (Morgenthaler *et al.*, 1975).

B.   Composition and Handling of Silica-Containing Media

Owing to their high particle weight silica sols are osmotically largely inactive. In subcellular fractionation, they generally require an osmoticant to prevent bursting of the organelles. Since Ludox is miscible with a wide variety of water-soluble compounds, this poses no serious difficulty, unless gelation or precipitation occurs. Usually, the media used in isolating and resuspending organelles also form the basis of silica sol gradients. For example, the media of Suyama and Bonner (1966), Honda *et al.* (1966), and Jensen and Bassham (1966) have been used successfully in the preparation of silica sol gradients.

Ludox is stable as a basic sol (Ludox HS 40, pH 9.7). It can be sterilized by autoclaving. Freezing causes irreversible coagulation. The tendency of the sols to form gels or to precipitate is influenced by the composition, especially the ionic composition of the supporting medium. Neutralization of the sols, addition of bi- or trivalent cations, or of quaternary ammonium salts all accelerate gelation or even lead to an immediate precipitation of the sols. With physiological media comparable to those mentioned above, gelation times are usually in the range of days and should not affect experiments. Ludox-containing media ready for use should not be stored for prolonged periods unless knowledge about sol–gel transformation is available. The stability of Ludox AM is increased by neutralization or acidification.

The commercial silica sols apparently contain impurities detrimental to cells and cell components. Moreover, silica itself may interact with organelles. These difficulties may be overcome by purification of the sols and/or by addition of protective substances.

Purification can be achieved as recommended by Morgenthaler *et al.* (1974) with activated carbon, filtration through diatomaceous earth and treatment with a cation-exchange resin. In certain cases dialysis may be sufficient. Thereafter, the density should be checked gravimetrically or areometrically, and readjusted if necessary.

No general rules for the choice of protective agents can be given at the present state of knowledge. In addition to the usual antioxidants (thiol reagents, ascorbate), several polymers have been employed. Dextran (2–5%), Ficoll (1–2.5%), polyvinyl sulfate (0.2%), and bovine serum albumin (BSA; 0.05–0.1%) have been included in gradient media for the isolation of mitochondria (Gronebaum-Turck and Willenbrink, 1971), nuclei (Hendriks, 1972), and chloroplasts (Schmitt *et al.*, 1974; Salisbury *et al.*, 1975), but the authors did not always clearly state the consequences of omission. In this laboratory 5% dextran included in both Ludox HS 40- or Ludox AM-containing media did not preserve the ability of chloroplasts to fix CO_2 or preserve or restore the activity of some (Mg^{2+}-dependent) chloroplast enzymes. According to Morgenthaler *et al.* (1974) and Morgenthaler and Mendiola-Morgenthaler (1976), 10% polyethylene glycol added to media containing purified Ludox AM maintained protein and especially photosynthesizing, capacity. Sixteen percent polyethylene glycol in Ludox HS 40 was found to be toxic for HeLa cells (Wolff and Pertoft, 1972). BSA (0.05–0.1%) has been beneficial for preservation of organelle morphology. In the absence of albumin and/or at low pH, plastid envelopes exhibited a marked tendency to swell irreversibly and to form protuberances (Schmitt *et al.*, 1974). Higher concentrations of albumin increase the rate of silica gelation. Polyvinylpyrrolidone occasionally used as absorbant for tannins precipitates silica. Polyethylene glycol may exhibit similar effects, especially with aging sols. Where appropriate these compounds may be effective in removing silica precipitates from organelle surfaces after gradient separation (see Section IV).

C.   Preparation of Medium

Ludox-containing media are readily prepared. To make, for example, a 30% Ludox sol (v/v), mix 50 volumes of double concentrated medium with 30 volumes of Ludox. (The silica content of this solution depends on the sol type used. It is approximately 13%, w/v, in the case of Ludox HS 40.) Adjust the pH with the acidic component of the buffer under vigorous stirring. If concentrated acid causes precipitation, use dilute acid. The use of cation exchange resins, such as Bio-Rad AG 50 W-X 8, is equally effective (Morgenthaler *et al.*, 1974). Back titration is not possible, because alkali (NaOH) precipitates the sol. Fill up with water to 100 volumes; the sol is then ready for use. Alternatively, Ludox itself may be pH adjusted and mixed subsequently.

D.   Separation Methods

Cell fractionation in silica sols may be accomplished with any kind of rotor, in angle or swinging-bucket heads as well as in batch (continuous flow or zonal) or elutriator rotors, using their general principle of operation. The silica material can accommodate to differential centrifugation and isopycnic separations. It has, however, for reasons outlined below rarely been employed in velocity band centrifugation (Pertoft *et al.*, 1967).

Discontinuous and continuous gradients have been used for purification of organelles. Discontinuous gradients may be prepared in the usual manner by superimposing layers of decreasing density, that is, of decreasing silica concentration. Continuous gradients may be preformed either in the standard way with any instrument for gradient production at any desired shape, or established by centrifugal fields prior to or even simultaneously with particle separation (Schmitt *et al.*, 1974) In the case of preformed gradients, the sample to be separated is layered above the liquid column. In the case of gradients formed during fractionation, the sample is mixed with buffered silica sol of appropriate tonicity and concentration.

Silica gradients form quite rapidly by sedimentation of sol particles, unlike gradients of heavy salts, such as CsCl, where the density changes are produced during a rather long centrifugation period under the opposing influence of gravity field and diffusion until an equilibrium is reached. In silica gradients the diffusion parameter is practically negligible. Therefore, the gradient itself changes permanently during centrifugation and true equilibrium is not reached. Nevertheless, particles band at a density position within the gradient column that is equal to their own apparent density, that is, isopycnically (quasi-equilibrium gradient; Meselson and Nazarian, 1963). Consequently, a particle zone will isopycnically move along the tube or field axis with the time of centrifugation or as the rotor speed is increased. It may be seen from these brief considerations that the final shape of a gradient will be a function of the initial sol concentration, the speed and time of centrifugation, and its preformed shape. It can be seriously affected by various additives (see Section IV).

III.   Fractionation of Organelles

A.   Chloroplasts

1.   ORGANELLE PREPARATION

The chloroplasts to be used as starting material for gradient purification are collected from homogenates by differential centrifugation. With short centrifugation times, differences in density and friction between unbroken and broken chloroplasts can efficiently be used to enrich pellets in unbroken organelles. Moreover, the organelles are removed rapidly from the homogenate, and thus from injurious vacuolar contents. This method now constitutes one of the essential steps for studies with chloroplasts because it can provide functionally excellent organelles in a simple and rapid manner

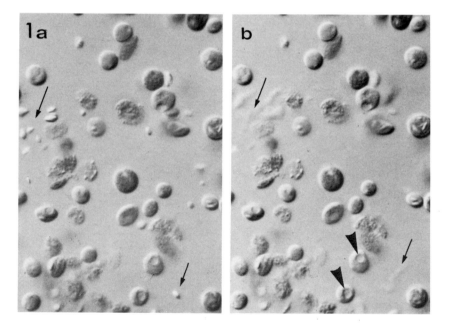

FIG. 1.   Interference contrast micrographs of a crude chloroplast preparation obtained by differential centrifugation (2000 g, 40 seconds) containing unbroken (smooth surface) and broken chloroplasts (grana structure visible) and particulate material, mostly starch grains. Several chloroplasts contain starch grains. Note also the partially deformed organelle surfaces (bold arrowheads). (a) Flashlight exposure (Leitz Orthomat) and (b) exposure of the same preparation for 2 seconds. Under the latter conditions, small particles are not visible or are visible as tracks only (small arrows). × 1200.

(Jensen and Bassham, 1966; Heber and Kirk, 1975). It has been described in detail by Walker (1971) and Heber (1973).

Although the resulting chloroplast fractions contain considerably fewer mitochondria than fractions prepared by the usual centrifugation at about 1000 g for 5–10 minutes (Herrmann et al., 1975), they still consist of a mixture of unbroken organelles and thylakoidal systems and contain starch grains, crystals, and nuclear remnants as contaminants (Fig. 1a).

In practice the leaf tissue (75–100 gm) is usually homogenized with a Waring blender (2 full-speed bursts, 3 seconds each) in about 4 volumes (w/v) of an appropriate medium (Jensen and Bassham, 1966; Honda et al., 1966; Walker, 1971; Heber, 1973; Herrmann et al., 1975). The homogenate is strained through one layer, then filtered through 2–4 layers of nylon gauze of defined pore size (20 μm; Schweizer Seidengaze AG, Zürich, Switzerland). This cloth type is available in several pore size gradations and was found to be more effective in retaining cells and tissue fragments than other clothes tested in our laboratory. The chloroplasts are spun down by a 60-second centrifugation at 2000 g in a swinging-bucket head (6 × 100 ml). Acceleration and braking should be as fast as possible. The sediment is immediately taken up in fresh medium (200 ml). A first chloroplast pellet is collected by a 45-second centrifugation at 2000 g, a second one from the resulting supernatant by a subsequent 40-second centrifugation. The chloroplast fractions are suspended in a 4-fold volume of medium, combined, and used for gradient centrifugation (Fig. 1a). A centrifugation with low gravity fields, for example at 100 g for 1–2 minutes, often included to remove starch, crystals, and nuclei, does not significantly improve purity.

2. Gradient Centrifugation

Discontinuous silica sol gradients preformed by overlayering, or continuous gradients formed by mixing or by gravity fields may be used to further purify chloroplasts and to separate broken and unbroken organelles.

In the case of separation by means of a field-formed gradient (Heber and Santarius, 1970; Schmitt et al., 1974), the sol, adjusted to an appropriate Ludox concentration, is centrifuged at 2°C for 20 minutes at 40,000 g in a swinging-bucket rotor. Tubes of 5-ml or 34-ml volumes have been used in our laboratory, but others should do as well. The resulting gradients are continuous and shallow. They are overlayered with chloroplast suspension, corresponding to 300–500 μg chlorophyll per 34-ml tube, and centrifugated again under the same conditions. Two well separated zones about 5–6 cm apart will be obtained (Fig. 2b). Because of their great distance they can conveniently be collected by a wide-orificed pipette. If fractionation

FIG. 2.   Silica sol gradients in 5-ml centrifuge tubes. (a and b) Continuous gradients demonstrating isodensity banding of (a) mitochondria of *Acanthamoeba castellanii* and (b) chloroplasts of *Spinacia oleracea*. Original sol concentration 32% (Ludox HS 40), isotonic in Jensen and Bassham's medium C (1966). The gradients were preformed, and the organelles were separated at 40,000 g for 20 minutes at 0°C. Unbroken organelles band at the top, broken ones at the bottom, of the gradient (arrows). The plastids of the top band are shown in Fig. 3a. (c) Separation of mutant plastids from *Oenothera grandiflora* plastome mutant III β in a discontinuous silica gradient. Ludox HS 40 concentrations are 40, 35, 32, and 28%, 1 ml each, isotonic in medium C of Jensen and Bassham (1966). Centrifugation was at 30,000 g for 20 minutes at 0°C. Two bands with unbroken organelles were recovered (arrows). A yellowish band at the 28%/medium interphase (upper arrow) contained plastids largely lacking internal membrane structures. The plastids banding at the 28/32% Ludox interphase (faint green; lower arrow) are shown in Figs. 7 and 8.

by piercing is necessary, the needle of the fractionator should be long enough to penetrate a small Ludox pellet at the tube bottom.

The upper band contains unbroken plastids, which are sometimes slightly clumped. They look refractive and sharp-edged under the phase-contrast microscope and show a smooth surface under the interference microscope (Fig. 3a). They have retained both the outer and the inner envelope (Figs. 6 and 8 in Schmitt *et al.*, 1974). The lower band contains broken plastids that look brownish with visible grana under phase contrast and exhibit a rough surface in the interference contrast microscope. Some unbroken plastids may be present, too, but they appear smaller than those of the top layer.

The outlined procedure is effective only when nuclei or multiorganelle complexes are absent in the plastid fractions. Both contaminants would migrate with the intact chloroplasts and collect in the top band. Multiorganelle complexes (peak III chloroplasts according to Larsson *et al.*, 1971) consist of a varying number of chloroplasts, mitochondria and even nuclei surrounded by cytoplasm and by a common membrane. They form during homogenization, particularly when young tissue is used, and in media

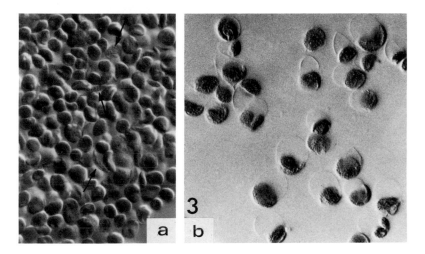

FIG. 3. Interference-contrast micrographs of (a) unbroken chloroplasts obtained from the top layer of the silica sol gradient shown in Fig. 2b. Arrows: A few broken organelles with visible grana structures (surface or side view) are present. (b) Chloroplasts with extremely swollen envelopes derived from a continuous silica gradient (sol concentration 25%) devoid of albumin, at pH 6.3. × 1000.

that increase the flexibility of membranes, e.g., in those containing polyvinyl-pyrrolidone. Contrary to plastids, such complexes exhibit considerable stability against tonicity changes. Since these structures may escape detection under phase or interference contrast microscopy even after mildly hypotonic treatment, the most reliable method for demonstrating them is electron microscopy (Fig. 4). There is no satisfactory centrifugation procedure at present which *quantitatively* separates nuclei and multiorganelle complexes from unbroken plastids. However, since nuclei and multiorganelle complexes band at a somewhat lower density in silica than do plastids, plastids essentially free from these contaminants can be prepared by interposing a layer of lower silica concentration (range 20–27%) before the gradient is formed by the centrifugal field. Otherwise the procedure is identical to that described above. The material on this layer consists of multiorganelle complexes and nuclei, and contains 10–15% of the unbroken plastids (Fig. 4). It should be noted that the morphology of these plastids is generally less maintained than that of plastids collected from the 27/32% Ludox interphase. The former tend to be somewhat swollen.

A related class of chloroplasts, unbroken but with swollen and seemingly detached envelopes, can be separated in Ludox concentrations below 25%. This organelle class occurs in significant proportion in the preparation

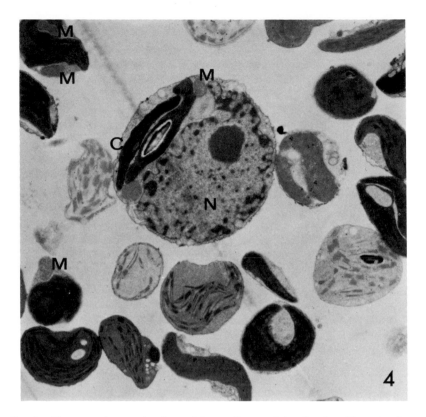

FIG. 4. Electron micrograph of a multiorganelle-containing fraction (*Spinacia oleracea*) banding at the top of a continuous silica gradient isotonic in Jensen and Bassham's medium C (1966). Original Ludox concentration 25%; centrifugation was performed as described in Fig. 2a,b. The membrane-covered multiorganelle complexes contain plasm and varying numbers of broken or unbroken plastids (C), mitochondria (M), and nuclei (N). Complexes containing nuclei are rarely found. Note also the somewhat swollen appearance of the unbroken chloroplasts. × 6300.

when a small percentage of Ludox is included into the grinding medium, the pH is kept below 6.5 (Fig.3b), and medium and gradient are devoid of BSA.

With chloroplasts containing large amounts of starch or with those derived from mature tissue, a quantitative separation of plastid classes is sometimes not possible (Schmitt *et al.,* 1974). In order to deplete the starch, plants should be kept in dark or low light before leaves are harvested.

Antirrhinum, Beta, and *Spinacia* tissue yielded good results with sols containing concentrations of between 30 and 35% of Ludox HS 40. If new species or material grown under different conditions are to be used, the best procedure must be found empirically. The gradient should be adapted to the material by varying the sol concentration in steps of 5 % in a range from 25 to 40 %. The resulting bands should be examined visually for an optimum amount of chlorophyll combined with a satisfactory percentage of unbroken chloroplasts in the top layer. The separation may subsequently be refined with a narrower Ludox gradation.

Alternative procedures have been reported in the literature. Lyttleton (1970) and Lyttleton et al., (1971) have separated Spinacia, Chenopodium, and Amaranthus chloroplasts on discontinuous gradients of Ludox HS 40 preformed by overlayering. After centrifugation for 15 minutes at 2500 g in a swinging-bucket rotor, unbroken chloroplasts were found at the interphase between the densities 1.08 and 1.13, the broken ones banded between those of 1.17 and 1.22 gm·cm^{-3}. A mixed population was found at the 1.13/1.17 gm·cm^{-3} Ludox junction.

Morgenthaler et al. (1974) have used continuous 10–80 % gradients of purified Ludox AM preformed with a gradient mixer. Centrifugation was 15 minutes at 8200 g in a swinging-bucket head. Unbroken plastids banded at higher densities than broken ones. This effect was shown to be dependent on the presence of polyethylene glycol, which was included as a protective agent in the gradient by these authors (Morgenthaler et al., 1975) (see Section IV).

B.   Mutant Plastids

The isolation of mutant organelles presents special problems, as they are often mechanically labile and differ in size and density from wild-type organelles. Much knowledge in molecular genetics of viral and prokaryotic systems has been gained from the study of mutant phenotypes. This makes it likely that the study of eukaryotic mutants will increase in importance in future. We will consider briefly problems encountered in the study of organelles from mutants of higher plants, using chlorophyll-deficient plastids of Oenothera as an example.

1.   DIFFICULTIES ASSOCIATED WITH THE PREPARATION OF MUTANT PLASTIDS

Color variation in plastids caused by mutation is generally accompanied by changes in organelle structure. Phenotypically, mutated plastids may resemble wild-type organelles, but more often they appear in the light microscope as tiny, more or less spherical, colorless organelles, regardless of whether their expression is arrested by chromosomal or extrachromosomal mutation (Walles, 1971). Similarly stunted organelles, but of wild type, may result when incompatible genomes and plastomes are combined (Kutzelnigg and Stubbe, 1974). At the electron microscopic level, plastids of different mutants show extraordinary variability in detail. They may still develop a structurally normal internal membrane system, or form some thylakoids, but no grana, or, in extreme cases consist merely of envelopes surrounding the matrix, a few vesicles, tubules, or other kinds of structures (Kutzelnigg *et al.,* 1975). Typical examples of such plastids are presented in Fig. 8.

Mutant plastids may undergo drastic transformations, including degeneration during leaf development. They are generally more sensitive to environmental changes, especially to changes of light, than wild-type plastids. Considerable difference in organelle architecture may therefore be observed not only between different mutants but also in different tissues of a given mutant and even within single cells. Thus, heterogeneity of plastids may cause problems in cell fractionation.

Many higher plants carrying mutant plastids are unable to photosynthesize on soil. These plastids usually are light-sensitive and exhibit varying degrees of photobleaching of chlorophyll. The classical maintenance of chlorophyll mutants requires therefore the cultivation of plants, whose tissues contain two types of plastids, wild-type and mutant ones. These plastids will segregate during ontogenesis and produce variegation. According to the succession of cell lineages in dicotyledons, genetic homogeneity is guaranteed only in pure bleached shoots or in bleached halves of leaves, completely devoid of green sectors.

The problem concerning cell fractionation with mutant phenotypes of higher plants then resolves itself into three main parts: (1) how to obtain sufficient and developmentally uniform tissue for biochemical analyses; (2) how to break the cells without destroying the organelles; and (3) how to separate and fractionate the labile mutant plastids, which would scarcely resist the tonicity stresses of sucrose gradient centrifugation, from other subcellar particles.

2. MASS CULTIVATION OF MUTANT MATERIAL

Concerning the first topic, material grown *in vitro* offers considerable technical flexibility and would appear to be most appropriate from a practical viewpoint. We have therefore tested the possibility of growing plastome (extrakaryotic) mutants as heterotrophs in tissue culture. *Oenothera* was chosen because of its special genetic advantages (Kutzelnigg and Stubbe, 1974). We found that plastome mutants proliferate vigorously on defined media without remarkable periods of adaptation (Mehra-Palta and Herrmann, 1975). About 30 mutant lines have so far been cultured serially as stable plants. Fine stuctural details of plastids and karyotype analysis suggest that this material does not undergo genotypic alteration during *in vitro* growth.

3. TISSUE DISINTEGRATION

One of the most gentle procedures used for tissue disintegration is manual or mechanical chopping with razor blades. Although many plastids may be released morphologically unbroken, this method is not recommended, for only low organelle yields are generally obtained, and organelles are exposed for a long time to vacuolar contents which may be destructive. These drawbacks can be avoided if viable protoplasts are used as source material. Protoplasts are exceedingly fragile to mechanical stress and can easily be broken. Their plastids can be recovered from the homogenate within less than 3–4 minutes after breaking the cells.

Protoplasts are most conveniently prepared from the leaf tissue by a several hours' enzymic digestion with cellulases and pectinases. Because of the ease of protoplast release, leaf material from well developed plants grown under axenic conditions should be preferred. If greenhouse or field-grown material is to be used, leaves must be surface sterilized in order to prevent bacterial growth during the generation of protoplasts. All subsequent procedures may be carried out under aseptic conditions.

The leaves, immersed in the medium described by Frearson et al. (1973) containing 0.4 M mannitol and 0.1% polyvinylpyrrolidone, were cut into 1–2 mm squares with a stainless steel scalpel or a razor blade. It may sometimes be useful to peel off the lower epidermis with a forceps. After 30 minutes of incubation, the pieces were transferred into petri dishes, lower side down if the epidermis was removed, onto the enzyme solution, which was sterilized by ultrafiltration before use. The solution contained a mixture of 3% cellulase Onozuka SS and 2.5% Mazerozyme (both Yakult Biochemicals, Japan) in the above medium containing 0.6 M mannitol (pH 5.2). Other basic media and enzymes may be used as well. The leaf pieces, 1 or 2 gm per 10 ml of solution, were incubated for about 10 hours at 25°–30°C with occasional gentle shaking. During this time the tissues largely disintegrated. For further details of protoplast preparation, the reader is referred to the article of Constabel (1975).

The protoplast suspension is then filtered through stainless steel sieves (one layer of each 400 and 100 μm pore size), and the protoplasts are collected from the filtrate by low-speed centrifugation at 20–50 g for 3–5 minutes. In this step, most of the damaged protoplasts and protoplast fragments should remain in the supernatant. The sedimented protoplasts are resuspended very gently, washed twice in the above solution free of enzyme, and then once in the same medium but of lower tonicity (0.4 M mannitol, pH 7.8).

The hypotonic treatment increases the percentage of viable protoplasts in the final pellet (Fig. 5) and consequently that of unbroken organelles in the

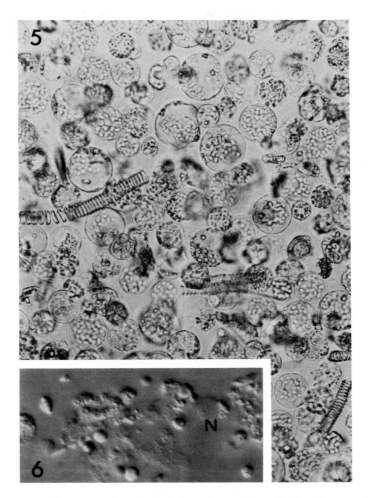

FIG. 5. Protoplast suspension obtained from the extrakaryotic plastid mutant III β of *Oenothera grandiflora* grown in tissue culture. The protoplasts were generated by enzymic treatment and prepared by several low-speed centrifugations for biochemical investigation as described in Section IV. × 350.

FIG. 6. Interference contrast micrograph of a crude homogenate derived from the protoplast suspension shown in Fig. 5. N, nucleus. × 1000.

homogenate. Within several minutes, the washed protoplasts should be broken in a loose-fitting Dounce homogenizer. Two strokes are usually sufficient to open the cells; it is recommended to monitor success with the phase or interference microscope.

4. PREPARATION OF PLASTIDS

The homogenate is filtered through 4 layers of nylon gauze (pore size 5, 7, or 10 μm; see Section III,A). This step eliminates larger protoplast fragments, if present, and part of the nuclei, which are easily discernible in the homogenate along with plastids and mitochondria (Fig. 6). Depending on the mutant phenotype, the supernatant is centrifuged 1–2 minutes at 3000 g. A swing-out rotor is favored in order to reduce mechanical stress due to sliding of organelles along the tube wall. The supernatants, containing the broken plastids, most of the mitochondria, and smaller particles, are decanted from the pellet. This operation may be repeated in medium without polyvinylpyrrolidone. Remaining nuclei are removed by two low-speed centrifugations for 2 minutes at 250 g. It is usually much easier to prepare plastids free of nuclei than it is to prepare chloroplasts free of nuclei because of greater differences in organelle size and density. The final wet-packed plastid pellet is resuspended in five volumes of medium.

5. GRADIENT CENTRIFUGATION

To further purify mutant plastids by silica sol gradient centrifugation, the methods outlined in the foregoing paragraphs are applicable. One-milliliter aliquots of the plastid suspension are layered on top of 3.5-ml gradients or columns and centrifuged isopycnically in a rotor of the swinging-bucket type. It is necessary to establish the appropriate sol concentration for each tissue or mutant. As a rule, more impaired, unbroken organelles collect generally at a lower density, corresponding to 24–28 % Ludox HS 40.

In some instances a purer preparation may be achieved with discontinuous gradients. An example of a discontinuous gradient with the silica solutions of 40, 35, 32, and 28 %, 1.0 ml of each, is shown in Fig. 2c. It was overlayered with a plastid fraction of the plastome mutant III β from *Oenothera grandiflora* (see Kutzelnigg *et al.*, 1975) obtained by differential centrifugation, and spun for 15 minutes at 20,000 g. This resulted in the isolation of two bands with unbroken plastids. Yellowish organelles were concentrated at the junction of medium to 28 % silica; a faint green plastid band was recovered at the interface of the 28/32 % silica layers (Fig. 2c). Interference and electron micrographs (Figs. 7 and 8) of these fractions demonstrated the morphological integrity of the mutant organelles as well as the purity and heterogeneity of the fractions. Moreover, considerable difference in organelle morphology was noted between the two organelle populations. Thus the gradient, when used with finer gradation, might work adequately to subfractionate mutant organelles into morphologically distinct classes. This kind of application has not yet been studied extensively. As expected, the lower layers contained only broken organelles.

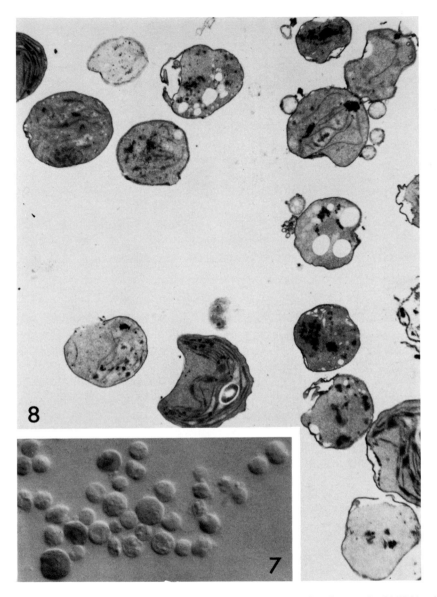

Fig. 7. Interference contrast micrograph of mutant plastids banding at the 28/32% sol interphase of the discontinuous isopycnic silica gradient shown in Fig. 2c. Note the considerable size heterogeneity of the mutant organelles (cf. Fig. 3a) and the degree of purification (cf. Fig. 6). The plastid band contains almost exclusively unbroken organelles (smooth surface). × 1400.

Fig. 8. Electron micrograph of the fraction of mutant plastids as shown in Fig. 7. All organelles shown are envelope-covered. Note the considerable heterogeneity in the interior of different plastids. See text of Section III,B. × 6200.

C. Nuclei

The procedure described here for the isolation of plant nuclei is based on a modified version of Kuehl's method (1964) and may involve differential Triton X-100 solubilization (Spencer and Wildman, 1963) in order to facilitate the removal of contaminating plastids.

Generally, nuclei are purified by differential centrifugation or by combinations of differential and isopycnic sucrose gradient centrifugations. However, even if quantitative removal of chlorophyll is achieved, the nuclear fractions isolated by these techniques are generally impure. Moreover, enzyme measurements and composition studies are often of doubtful significance, and there is loss of biological function under the influences of hypertonic sucrose or/and detergent. Nuclei obtained after differential centrifugation may, however, be purified in silica gradients (Hendriks, 1972) taking advantage of the fact that unbroken organelles may band at the top, and broken ones at the bottom, of a gradient.

Best yields of plant nuclei are generally obtained with young tissue. The isolation of nuclei from mature tissue is complicated by the lability of the nuclei, the rigidity of the cell wall, and the presence of injurious vacuolar contents, such as tannins. Together with the shearing forces that act during homogenization, tonicity differences between even adjacent cells may cause considerable disruption of nuclei. Incubation of tissue in medium equilibrates tonicities and probably effects other kinds of adaptions as reflected in an increased organelle yield upon prolonged infiltration (4 vs 14 hours). The plastid-nuclei ratio is a convenient index to assess the maintenance of nuclei during a homogenization, especially if the average plastid numbers per cell are known (Butterfass, 1973). It is easily estimated with aldehyde-fixed samples.

1. PREPARATION OF NUCLEI

Seedlings or leaves of *Spinacia, Beta,* or *Antirrhinum* soaked at 0°C for 10 minutes in 0.5 % sodium hypochlorite containing 0.5 % Tween 80 are washed with sterile demineralized water. After evacuation in a 3-fold volume of sterile medium at room temperature, the sample is ventilated slowly. The medium consists of 0.2 M sucrose, 10 mM MES, 1 mM EDTA, 2 mM $MgCl_2$, 5 mM mercaptoethanol, 0.1 % BSA, pH 6.7. Other media may be used instead. The tissue is then chilled and incubated for several hours or overnight. All following steps are conducted at 0°–4°C. The infiltration solution replaced by approximately 3 volumes of fresh medium, and the tissue is homogenized by two to three bursts of 2–3 seconds each in a Waring blender, by hand chopping with razor blades, or at a rate of 120 chops per minute using a motor-driven device for about 5 minutes. Chopping is generally found to be superior to blending. The homogenate is strained and filtered, respectively, through nylon gauzes (pore sizes 50 or 20 μm; see Section III, A). This step efficiently removes whole cells and tissues. The residual material is chopped again, and the passages through gauzes are repeated. The nuclear fractions are prepared by low-speed centrifugation for 3 minutes at 250 g in a swinging-bucket head. The centrifuge is decelerated slowly to avoid whirling of the pellet. The supernatant, which is largely devoid of nuclei, is discarded. The sediment is washed once after resuspension by gentle agitation. The final faintly green pellet is resuspended in a 10- to 20-fold volume of medium and treated with detergent, or is directly transferred onto continuous silica gradients.

To solubilize chloroplasts, Mg^{2+} is added, followed by 0.05 to 0.1 volume of cooled 10% Triton X-100 in medium. In the presence of Mg^{2+}, the detergent solubilizes chloroplast membranes but does not destroy nuclear envelopes. The optimum concentration of Mg^{2+} (5–20 mM is suitable) and of the nonionic detergent may vary with the tissue. After a few minutes, the lysate is centrifuged at 250 g for 3 minutes. The green supernatant is discarded, and the crude nuclear pellet is washed twice in an excess of Mg^{2+}-containing medium. The final fraction should be completely free of chlorophyll, but could still contain some partially lysed, colorless plastid fragments (including probably membrane-associated plastid DNA; Herrmann et al., 1974), especially if the concentration of divalent cations was high.

An alternative possibility, probably more sparing to the functionality of nuclei and therefore preferred in our laboratory, is to destroy the plastid envelopes by a brief exposure to traces of Triton and to make use of the advantage of silica gradients to separate broken plastids from unbroken nuclei.

2. GRADIENT CENTRIFUGATION

The crude nuclear fractions are layered onto buffered discontinuous gradients of correct tonicity, consisting of 35, 32, 28, 24, and 20% silica in a ratio to sample of 0.5, 1.0, 1.0, 1.0, and 0.5, followed by centrifugation at 2000 g for 15 minutes. The nuclei band at the upper third of the gradient predominantly at the 20/24 and 24/28% silica interphases. Triton-treated nuclei sometimes tend to swell and to lose their internal structures. In extreme cases (absence of Mg^{2+}), they assume the shape of large membrane-surrounded sacs. Untreated nuclei are still contaminated by some chloroplasts and small particulate material (Fig. 9). The latter may be removed by higher gravity fields during gradient centrifugation. Further purification of nuclei may also be accomplished by differential centrifugation in silica-containing media (10–20%) at 500–1000 g for 2–4 minutes. The final suspension of twice- or thrice-sedimented nuclei is reasonably pure.

The preparations of leaf nuclei inevitably reflect the heterogeneity of

FIG. 9. Interference contrast micrograph of nuclei of *Spinacia oleracea* purified in a discontinuous Ludox gradient (sol concentrations 35, 32, 28, 24, and 20%; nuclei were taken from the 28/24% junction). Note the considerable differences in size and shape. × 1400.

plant tissues. They show very different nuclear shapes (including sickle- and spindle-shaped ones) and sizes, reflecting variation in cell type, ploidy, and physiological state.

D.   Mitochondria and Other Organelles

Nordhorn and Willenbrink (1972) and Gronebaum-Turck and Willenbrink (1971) prepared functionally active mitochondria of *Spinacia* and *Beta* in continuous Ludox HS 40 gradients containing 2% dextran. The gradient was preformed by centrifugation in a swinging-bucket head at 50,000 g for 20 minutes. Organelle separation was achieved under the same conditions of centrifugation. Unbroken mitochondria still contaminated by a few thylakoids banded at the top of the column (Turck and Willenbrink, 1969). A comparable gradient shown in Fig. 2a also separated mitochondria of *Acanthamoeba castellanii* from contaminating material.

Duffus and Rosie (1975) have purified amyloplasts in discontinuous gradients consisting of 90, 88, and 84% Ludox TM in water. Centrifugation was performed at 2500 g for 10 minutes, and the fractions obtained were subsequently characterized by assaying starch synthetase.

IV.   Purity and Functionality of Silica-Purified Organelles

We have outlined in the preceding sections methods for the purification of several subcellular particles in silica-containing media and discussed experimental problems. It is necessary to characterize the fractions obtained with regard to their purity, homogeneity, and pertinent physiological functions. Obviously, criteria for assessing organelle quality will depend upon the research interests of the investigator.

There is no doubt that on the basis of *cytological criteria* silica sols can be used with excellent results in small- and large-scale organelle preparations. The most important positive aspects of silica sol fractionation are that unbroken and broken organelles are efficiently separated from each other and that unbroken organelles can often be fractionated into morphologically distinct subpopulations. As already mentioned, this advantage can be exploited fully only if silica sol density gradient centrifugation is preceded or followed by careful differential centrifugation. Nevertheless, it has repeatedly been demonstrated by cytological and biochemical criteria that individual organelle bands may display a high degree of both homogeneity and purity.

Phase contrast microscopy provides a convenient method of distinguishing between broken and unbroken organelles (Kahn and von Wettstein, 1961), while interference contrast microscopy permits the detection of fine-structural changes, such as swelling of membrane areas and the detection of contaminating vesicles (Figs. 1, 3, 6, 7, and 9). Photographic analysis and documentation of cell fractions with the light microscope should generally be made with a flash unit. Otherwise, contaminating small particles will escape detection because of their Brownian movement (Fig. 1). Electron microscopy should be regarded as an indispensable technique in the qualitative assessment of organelle fractions. It has been used to estimate the degree of structural preservation and to demonstrate the absence of even minute particulates and of small quantities of cytoplasm that may adhere to organelle envelopes (Figs. 4 and 8). In order to obtain a quantitative estimate of purity, the cell fractions should be homogeneously mixed and surrounded by agar according to Preston *et al.* (1972) after aldehyde fixation. This procedure not only facilitates the preservation of labile structures, but also leads to the retention of small particles, which might otherwise be lost during dehydration and monomer infiltration before embedding.

Electron microscopic examination of plastid, nuclear, and mitochondrial fractions (including preparations of mutant plastids) from several plant species emphasizes that silica-treated organelles closely resemble the organelles of whole tissue sections. Furthermore, the purity and homogeneity of the organelles are far superior to those of organelles prepared by differential and sucrose density gradient centrifugation. For example, on scanning many sections through purified plastid preparations, such as those shown in the Figs. 3a and 8, almost no particulate material was detected, and less than one mitochondrion was seen for every 500 to 900 plastids.

Mutant plastids are presently identified exclusively on the basis of their morphology, which often is so characteristic that it serves as a reliable marker. We have demonstrated that the silica gradient is able to resolve unbroken organelles of a given mutant into several morphologically distinct classes. However, until a more complete study of mutant material is made, it will be impossible to state whether these classes are also distinct biochemically.

Though morphological analysis constitutes an essential element of judging the purity of subcellular fractions, it is evident that absence of visible contamination is not in itself a conclusive proof of purity of subcellular fractions. *Chemical* or *biochemical analysis* frequently gives better evidence of the presence or the absence of impurities. In turn, morphological integrity and purity bear no strict relationship to the *functional competence* of an organelle.

The biochemical characterization of silica-purified organelles has been poor until very recently; and tests for function have frequently been negative. For example, while some enzymic activities were not affected by the gradient material (Lyttleton, 1970; Schmitt *et al.,* 1974; Duffus and Rosie, 1975), others were inhibited to various degrees, some severely (Nilsson and Ronquist, 1969; Schmitt *et al.*, 1974). Similarly, whereas mitochondria prepared in Ludox HS 40 were apparently capable of high rates and control of respiration (Gronebaum-Turck and Willenbrink, 1971), photosynthetic reduction of phosphoglycerate or of carbon dioxide was found to be inhibited in chloroplasts purified in *commercial* Ludox HS 40 (Schmitt *et al.,* 1974). In some instances it had been possible to restore or to maintain fully or partly the activities by protective agents (see Section 11 and below). Several reasons have been proposed to account for these results. The exposure of an organelle to the colloid causes silica precipitation on organelle membranes (Schmitt *et al.,* 1974; see also Pertoft *et al.,* 1967). This precipitate, though providing an effective means of protecting more delicate particles against mechanical stress due to its stabilizing effect, can interfere in the extractability of compounds and in photometric (UV) analyses. It can also affect membrane properties. In fact, chloroplasts that had been exposed to Ludox HS 40 exhibited an increased permeability for pyridine nucleotides (Schmitt *et al.,* 1974). Silica may also interact with cofactors or enzymes. The inhibition of NAD-dependent phosphoglycerate kinase/triosephosphate dehydrogenase may well be explained by competition for magnesium (Schmitt *et al.,* 1974). Recent developments indicate, however, that it is possible to overcome these difficulties. Two examples may illustrate this.

The identification and determination of type-purity of plastid DNA has been the subject of considerable controversy during the past years. Silica sol gradient centrifugation permits to determine unequivocally the intracellular localization of DNA components. In certain species, such as *Antirrhinum majus,* chloroplast and mitochondrial DNA have base compositions and thus buoyant densities in CsCl different from each other and from that of nuclear DNA. Pycnographic analysis of DNA in CsCl equilibrium gradients is therefore a valid criterion to assess the type-purity of isolated organelles. Using this approach we have recently demonstrated that purification of chloroplasts in sucrose gradients does not yield type-pure chloroplast DNA. In contrast, silica sols efficiently remove the main sources of contaminating nuclear DNA (multiorganelle complexes, nuclei, nuclear fragments, broken plastids that entrap nuclear remnants; see Figs. 3a and 4). Only one unimodally banding DNA species was found in DNA extracts from silica-purified chloroplasts. The plastidic origin of this DNA is unquestionable since it was derived from a fraction of unbroken organelles (Herrmann et al., 1975).

As far as retention of functionality is concerned, Morgenthaler et al. (1975) have described a method for the isolation in silica gradients of chloroplasts that remain functionally active. These authors have found that carbon assimilation and associated oxygen evolution were apparently unaffected after purification of the organelles in Ludox AM. Photosynthesis rates as high as 100 μmoles of CO_2 reduction $\times$ mg chlorophyll^{-1} $\times$ hr^{-1} were obtained. The maintenance of complete photosynthesis can be taken as a sensitive indicator of the metabolic competence of the chloroplasts, as it involves the complete machinery of the carbon reduction cycle. Recently, the capacity of Ludox-purified organelles to incorporate radioactive amino acids into defined polypeptides was successfully used as another test for function (Morgenthaler and Mendiola-Morgenthaler, 1976). The Ludox media employed by Morgenthaler and co-workers differ from those used by others in that the sol, Ludox AM, was freed from yet unidentified toxic substances (see Section II) and that polyethylene glycol was added as protective agent. Presumably these modifications help to preserve the inner envelope since polyethylene glycol does not prevent sol precipitation on organelle surfaces (see Fig. 6 in Morgenthaler et al., 1975). This precipitate has been shown by electron microscopy to adhere exclusively to the outer envelope of unbroken organelles (Schmitt et al., 1974).

Morgenthaler's results indicate that the initial failure to obtain functional organelles after silica sol fractionation was due to avoidable damage. In view of the remarkable separation qualities of silica sols, the search for compounds that prevent adverse sol–organelle interactions without affecting separation therefore seems to be a rewarding task.

ACKNOWLEDGMENTS

The authors thank Professor Dr. U. Heber (University of Düsseldorf) and Dr. J. Bennett (University of Warwick) for critically reading the manuscript. We are greatly indebted to Mrs. B. Pietsch, Miss B. Schiller, and Miss M. Streubel for expert technical assistance. This work was supported by the Deutsche Forschungsgemeinschaft (Grant He 693/5).

REFERENCES

Butterfass, Th. (1973). *Protoplasma* **76**, 167–195.
Constabel, F. (1975). *In* "Plant Tissue Culture Methods" (O. L. Gamborg and L. R. Wetter, eds.), p. 109. Natl. Res. Council of Canada, Ottawa, Canada.
Duffus, C. M., and Rosie, R. (1975). *Anal. Biochem.* **65**, 11–18.
Frearson, E. M., Power, J. B., and Cocking, E. C. (1973). *Dev. Biol.* **33**, 130–137.
Gronebaum-Turck, K., and Willenbrink, J. (1971). *Planta* **100**, 337–346.
Heber, U. (1973). *Biochim. Biophys. Acta* **305**, 140–152.
Heber, U., and Kirk, M. R. (1975). *Biochim. Biophys. Acta* **376**, 136–150.
Heber, U., and Santarius, K. A. (1970). *Z. Naturforsch. Teil B* **25**, 718–728.
Hendriks, A. W. (1972). *FEBS Lett.* **24**, 101–105.
Herrmann, R. G., Kowallik, K. V., and Bohnert, H. J. (1974). *Port. Acta Biol., Ser. A* **14**, 91–110.
Herrmann, R. G., Bohnert, H. J., Kowallik, K. V., and Schmitt, J. M. (1975). *Biochim. Biophys. Acta* **378**, 305–317.
Honda, S. I., Hongladarom, T., and Laties, G. G. (1966). *J. Exp. Bot.* **17**, 460–472.
Jensen, R. G., and Bassham, J. A. (1966). *Proc. Natl. Acad. Sci. U.S.A.* **56**, 1095–1101.
Kahn, A., and von Wettstein, D. (1961). *J. Ultrastruct. Res.* **5**, 557–574.
Kuehl, L. R. (1964). *Z. Naturforsch. Teil B* **17**, 525–532.
Kutzelnigg, H., and Stubbe, W. (1974). *Sub-Cell Biochem.* **3**, 73–89.
Kutzelnigg, H., Meyer, B., and Schötz, F. (1975). *Biol. Zentralbl.* **94**, 513–526.
Larsson, C., Collin, C., and Albertsson, P. A. (1971). *Biochim. Biophys. Acta* **245**, 425–438.
Lyttleton, J. W. (1970). *Anal. Biochem.* **38**, 277–281.
Lyttleton, J. W., Ballantine, J. E. M., and Forde, B. J. (1971). *In* "Autonomy and Biogenesis of Mitochondria and Chloroplasts" (N. K. Boardman, A. W. Linnane, and R. M. Smillie, eds.), pp. 447–452. North-Holland Publ., Amsterdam, New York.
Mateyko, G. M., and Kopac, M. J. (1963). *Ann. N.Y. Acad. Sci.* **105**, 219–284.
Mehra-Palta, A., and Herrmann, P. G. (1975). *Proc. Int. Bot. Congr., 12th, 1975* Vol. p. 301.
Meselson, M., and Nazarian, G. M. (1963). *In* "Ultracentrifugal Analysis in Theory and Experiment" (J. W. Williams, ed.), pp. 131–142. Academic Press, New York.
Morgenthaler, J. J., and Mendiola-Morgenthaler, L. (1976). *Arch. Biochem. Biophys.* **172**, 51–58.
Morgenthaler, J. J., Price, C. A., Robinson, J. M., and Gibbs, M. (1974). *Plant Physiol.* **54**, 532–534.
Morgenthaler, J. J., Marsden, M. P. F., and Price, C. A. (1975). *Arch. Biochem. Biophys.* **168**, 289–301.
Nilsson, O., and Ronquist, G. (1969). *Biochim. Biophys. Acta* **183**, 1–9.
Nordhorn, G., and Willenbrink, J. (1972). *Planta* **103**, 147–154.
Pertoft, H. (1966). *Biochim. Biophys. Acta* **126**, 594–596.
Pertoft, H. (1967). *Exp. Cell Res.* **46**, 621–623.
Pertoft, H., and Laurent, T. C. (1969). *In* "Modern Separation Methods of Macromolecules and Particles" (T. Gerritsen, ed.), pp. 71–90. Wiley (Interscience), New York.
Pertoft, H., Philipson, L., Oxelfelt, P., and Höglund, S. (1967). *Virology* **33**, 185–196.
Preston, J. F., Parenti, F., and Eisenstadt, J. M. (1972). *Planta* **107**, 351–367.
Salisbury, J. L., Vasconcelos, A. C., and Floyd, G. L. (1975). *Plant Physiol.* **56**, 399–403.
Schmitt, J. M., Behnke, H. D., and Herrmann, R. G. (1974). *Exp. Cell Res.* **85**, 63–72.
Spencer, D., and Wildman, S. G. (1963) *Aust. J. Biol. Sci.* **15**, 599–610.
Suyama, Y., and Bonner, W. D. (1966). *Plant Physiol.* **41**, 383–388.
Turck, K., and Willenbrink, J. (1969). *Naturwissenschaften* **56**, 222.

Walker, D. A. (1971). *In* "Methods in Enzymology" (A. San Pietro, ed.), Vol. 23, Part A, pp. 211–220. Academic Press, New York.

Walles, B. (1971). *In* "Structure and Function of Chloroplasts" (M. Gibbs, ed.), pp. 51–88. Springer-Verlag, Berlin and New York.

Wolff, D. A., and Pertoft, H. (1972). *J. Cell Biol.* **55**, 579–585.

Chapter 12

Preparation of Photosynthetically Active Particles from Synchronized Cultures of Unicellular Algae[1]

HORST SENGER AND VOLKER MELL

Botanisches Institut der Universität Marburg,
Marburg, West Germany

I. Introduction

The preparation of chloroplasts and chloroplast particles from higher plants, as introduced by Hill (1937), provides more homogeneous experimental material than do whole cells, abolishes the effects of the stomatal

[1] This investigation was supported by the Deutsche Forschungsgemeinschaft.

barrier, transport, and permeation problems. The methods of preparation and the classification of chloroplasts have, however, become very elaborate (Hall, 1972).

Unicellular algae, introduced into photosynthetic research by Warburg (1919), provide most of the advantages of isolated chloroplasts. Only the problems of permeability of several photosynthetic inhibitors (Frickel-Faulstich and Senger, 1972), cofactors, electron acceptors and donors, as well as the demand to prepare particles of single photosystems (Grimme and Boardman, 1972, 1973) make it necessary to disrupt algae cells.

Several methods for breaking algae cells have been reported. Some of them are successful for only one species. Cells of *Chlorella* type are resistant to most of the methods.

The sudden expansion of *Chlorella* cells when passed through a capillary under high pressure (French press) has been used successfully for particle preparation (French and Milner, 1955). This procedure is still widely used, but it yields satisfactory and reproducible amounts of disruption only for a few organisms.

Breaking by ultrasonic treatment is successful only for cells with walls of limited resistance, like *Chlamydomonas* (Levine and Volkmann, 1961; Boschetti and Grob, 1969). *Chlamydomonas* and *Euglena* cells can even be disrupted with sand in a mortar (Gorman and Levine, 1965; Price and Hirvonen, 1967).

Cell walls of *Euglena* have been dissolved enzymically by trypsin and the whole protoplast was obtained (Rawson and Stutz, 1969). Böger (1971) successfully isolated whole protoplasts from three species of blue-green algae by enzymic treatment. Our similar experiments with *Chlorella* were unsuccessful.

A Waring blender was used to isolate chloroplasts from *Euglena* (Kahn, 1966). The method of shaking cell suspensions with glass, steel, or polyethylene beads has been the most successful and widely used procedure. Of the apparatus employed, the Nossal Shaker, the Braun-Merkenschlager homogenizer, and the Vibrogen Zellmühle are the best known. The Nossal shaker has been applied on *Scenedesmus* (Kok and Datko, 1965) and *Chlorella* cells (Grimme and Boardman, 1973). The Braun Merkenschlager homogenizer yielded good disruption of *Chlorella* (Böger, 1964) and *Bumilleriopsis filiformis* (Böger, 1969).

Successful application of the Vibrogen Zellmühle was reported for *Chlorella, Cyanidium, and Anacystis* (Richter and Senger, 1964) and for *Scenedesmus* (Pratt and Bishop, 1968; Berzborn and Bishop, 1973).

More comprehensive literature of chloroplast preparation can be found in San Pietro (1971) and Jacobi (1972, 1974).

Many of the methods reported do not work on *Chlorella*- type cells, and the different parameters influencing the percent disruption have not been systematically investigated. The current investigation was carried out to establish a method fulfilling the following requirements: (1) fast disruption with high yield, (2) high activity of the particles, (3) applicability at different stages of synchronous cultures of different algae, (4) high reproducibility.

II. Organisms, Growth Conditions, Synchronization

In the study to be reported here, the unicellular green algae *Chlorella pyrenoidosa,* strain 211-8b (Algae Collection, Göttingen) and *Scenedesmus obliquus,* strain D3 (Gaffron) were used. Additional experiments (not extensively reported here) have been carried out with *Chlamydomonas reinhardii,* strain 11–32 (Algae Collection, Göttingen).

The culture medium employed for *Chlorella* and *Scenedesmus* was described by Bishop and Senger (1971) and that for *Chlamydomonas* by Kuhl (1962). All cultures were grown at 28°C in a light-thermostat (Kuhl and Lorenzen, 1964) and aerated with 3% CO_2 in air (20 liters/hour).

The synchronization was obtained by a light–dark regime of 14 : 10 hours combined with a photoelectrically controlled automatic dilution at the beginning of each life cycle (Senger *et al.,* 1972). The number of starting cells was 3.12×10^6 cells/ml for *Chlorella* and *Scenedesmus* and 156×10^6 cells/ml for *Chlamydomonas.* To obtain enough cells for the experiments, cells were subcultured from an automatically controlled culture into many culture tubes.

The synchronization achieved for the three organisms met the standards proposed by Senger and Bishop (1969). A comparative characterization of the photosynthetic behaviour and the changes in the photosynthetic apparatus of synchronized cultures of *Chlorella, Scenedesmus,* and *Chlamydomonas* has been reported (Senger, 1975).

III. Particle Preparation

A. The Vibrogen Zellmühle

The Vibrogen Zellmühle was applied for disruption of the cells in all experiments reported here (Fig. 1). The principle is a vertical shaking of a stainless steel cup with a frequency of 75 Hz and an amplitude of 7 mm. Cups of 7-, 18-, 52-, or 108-ml volume could be inserted in the shaker. The

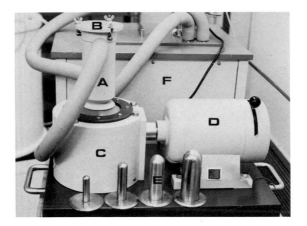

FIG. 1.   The Vibrogen Zellmühle. A, Compartment; B, lid; C, eccentric shaker; D, motor; E, stainless steel cups; F, cryostat. Manufactured by Firma E. Bühler, 7400 Tübingen, Germany.

cups were filled with glass beads, algae, and buffer (Fig. 2). As it turned out, the rate of disruption was dependent on the size of the glass beads, the ratio of beads to algae, the free space above the liquid and the shaking time. The flow-through attachment for continuous breaking was not used in these experiments.

After disruption of the cells, each sample was microscopically examined for the success of breaking. During the shaking of the cells with glass beads, the cell walls split open and the protoplasts are squeezed out. The remaining cell walls ("ghosts") can readily be seen under the microscope, especially with the phase contrast method. To estimate the percentage of disrupted cells the ratio of cell number before breaking to unbroken cells after the suspension had been separated from the glass beads was not satisfactory, since some cells were lost and the suspension was diluted during the washing procedure. We thus used the ratio of the number of "ghosts" to the number of

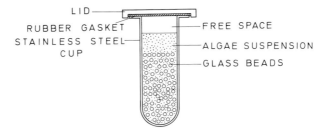

FIG. 2.   Schematic diagram of a stainless steel cup with glass beads and algal suspension.

"ghosts" plus unbroken cells and expressed this ratio in percent. The counting was done in a hemacytometer under the microscope.

B. Temperature Control

The stainless steel cup of the Vibrogen Zellmühle is surrounded by a jacket through which a cooling liquid can be pumped. For cooling we used a cryostat (Type MC 3, Colora Messtechnik, Lorch), and water at 0.5°C was constantly pumped through the jacket.

Since it is very important for the maintenance of the activity of the particles to keep the temperature low, we tested the temperature in the beaker after different shaking times. After 5 or 10 minutes of shaking, the temperature in the beaker increased, respectively, 1° or 2°C above the cooling temperature.

C. Cell Disruption and Resuspension Media

For the preservation of the activity of the particles, several media were tested during disruption. The best results were obtained with the medium reported by Berzborn and Bishop (1973), modified after Selman and Bannister (1972) and Garnier and Maroc (1972) (medium I): NaCl, 10^{-2} M; KCl, 10^{-2} M; $MgCl_2$, 2.5×10^{-3} M; EDTA, 10^{-5} M; Tricine/KOH buffer, pH 7.5, 2×10^{-2} M.

For photosystem I reactions, the buffer was replaced by a phosphate buffer, pH 6.5, 5×10^{-2} M (medium II); this medium was also used as the resuspension medium.

The glass beads were separated from the cell debris by filtration through a coarse fritted-glass funnel into a suction flask and washed twice with a total of 30 ml of resuspension solution. The resuspension medium contained in addition 0.56 M sucrose (medium III) to establish a final concentration of the combined suspensions of 0.34 M sucrose.

D. Dependence on Diameter and Amount of Glass Beads

It is demonstrated in Figs. 3 and 4 that achievement of disruption is extremely dependent on the size of the glass beads. It is surprising that glass beads with a diameter about 100 times that of the cells cause the best breaking. In this aspect it is understandable that the relatively small differences in size of cells of the different stages of the synchronous culture have little effect on the rate or percentage of disruption. For the different organisms and the various developmental stages investigated, glass beads with a diameter of 0.7 mm proved to be the most effective. The same was true for *Chlamydomonas*. Thus in all further experiments, glass beads of 0.7 mm were used.

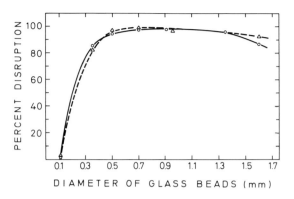

FIG. 3.   Percent disruption of cells from the 8th (O———O) or 16th (△---△) hour of a synchronous culture of *Chlorella pyrenoidosa* in relation to the diameter of the glass beads; 52-ml cup, 5 minutes shaking time.

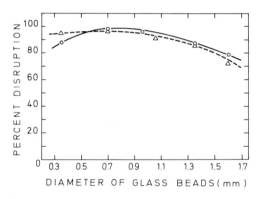

FIG. 4.   Percent disruption of cells from the 8th (O———O) or 16th (△---△) hour of a synchronous culture of *Scenedesmus obliquus* in relation to the diameter of the glass beads; 52-ml cup, 5 minutes shaking time.

   The ratio of the volume of glass beads to the volume of algal suspension influenced the rupture of the cells, as demonstrated in Table I. The best percentage of broken cells was obtained when the stainless steel cup (52 ml) was filled within 2 mm of the surface with beads (50 ml) and the interspace between the beads was filled with the algal suspension (19.5 ml). This free space between the surface of the cup filling and the lid was quite important, as can be concluded from the data of Table II. Since an increase of free space of more than 2 mm did not improve the breakage very much, and additional air could possibly increase oxidation processes, we left a free space of 2 mm as a standard procedure for all following experiments.

TABLE I

EFFECT OF THE VOLUME RATIO OF GLASS BEADS TO ALGAL
SUSPENSION ON PERCENT DISRUPTION[a]

Glass beads (ml)	20	30	40	45	50
Algal suspension (ml)	39	32	26	23	19.5
Percent disruption	18	30	52	67	80

[a] Algal suspensions (25 μl of PCV/ml) were shaken with glass beads 0.7 mm in diameter for 3 minutes in a 52-ml stainless steel cup.

TABLE II

EFFECT OF FREE SPACE AT THE SURFACE OF THE ALGAL/GLASS
BEADS MIXTURE[a]

Free space (mm)	0	2	5	10
Glass beads (ml)	51.5	50.2	48.5	44.5
Algal suspension (ml)	20	19.5	18.2	16.2
Percent disruption	65	90	95	95

[a] Algal suspensions (25 μl of PCV/ml) were shaken with glass beads 0.7 mm in diameter for 5 minutes in a 52-ml stainless steel cup.

E. Dependence on Disruption Time

For the activity of chloroplasts or particles, it is most important to keep the preparation time as short as possible (Kulandaivelu and Hall, 1976). On the other hand, the shaking time has to be long enough to guarantee a high percentage of disruption. The dependence of disruption on time is shown in Figs. 5 and 6. The *Chlorella* cells reach the optimal disruption within

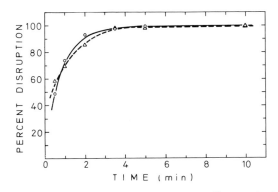

FIG. 5. Percent disruption of cells from the 8th (O———O) or 16th (△‑‑‑△) hour of a synchronous culture of *Chlorella pyrenoidosa* in relation to the shaking time; 52-ml beaker, 50 ml of glass beads 0.7 mm in diameter, 19.5 ml of algal suspension of 25 μl of PCV/ml.

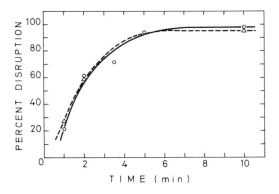

FIG. 6.   Percent disruption of cells from the 8th (O———O) or 16th (△---△) hour of a synchronous culture of *Scenedesmus obliquus* in relation to the shaking time; 52-ml beaker, 50 ml of glass beads 0.7 mm in diameter, 19.5 ml of algal suspension of 25 μl of PCV/ml.

3 minutes. The *Scenedesmus* cells demonstrate even more resistance against shaking, and it takes about 5 minutes to get optimal disruption. For both organisms the developmental stage of the synchronous cultures is of minor influence. Allowing 5 minutes of shaking time guaranteed at least 90% disruption for both strains and all developmental stages. The same was true for *Chlamydomonas reinhardii*.

F.   Standardized Procedure

A sample equivalent to 500 μl of PCV was drawn from the culture, centrifuged, washed with medium I, and resuspended in 20 ml of medium I. The 52 ml stainless steel cup was filled with 50 ml of glass beads 0.7 mm in diameter and 19.5 ml of the cell suspension. Glass beads and cell suspension were homogeneously mixed with a glass rod. The lid was closed, and the mixture was cooled to 1.4°C, then shaken for 5 minutes. The cell debris was separated from the glass beads by sucking through a coarse fritted-glass funnel. The glass beads were washed with 30 ml of medium III, and the suspensions in medium I and medium III were combined. To separate the whole cells from the particles, the suspension was centrifuged for 5 minutes at 600 g. The supernatant was checked microscopically for whole cells and used for the experiments.

This method works for *Chlorella*, *Scenedesmus*, and *Chlamydomonas* during the synchronous growth, for homocontinuous and fast growing autotrophic and heterotrophic batch cultures. With aging of a batch culture, the resistance of its cells to disruption increases considerably.

IV.   Particle Characterization

A.   Size Analysis

To ensure that the cells were completely broken and to estimate the particle size, whole cells and particles were measured with a Coulter Counter (Model Z_B with Channelyzer, Coulter Electronics). The volume distribution of whole cells of the 8th and 16th hour of synchronous cultures of *Chlorella* and *Scenedesmus* are shown in Figs. 7 and 8. The cells differ significantly in size in relation to age and species. It is more surprising that particles yielded in all 4 cases are of more or less the same size and distribution (Figs. 9 and 10). Since the cells of these algae contain only one chloroplast, the ratio in size of whole cells to particles (Table III) clearly demonstrates that there are only small particles of chloroplasts in the supernatant. Microscopic and size distribution controls clearly proved that there were no whole cells in the supernatant used in the experiments.

B.   Pigment Analysis

The disruption of the cells into relatively small particles might have caused a fractionation in different pigment groups. In order to test whether this might be the case, low-temperature spectra of whole cells, crude extracts, and particles were recorded with a Shimadzu MPS 50L spectrophotometer (Figs. 11 and 12). There is no change in the position of peaks and shoulders. Changes in the heights of the extinction in the blue region might be due to stray light effects of the frozen sample. From the similarity of the spectra, we draw the conclusion that the pigment composition of the particles is the same as that of whole cells.

V.   Photosynthetic Reactions of the Particles

The effectiveness of the procedure in the preparation of photosynthetically active particles can be evaluated only in photosynthetic experiments. Thus we tested the particles for their capacity for photosystem II and photosystem I reactions. Another interesting question was whether the particles of the 8th and 16th hour of the synchronous life cycles would show the same differences as seen in photosystem II reactions of whole cells. Since changes in the photosynthetic apparatus of synchronous cultures of *Scenedesmus* are best known, we concentrated on tests with this organism. Reactions with particles of

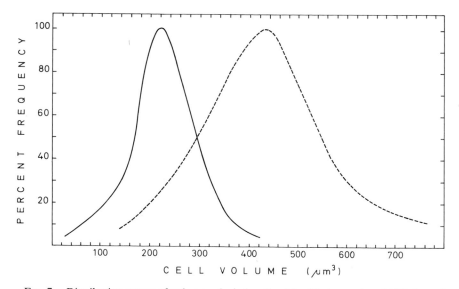

FIG. 7.   Distribution curves of volumes of whole cells of the 8th (———) and 16th (– – – –)
hour of a synchronous culture of *Chlorella pyrenoidosa*. Volumes were determined with the
Coulter Counter. Synchronization as described in Section II.

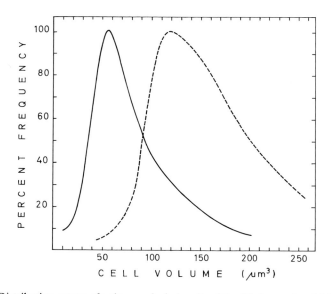

FIG. 8.   Distribution curves of volumes of whole cells of the 8th (———) and 16th (– – – –)
hour of a synchronous culture of *Scendesmus obliquus*. Volumes were determined with the
Coulter Counter. Synchronization as described in Section II.

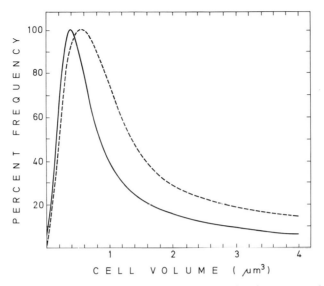

FIG. 9.   Distribution curves of volumes of particles of cells of the 8th (———) and 16th (– – – –) hour of a synchronous culture of *Chlorella pyrenoidosa*. Volumes were determined with the Coulter Counter. Particles were obtained as described in Section III,F.

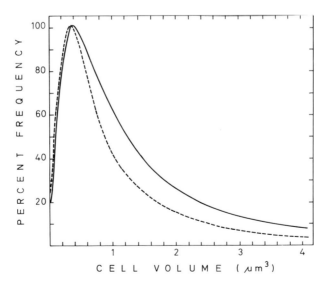

FIG. 10.   Distribution curves of volumes of particles of cells of the 8th (———) and 16th (– – – –) hour of a synchronous culture of *Scenedesmus obliquus*. Volumes were determined with the Coulter Counter. Particles were obtained as described in Section III,F.

TABLE III

Most Frequent Volumes of Cells and Particles of *Chlorella* and *Scenedesmus*[a]

	Chlorella		Scenedesmus	
	8th Hr	16th Hr	8th Hr	16 Hr
Whole cells	65	120	220	430
Particles	0.4	0.6	0.4	0.4

[a] Volumes of cells and particles measured with a Coulter Counter are given in μm^3. Cells were disrupted as described in Section III,F.

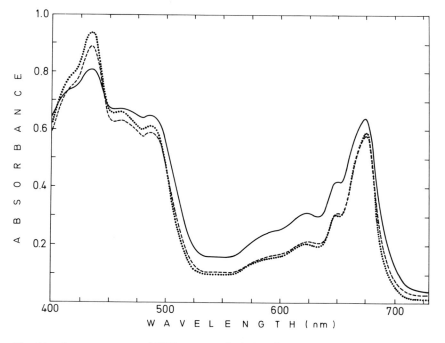

Fig. 11. Low temperature (77°K) spectra of whole cells (———), crude extract (————), and particles (. . . .) of cells of the 8th hour of a synchronous culture of *Scenedesmus obliquus*. Spectra were recorded with a Shimadzu MPS 50 L spectrophotometer. Cells were disrupted as described in Section III,F.

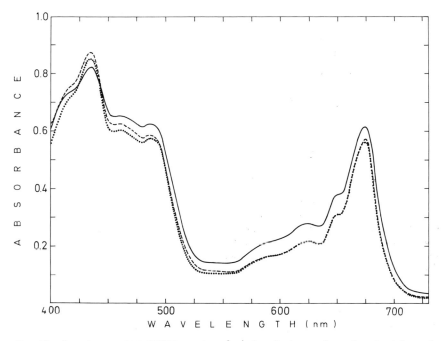

FIG. 12.   Low-temperature (77°K) spectra of whole cells (———), crude extract (– – – –), and particles (. . . .) of cells of the 16th hour of a synchronous culture of *Scenedesmus obliquus*. Spectra were recorded with a Shimadzu MPS 50 L spectrophotometer. Cells were disrupted as described in Section III,F.

Chlorella and *Chlamydomonas* have been reported elsewhere (Senger, 1975; Senger and Frickel-Faulstich, 1975).

A.   Photosystem II Reactions

To evaluate the activity of photosystem II and the water-splitting system, Hill reactions with *p*-benzoquinone or dichlorophenolindophenol (DCPIP) as electron acceptors were carried out. Since *p*-benzoquinone easily penetrates the cell membranes, it was most suitable to compare Hill reaction rates in whole cells and particles. Whole cells were suspended in phosphate buffer (0.05 M, pH 7); *p*-benzoquinone at a final concentration of 5×10^{-3} M was added, and oxygen evolution upon illumination with different intensities of red light above 620 nm was measured polarographically (Oxygraph, Gilson Medical Electronics). Particles were suspended in the combination of media I and III. The measuring procedure was the same as for whole cells. The light-dependent oxygen evolution of Hill reaction in particles of cells

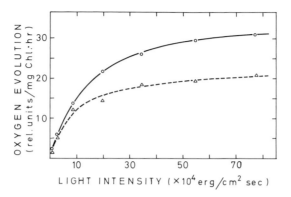

FIG. 13.   Light intensity-dependent oxygen evolution of Hill reaction with *p*-benzoquinone (final conc. 5×10^{-3} *M*) in particles of cells of the 8th (O ———O) and 16th (△---△) hour of a synchronous culture of *Scenedesmus obliquus*. Oxygen evolution was measured polarographically under red light above 620 nm and calculated per milligram of chlorophyll per hour. Cells were synchronized as described in Section II, and particles were obtained as described in Section III,F.

from the 8th and 16th hour of a synchronous culture is shown in Fig. 13. These curves are very similar to those obtained for Hill reaction or photosynthetic oxygen evolution in whole cells (cf. Senger, 1975). The relative rates per chlorophyll obtained in the saturated light region are given in Table IV. The oxygen evolution of Hill reaction with whole cells reached about 75 % of that of total photosynthesis. The activity of particles is almost identical.

Another method of measuring photosystem II activity is the Hill reaction with DCPIP. The light-dependent reduction of DCPIP (final concentration

TABLE IV

COMPARISON OF PHOTOSYNTHESIS AND HILL REACTION IN WHOLE CELLS WITH
HILL REACTION IN PARTICLES OF THE 8TH AND 16TH HOUR OF A
SYNCHRONOUS CULTURE OF *Scenedesmus obliquus*[a]

	8th Hr	16th Hr
Photosynthesis	41.4	30.0
Hill reaction, whole cells	30.8	22.8
Hill reaction, particles	31.1	20.8

[a] Oxygen evolution was measured polarographically in saturated red light above 620 nm and expressed in arbitrary units per chlorophyll per hour. For further conditions see Fig. 13.

4×10^{-5} M) in the presence of active particles was determined spectrophotometrically as absorbance change at 578 nm in a Shimadzu MPS 50 L spectrophotometer (see Senger and Frickel-Faulstich, 1975, for details). The result with particles from cells of the 8th and 16th hour of a synchronous culture of *Scenedesmus obliquus* are shown in Fig. 14. These curves represent the same pattern as those of the Hill reaction with *p*-benzoquinone. The difference in activity between cells of the 8th and 16th hour is the same as in total photosynthesis and *p*-benzoquinone Hill reaction. Oxygen evolution of total photosynthesis, measured manomentrically in Warburg buffer No. 9, yielded 249 and 159 μmoles of O_2 per milligram of chlorophyll per hour for the 8th and 16th hour, respectively, of the synchronous life cycle. Theoretically, the rates of oxygen and reduced DCPIP produced should have a ratio of $1:2$. The observed rates of 485 and 293 μmoles of DCPIP red per milligram of chlorophyll per hour, respectively, approach this theoretical value.

These results indicate that the particles are highly active in photosystem II activity, that the water splitting system is still intact, and that the changes in the photosynthetic capacity of whole cells are manifested to the same extent in the particles.

B.   Photosystem I Reactions

The activity of photosystem I was measured as reduction of methylviologen with artificial electron donors. The reduction of methylviologen (final concentration 2×10^{-4} M) was assayed polarographically as oxygen uptake

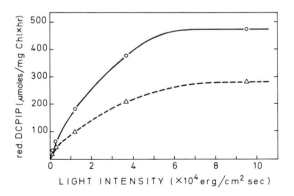

FIG. 14.   Light intensity-dependent reduction of the Hill reagent dichlorophenolindophenol (DCPIP) (final conc. 4×10^{-5} M) in particles of the 8th (O———O) and 16th (△---△) hour of a synchronous culture of *Scenedesmus obliquus*. DCPIP reduction was measured spectrophotometrically at 578 nm under actinic illumination with red light above 620 nm and calculated per milligram of chlorophyll per hour. For further conditions see Fig. 13.

with the oxygen under different intensities of red light above 620 nm. As electron donor the following systems were used: ascorbate $(5 \times 10^{-3}$ M/DCPIP, 10^{-4} M) (cf. Strotmann and von Gösseln, 1972) and ascorbate 5×10^{-3} M/2, 3, 5, 6-tetramethylphenylenediamine (DAD) 10^{-4} M (cf. Trebst and Pistorius, 1965; Izawa et al., 1966).

To measure the photosystem I reaction exclusively and to avoid any electron flow from photosystem II 2 μM 3-(3, 4-dichlorophenyl)-1, 1-dimethylurea (DCMU) was added to the reaction mixture. Particles were prepared from cells of the 8th hour of a synchronous culture of *Scenedesmus obliquus*. The results are demonstrated in Fig. 15.

The curves of light-dependent oxygen uptake are very similar to those of the Hill reaction. The lower values of the photosystem I reaction with ascorbate/DCPIP as electron donors might be due to the different sites where the electrons are fed into the electron transport chain.

A comparison of the maximal values of methylviologen reduction, assayed as O_2 uptake, is presented in Table V. The lower activity obtained with ascorbate/DCPIP as electron donor is most probably due to the fact that the electrons are fed into the transport chain before plastocyanine and that plastocyanine is partially washed out during the disruption process. The data obtained with the donor system ascorbate/DAD were very good in comparison to those of spinach chloroplasts.

Experiments with *Chlamydomonas* support these results. Particles pre-

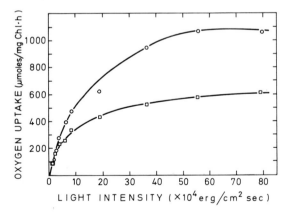

FIG. 15. Photosystem I-mediated reduction of methylviologen assayed as O_2 uptake in relation to different intensities of actinic red light above 620 nm. Ascorbate/DCPIP (□———□) or ascorbate/DAD (O———O) served as artificial e⁻ donors. Photosystem II was blocked by DCMU (0.2 μM). O_2 uptake was measured polarographically in the oxygraph. Particles were prepared (see Section III,F) from cells of the 8th hour of a synchronous culture of *Scenedesmus obliquus* (see Section II).

TABLE V

COMPARISON OF MAXIMAL VALUES OF PHOTOSYSTEM I-MEDIATED REDUCTION OF
METHYLVIOLOGEN ASSAYED AS O_2 UPTAKE

Author	Subject	O_2/Chlorophyll per hour (μmoles/mg)	e$^-$ donor[a]
Epel and Neumann (1973)	*Lactuca* Class II chloroplasts	1087	Asc/DCPIP
Postius (1972)	*Spinacia* chloroplasts	1220	Asc/DCPIP
Own measurements	*Scenedesmus* particles	630	Asc/DCPIP
Bar-Nun and Ohad (1975)	*Chlamydomonas* PS I particles	3000	Asc/DCPIP
Hauska (1975)	*Spinacia* chloroplasts	815	Asc/DAD
Ort and Izawa (1973)	*Spinacia* chloroplasts	1560	Asc/DAD
Own measurements	*Scenedesmus* particles	1130	Asc/DAD

[a]Asc, ascorbate; DCPIP, dichlorophenolindophenol; DAD, 2,3,5,6-tetramethylphenylene-diamine.

pared with the same method from *Scenedesmus* and *Chlamydomonas* provided excellent material to determine the concentrations of different cytochromes (Senger and Frickel-Faulstich, 1975).

VI. Evaluation of Reproducibility of Particle Preparations and Activity

High reproducibility of experimental results are most desirable. Synchronous cultures are very well standardized and present therefore a good initial experimental material. For such synchronous cultures, the reproducibility

TABLE VI

MEAN VALUES AND STANDARD DEVIATION OF PERCENT DISRUPTION
OF CELLS FROM SYNCHRONOUS CULTURES OF *Chlorella pyrenoidosa*
AND *Scenedesmus obliquus*[a]

Culture	8th Hour	16th Hour
Chlorella	97.5 ± 2.1	98.9 ± 1.4
Scenedesmus	97.1 ± 4.3	95.9 ± 4.0

[a] For details see Sections II (synchronization) and III,F (disruption).

TABLE VII

MEAN VALUES AND STANDARD DEVIATION OF HILL REACTIONS WITH
p-BENZOQUINONE AND DCPIP AS ACCEPTORS IN PARTICLE
PREPARATIONS OF CELLS FROM SYNCHRONOUS CULTURES OF
Scenedesmus obliquus[a]

Hill reagent	8th Hour	16th Hour
p-Benzoquinone	31.1 ± 5.2	21.0 ± 5.1
DCPIP	481 ± 116	293 ± 32

[a] Data are expressed as arbitrary units of O_2 per milligram of chlorophyll per hour (*p*-benzoquinone) or micromoles of O_2 per milligram of chlorophyll per hour (dichlorophenolindophenol). For details see Sections II (synchronization), III,F (disruption, and V,A (Hill reactions)).

of disruption has a standard deviation of less than 5% (Table VI). This accuracy has hardly been reached by any other method of cell breaking.

Many different parameters do influence the activity of chloroplast or particle preparations. In the literature reports on deviations of 2- and 3-fold in parallel experiments are quite common. Thus the deviation in photosystem II activity (Table VII) with maximal 40% is very good in comparison to other results.

Thus the demands set up for particle preparation in the Introduction are mostly met by the results.

The disruption is fast with a high yield and an efficient cooling system. The particles demonstrate high activities of photosystem II and I reactions. The method can be applied with high reproducibility on several organisms and their different developmental stages in synchronized cultures.

ACKNOWLEDGMENTS

We thank Mrs. A. Altmann, Mrs. G. Müller, and Mr. H. Hohl for technical assistance and Mrs. I. Caspar and Mr. H. Becker for valuable help in preparing the manuscript.

REFERENCES

Bar-Nun, S., and Ohad, J. (1975). *Proc. Int. Congr. Photosynth. 3rd, 1974* pp. 1627–1638.
Berzborn, R. J., and Bishop, N. I. (1973). *Biochim. Biophys. Acta* **292**, 700–714.
Bishop, N. I., and Senger, H. (1971). *In* "Methods in Enzymology", (A. San Pietro, ed.), pp. 53–66. Vol. 23, Part A, Academic Press, New York.
Böger, P. (1964). *Flora (Jena)* **154**, 174–211.
Böger, P. (1969). *Z. Pflanzenphysiol.* **61**, 85.
Böger, P. (1971). *In* "Methods in Enzymology" (A. San Pietro, ed.), Vol. 23, Part A, pp. 242–248. Academic Press, New York.
Boschetti, A., and Grob, E. C. (1969). *Prog. Photosynth. Res., Proc. Int. Congr. 1st, 1968* Vol. I, pp. 245–249.

Epel, B. L., and Neumann, J. (1973). *Biochim. Biophys. Acta* **325**, 520–529.

French, C. S., and Milner, H. W. (1955). *In* "Methods in Enzymology" (S. P. Colowick and N. O. Kaplan, eds.), Vol. 1, pp. 64–67. Academic Press, New York.

Frickel-Faulstich, B., and Senger, H. (1972). *Ber. Deusch. Bot. Ges.* **85**, 401–408.

Garnier, J., and Maroc, J. (1972). *Proc. Int. Congr. Photosynth. Res., 2nd, 1971* pp. 669–675.

Gorman, S., and Levine, R. P. (1965). *Proc. Natl. Acad. Sci. U.S.A.* **54**, 1665.

Grimme, L. H., and Boardman, N. K. (1972). *Biochem. Biophys. Res. Commun.* **49**, 1617.

Grimme, L. H., and Boardman, N. K. (1973). *Hoppe Seyler's Z. Physiol. Chem.* **354**, 1499.

Hall, D. O. (1972). *Nature (London) New Biol.* **235**, 125–126.

Hauska, G. A. (1975). *Proc. Int. Congr. Photosynth. 3rd, 1974* pp. 689–696.

Hill, R. (1937). *Nature (London)* **139**, 881–882.

Izawa, S., Conolly, I. N., Winget, G. D., and Good, N. E. (1966). *Brookhaven Symp. Biol.* **19** 169–127.

Jacobi, G. (1972). "Chloroplast Fragments." Eigenverlag, Göttingen.

Jacobi, G. (1974). "Biochemische Cytologie der Pflanzenzelle." Thieme, Stuttgart.

Kahn, J. S. (1966). *Biochem. Biophys. Res. Commun.* **24**, 329.

Kok, B., and Datko, E. A. (1965). *Plant Physiol.* **40**, 1171–1177.

Kuhl, A. (1962). *In* "Beiträge zur Physiologie und Morphologie der Algen," pp. 157–164. Fischer, Stuttgart.

Kuhl, A., and Lorenzen, H. (1964). *Methods Cell Physiol.* **1**, 159–187.

Kulandaivelu, G., and Hall, D. O. (1976). *Z. Naturforsch.* **31c**, 82–84.

Levine, R. P., and Volkmann, D. (1961). *Biochem. Biophys. Res. Commun.* **6**, 264.

Ort, D. R., and Izawa, S. (1973). *Plant Physiol.* **52**, 595–600.

Postius, S. (1972). *Proc. Int. Congr. Photosynth. Res., 2nd, 1971* Vol. 1, pp. 739–744.

Pratt, L. H., and Bishop, N. I. (1968). *Biochim. Biophys. Acta* **153**, 664–674.

Price, C. A., and Hirvonen, A. P. (1967). *Biochim. Biophys. Acta* **148**, 531–538.

Rawson, J. R., and Stutz, E. (1969). *Biochim. Biophys. Acta* **190**, 368–380.

Richter, G., and Senger, H. (1964). *Biochim. Biophys. Acta* **87**, 502–505.

San Pietro, A., ed. (1971). "Methods in Enzymology," Vol. 23, Part A, Sect. III. Academic Press, New York.

Selman, B. R., and Bannister, T. T. (1972). *Proc. Int. Congr. Photosynth. Res., 2nd, 1971* pp. 577–586.

Senger, H. (1975). "Les Cycles Cellulaires et leur Blocage chez Plusieur Protistes" Colloques Intern. CNRS, No. 240, 1974, pp. 101–108.

Senger, H., and Bishop, N. I. (1969). *In* "The Cell Cycle" (G. M. Padilla, G. L. Whitson, and I. L. Cameron, eds.), pp. 179–202. Academic Press, New York.

Senger, H., and Frickel-Faulstich, B. (1975). *Proc. Int. Congr. Photosynth., 3rd, 1974* pp. 715–727.

Senger, H., Pfau, J., and Werthmüller, K. -J. (1972). *Methods Cell Physiol.* **5**, 301–323.

Strotmann, H., and von Gösseln, C. (1972). *Z. Naturforsch., Teil B* **27**, 445–455.

Trebst, A., and Pistorius, E. (1965). *Z. Naturforsch., Teil B* **20**, 143–147.

Warburg, O. (1919). *Biochem. Z.* **100**, 230–270.

Chapter 13

Rapid Isolation of Nucleoli from Detergent-Purified Nuclei of Tumor and Tissue Culture Cells

MASAMI MURAMATSU AND TOSHIO ONISHI

Department of Biochemistry,
Tokushima University School of Medicine,
Tokushima, Japan

I. Introduction

In a previous review, one of the authors described general procedures for isolation of nuclei and nucleoli from various mammalian cells including tumor cells (Muramatsu, 1970). Techniques for isolation of nucleoli could be divided into three categories depending upon their basic principles of nuclear disruption: (1) sonication (Monty *et al.*, 1956; Rees *et al.*, 1962; Maggio *et al.*, 1963; Muramatsu *et al.*, 1963), (2) compression and decompression (Desjardins *et al.*, 1963; Liau *et al.*, 1965; Liau and Perry, 1969; McConkey and Hopkins, 1964), and (3) DNase treatment (Penman *et al.*, 1966; Willems *et al.*, 1968).

In each case, it is always preferable to isolate nuclei before disruption rather than to start with whole cells. However, it is known that most tumor cells are very resistant to homogenization, especially when divalent cations are used in the homogenization medium (Dounce, 1963; Muramatsu and Busch, 1967; Muramatsu, 1970). Since the inclusion of a critical amount of either Ca^{2+} or Mg^{2+} is the prerequisite for the isolation of intact nucleoli by the above procedures, nuclear preparations with various amounts of cytoplasmic tags have long been utilized for the isolation of nucleoli, except for a kind of tumor cell, such as Morris hepatoma 5123 D, for which relatively pure nuclear preparations could be obtained with a simple hypertonic sucrose procedure (Chauveau *et al.*, 1956; Muramatsu *et al.*, 1968). In fact, for usual purposes, small amounts of perinuclear cytoplasm of tumor cells did not hamper the isolation of nucleoli from these tissues because they were eventually broken up into small pieces by sonication and removed during the purification step.

However, isolation of nucleoli from cytoplasm-free nuclei is essential at least in two cases: when comparisons have to be made on certain nucleolar components directly with the nuclear counterparts without any interference from cytoplasm; and when there is a risk of damage of nucleolar components due to their contact with cytoplasmic components, such as lysosomal enzymes. In both cases, contamination of cytoplasm into the nuclear preparation may seriously affect the experimental results.

DNase treatment at a high ionic strength has been used to isolate nucleolar material from detergent-purified nuclei of HeLa cells (Penman *et al.*, 1966; Willems *et al.*, 1968). This method has been utilized successfully for the study of *in vivo* RNA and ribonucleoprotein metabolism, but could not be used in studies involving nucleolar DNA; e.g., *in vitro* RNA synthesis by isolated nucleoli. Small amounts of DNase are known to knock off the RNA synthetic activity of isolated nucleoli (Ro *et al.*, 1964). Incubation at a high ionic strength required for this method would inevitably extract considerable protein, including histones and other nuclear proteins as well as nucleolar RNA polymerase.

Therefore, it is highly desirable to work out a procedure to isolate nucleoli from "cytoplasm-free" nuclei of tumor cells without the use of DNase and high ionic solutions. In this chapter, we shall describe a method of isolation of nucleoli from detergent-purified nuclei from various tumor and tissue culture cells (Muramatsu *et al.*, 1974). The reason why tissue culture cells were treated like tumor cells is that most tissue culture cell lines are derived from tumors or transformed cells, so that they usually behave like a transplantable tumor cell against homogenization. The procedure is based on the fact that addition of a certain amount of Ca^{2+} or Mg^{2+} ions to detergent-purified nuclei of these cells can make the nucleoli sufficiently resistant to the sonication required for the complete destruction of nuclei. The nature of the nucleoli isolated with this procedure, including the pattern of RNA and RNA polymerase activity contained, will be described in detail.

II.   Isolation of Nuclei from Various Cells with Detergents

A.   Isolation Procedure

Unlike liver cells, most tumor and tissue culture cells resist ordinary homogenization. A combination of hypotonic shock and the use of detergents has been known to be most useful for efficient disruption of cells and the removal of perinuclear cytoplasm. Described below are the detailed steps for the preparation of cytoplasm-free nuclei as a starting material for isolation of nucleoli.

1.   Harvest the cells by washing the peritoneal cavity (in the case of ascites tumor) with a phosphate-buffered saline, PBS^- (0.14 M NaCl, 2.7 mM KCl, 8.1 mM Na_2HPO_4, 1.5 mM KH_2PO_4, pH 7.2) or by a policeman (in the case of a monolayer culture) also with PBS^-.

2.   Centrifuge at 500 g (1500 rpm) for 5 minutes.

3.   Suspend the cells in 20 volumes of ice cold RSB (Reticulocyte Standard Buffer, 0.01 M Tris-HCl, 0.01 M NaCl, 1.5 mM $MgCl_2$, pH 7.4).

4.   Stand the preparation in an ice bath for 10 minutes.

5.   Centrifuge at 600 g (2000 rpm) for 5 minutes to collect swollen cells.

6.   Resuspend the pellet in 10–20 volumes of RSB and add Nonidet P40 (Shell Chemical Co.) and sodium deoxycholate to defined final concentrations (Table I; also see Section II,B).

7.   Homogenize the suspension in a tightly fitting Potter–Elvehjem type of homogenizer with a Teflon pestle (0.4 mm clearance) giving 10 up and down strokes.

8.   Centrifuge at 1200 g (2500 rpm) for 5 minutes to sediment crude nuclei.

9.   Homogenize the pellet in 10 volumes (of the original cell volume) of 0.25 M sucrose, 3.3 mM $CaCl_2$, and transfer into a centrifuge tube.

10.   Underlay an equal volume of 0.88 M sucrose and centrifuge the two-layer system at 1200 g (2500 rpm) for 10 minutes.

The pellet contains purified nuclei (for morphology, see Section III,B and Figs. 1 and 3).

B. Selection of Detergents and Their Concentration

It is clearly desirable to isolate nuclei under conditions as mild as possible. However, the concentration of a detergent that could solubilize cell membrane is sometimes not enough to remove completely cytoplasmic tags that are bound tightly to the nuclear membrane. A higher concentration of detergents is usually required to obtain clean nuclei. When the hypotonicity of the homogenizing medium is fixed to RSB (see Section II,A), the kinds and the optimal concentrations of the detergents vary from one cell species to another (Table I). In some cases, mild treatment with a low concentration of a nonionic detergent is sufficient for this purpose as is the case for L cells or C3H2K cells. There appears to be a correlation between the tenacity with which the cytoplasm attaches to the nucleus and the malignancy of the cell. Thus, C3H2K cells are normal epithelial cells derived from the kidney of a newborn C3H/He mouse (Yoshikura et al., 1972), and L cells are usually nontumorigenic although their growth rate in culture is very high. For many other tissue culture cells as well as transplantable tumor cells, however, higher concentration of Nonidet P40 up to 1–2% is not effective to clean up perinuclear cytoplasm, but addition of certain amounts of deoxycholate, an ionic detergent, is essential. It may be mentioned that the concentrations listed in Table I are the *minimum* concentrations of detergents required for obtaining a cytoplasm-free nuclear preparation. However, it should be kept in mind that higher concentrations of either detergent tend to lyse the nuclei, at least in part. Also, the optimal concentration is in fact a function of the cell concentration in the lysate. To avoid this complication, it is recommended to use at least 10–20 volumes of the homogenizing medium always.

As to the kinds of the detergent, there are several nonionic detergents that

TABLE I

OPTIMAL CONCENTRATION OF DETERGENTS REQUIRED FOR THE COMPLETE REMOVAL OF PERINUCLEAR CYTOPLASM IN VARIOUS CELLS[a]

	Final concentration (%)	
Cell species	Nonidet P40	Deoxycholate
L cell	0.3	0
HeLa cell	0.3	0.2
Ehrlich ascites tumor	0.3	0.2
MH 134	0.3	0.3
C3H2K	0.2	0

[a]The indicated concentrations of both detergents were required for the isolation of cytoplasm-free nuclei from these cells under conditions described in the text. Nucleoli could be obtained from these purified nuclei by the procedure described.

are widely used in different laboratories, e.g., Nonidet P40, Triton X-100, Tween 80, Brij 58, etc. Although they are all alike, some difference in the potency does exist. For instance, we have obtained a clean nuclear preparation from a mouse erythroleukemic cell line T-3Cl-2 (Furusawa et al., 1971) with 0.2% of Triton X-100 alone (Sakai et al., 1976), whereas the same concentration of Nonidet P40 apparently destroyed some of the nuclei during purification. Therefore, different nonionic detergents may be tested for different cell species when no adequate conditions are found with one reagent. For ionic detergent, only deoxycholate has been tested up to the present time, and other reagents appear to be too drastic for the present purpose.

C.   Modifications with Magnesium Ions

There are occasions in which Mg^{2+} ions are preferable to Ca^{2+} ions, e.g., isolation of ribonucleoprotein particles and extraction of RNA polymerase. The procedure described in Section II,A could be modified for the use of Mg^{2+} as described previously (Higashinakagawa et al., 1972; Muramatsu et al., 1974). That is, in the purification step of detergent-treated nuclei, the crude nuclei are homogenized in 10 volumes of 0.25 M sucrose, 10 mM $MgCl_2$ instead of 0.25 M sucrose, 3.3 mM $CaCl_2$. A higher concentration of Mg^{2+} ions are required to protect nucleoli against sonication. A slight but definite difference is also present in the following procedure for isolation of nucleoli as described in Section III,B.

III.   Isolation of Nucleoli from Detergent-Purified Nuclei

A.   Method for Calcium Nuclei

It is relatively easy to isolate highly purified nucleoli from nuclei isolated by the procedure described in Section II,A. The method for nucleoli with Ca^{2+} ions is described below.

1. Suspend the nuclei in 10 volumes (of the original cell volume) of 0.34 M sucrose with a gentle homogenization.

2. Sonicate for 45–60 seconds to destroy all the nuclei: Almost any type of sonicator may be used provided that it has an output of 200 watts or more. Either 10 or 20 kHz will do. Longer sonication periods are required for sonicators with lower outputs. More than 99.9% of the nuclei could be disrupted without much damage to the nucleoli, which could be confirmed by examining a drop of sonicate under a light microscope after staining with Azure C or toluidine blue (Muramatsu et al., 1963). Usually, a 30-second

sonication is given first, followed by 15 seconds each for additional sonication, with light microscopic inspection at every interval.

3. Pour 15–20 ml of the sonicate into a 50-ml centrifuge tube and underlay an equal volume of 0.88 M sucrose.

4. Centrifuge at 2200 g (3300 rpm) for 15 minutes.

The pellet contains highly purified nucleoli (for morphology, see Section III,B, and Figs. 2 and 4–6).

B. Modifications for Magnesium Nuclei

When the nuclei are isolated with Mg^{2+} ions as described in Section II,C, the following modifications have to be made.

1. Use 0.34 M sucrose, 0.05 mM $MgCl_2$ instead of plain 0.34 M sucrose to suspend nuclei for sonication (Higashinakagawa *et al.*, 1972): This small

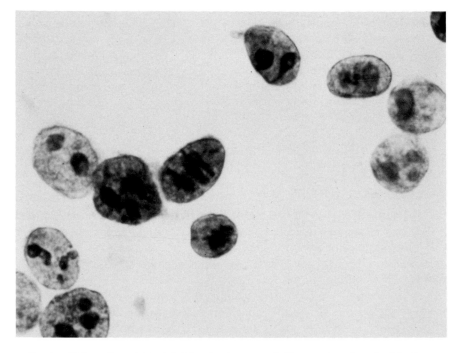

FIG. 1.   HeLa cell nuclei isolated by the detergent-Ca^{2+} procedure described in the text, stained with Azure C. × 1540.

amount of Mg^{2+} ions is indispensable for nuclei isolated by magnesium procedure. Without addition of Mg^{2+} at this step, many nucleoli are disintegrated during sonication. On the other hand, addition of a higher concentration of Mg^{2+} often tends to produce rigid nuclei that are never broken by a prolonged sonication.

2. Use 0.88 M sucrose, 0.05 mM $MgCl_2$ rather than plain 0.88 M sucrose as the underlaying solution for the purification of nucleoli: We did not describe this in the original paper (Higashinakagawa et al., 1972), but later it was found that addition of small mounts of Mg^{2+} could prevent the clumping of nucleoli in the pellet when they are resuspended in a buffer solution. The nucleoli appear to be leached with respect to Mg^{2+} during sedimentation through plain 0.88 M sucrose and become more gelatinous and sticky. It is possible that some nucleolar component is extracted in 0.88 M sucrose under these conditions.

In Figs. 1 and 2 are shown HeLa cell nuclei and nucleoli isolated by the

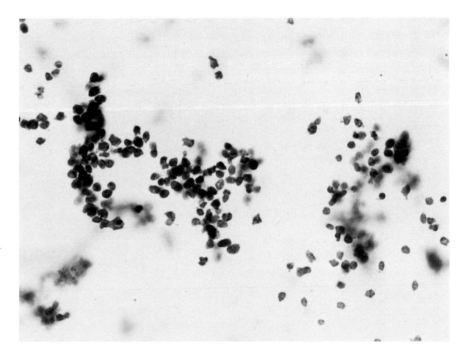

FIG. 2. HeLa cell nucleoli isolated from the detergent-Ca^{2+} nuclei shown in Fig. 1, stained with Azure C. Blurred spots are not contaminants, but nucleoli that are out of focus. × 600 (Muramatsu et al., 1974).

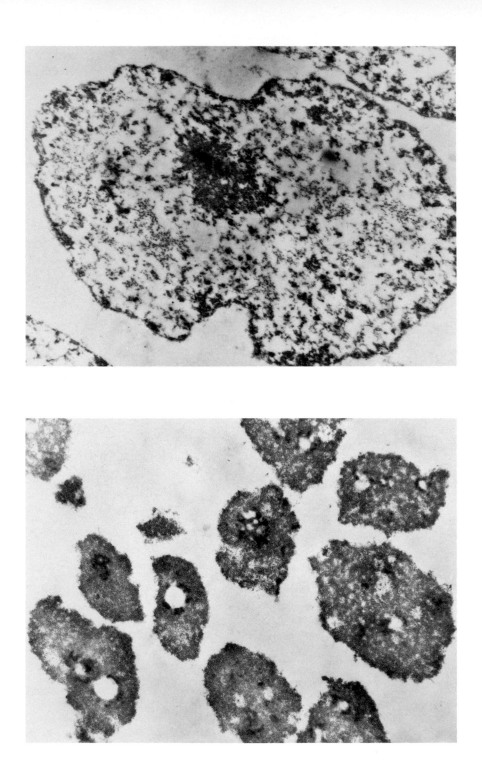

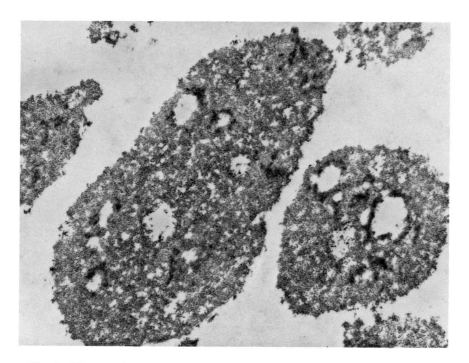

FIG. 5. Electron micrograph of HeLa cell nucleoli at a higher magnification. × 13,280.

Ca^{2+} procedure described in Sections II and III, which show both the purity and the integrity of the preparations. The preparations with Mg^{2+} procedure look essentially the same. Electron micrographs of these preparations are also shown in Figs. 3–6. It may be seen that outer membrane of the nuclei consisting of endoplasmic reticulum is mostly removed with this procedure. The ultrastructure of isolated nucleoli is very similar to that *in situ*, suggesting molecular integrity of this organelle, which is verified in the following biochemical analyses.

FIG. 3. Electron micrograph of HeLa cell nucleus isolated by the detergent-Ca^{2+} procedure. × 17,700 (Muramatsu *et al.*, 1974).
FIG. 4. Electron micrograph of HeLa cell nucleoli isolated from detergent-Ca^{2+} nuclei. × 12,300 (Muramatsu *et al.*, 1974).

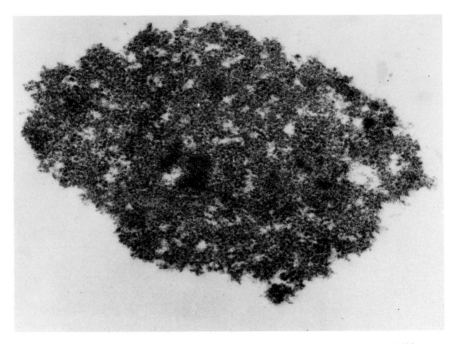

FIG. 6.   Higher magnification of an isolated nucleolus of the HeLa cell. × 26,600.

IV.   Applications

A.   Isolation and Properties of Nucleolar RNA

Intact nucleolar RNA could be extracted from these nucleoli with the conventional sodium sulfate–hot phenol procedure (Scherrer and Darnell, 1962; Muramatsu, 1973). A typical sucrose density gradient profile and a polyacrylamide gel pattern are shown in Figs. 7 and 8. Both patterns are similar to those obtained from other tumor cells without the use of detergents (Muramatsu and Busch, 1967; Muramatsu *et al.,* 1968; Nakamura *et al.,* 1967) and show no appreciable degradation of 45 S and other high-molecular-weight RNAs. Low-molecular-weight RNAs inherent in nucleoli do not seem to be lost either during this isolation procedure.

B.   *In Vitro* RNA Synthesis with Isolated Nucleoli

It has been shown that isolated nucleoli are capable of synthesizing RNA *in vitro* with added XTPs (Ro *et al.,* 1964). This reaction depends upon both

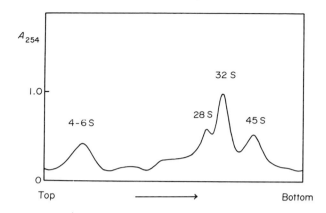

FIG. 7. Sucrose density gradient profile of nucleolar RNA of MH 134. Nucleoli were isolated from MH 134 cells by the Ca^{2+} procedure from detergent-purified nuclei as described in the text. RNA was extracted from nucleolar pellet with sodium dodecyl sulfate (SDS) hot phenol procedure. Sucrose gradient was linear 10–40% buffered with 0.1 M sodium acetate, pH 5.1, and centrifugation was continued for 17 hours at 30,000 rpm at 4°C. A similar pattern could be obtained for RNA from Mg^{2+}-prepared nucleoli.

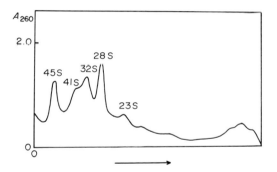

FIG. 8. Polyacrylamide gel electrophorogram of HeLa cell nucleolar RNA. RNA was extracted from nucleoli isolated by the Ca^{2+} procedure from HeLa cells. Electrophoresis was on a 2.2% gel (0.6 × 6.0 cm) at a constant current of 5 mA/tube for 2 hours. Essentially the same pattern was obtained for nucleoli isolated by the Mg^{2+} procedure (Muramatsu et al., 1974).

the template deoxyribonucleoprotein complexes and nucleolar RNA polymerase and also possibly upon some controlling factor(s) (Muramatsu et al., 1970; Higashinakagawa et al., 1972).

In Fig. 9, the RNA synthetic capacity of nucleoli isolated from detergent-purified nuclei is compared with that of nucleoli isolated from crude nuclei prepared without detergent. At 25°C, both activities continue linearly up to 15 minutes at the same rate. However, at 37°C, the activity of nucleoli

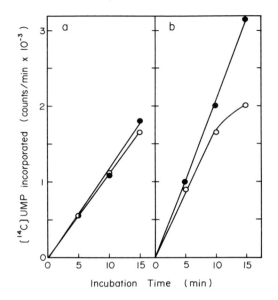

FIG. 9.   *In vitro* RNA synthetic activity of isolated nucleoli of MH 134. The nucleoli were isolated either from detergent-purified nuclei (○) (Mg^{2+} procedure) or from crude nuclei prepared without detergent (●). The reaction mixture contained in a final volume of 0.15 ml: 5 μmoles of Tris-HCl buffer, pH 7.9 (at 37°C), 1.0 μmole of dithiothreitol, 0.24 μmole of $MnCl_2$, 0.4 μmole of ATP, 0.1 μmole each of GTP and CTP, 7.5 μmoles of KCl, 0.1 μCi of [^{14}C]UTP (The Radiochemical Centre, Amersham, Bucks., England), 20 μl of rat liver RNase inhibitor (Sephadex G-100 fraction) and 4.5 μg of DNA-equivalent nucleoli from MH 134 cells. The reaction mixture was incubated for various periods at either 25°C (a) or 37°C (b) and then rapidly chilled in an ice bath. Then 1.0 ml of 7% trichloroacetic acid, 1% $Na_4P_2O_7$ was immediately added, and the mixture was allowed to stand for 10 minutes at 0°C. Acid-precipitable radioactivity was determined as described previously (Muramatsu *et al.*, 1974).

isolated from detergent-purified nuclei tends to level off earlier. Since the initial velocity of reaction is almost the same, this effect is probably due to the activation of RNase in this preparation. Whether or not this could be prevented by additional RNase inhibitors is now under study.

The recent demonstration by Grummt (Grummt and Lindigkeit, 1973; Grummt, 1975; Grummt *et al.*, 1975) that isolated rat liver nucleoli are able to synthesize 45 S RNA and possibly larger precursors to ribosomal RNA has prompted us to study the synthesis of 45 S RNA with tumor nucleoli isolated by the present procedure. The nucleoli isolated from MH 134/C cells by our Mg^{2+} procedure were incubated under standard conditions in the presence of an RNase inhibitor prepared from rat liver supernatant (Gribnau *et al.*, 1969), and the phenol-extracted RNA was then analyzed on a sucrose density gradient. It was found that nucleoli isolated from detergent-purified tumor nuclei could indeed synthesize RNA with a sedimentation coefficient of 45 S, although RNA with lower S values (28–41 S) was more abundant probably because of nucleolytic attack during incubation (Onishi and Muramatsu, 1976). In our hands, no RNA with S values larger than 45 S could be synthesized *in vitro*, even when rat liver nucleoli isolated without detergent were employed.

In any event, these results show that nucleoli isolated by the present procedure retain similar DNA template and RNA polymerase to those in the nucleoli isolated without detergent.

V. Commentary

The procedures described in this chapter proved to be effective for various tumor and tissue culture cells. In some cases, however, they may not work as well as for the cells described here. For instance, we could not obtain clean nucleoli from Morris 7288C, a cultured hepatoma, because nuclei could not be broken completely by sonication whether or not they were treated with detergents. For the application of these procedures to new cell types, it is recommended to test the optimal (minimum) concentrations of the detergents for individual cases.

As mentioned in the Introduction, these procedures could be applied for the comparison of nuclear and nucleolar content of RNA and protein, and also DNA, although some reservation may be needed for possible extraction of these components from the nucleus. This is especially true for nuclear proteins, some of which are known to be extremely soluble. Apart from the comparative studies, these procedures may be utilized for the preparation and analysis of nucleolar components, since they appear to be retained fairly well as judged from the ultrastructure of isolated nucleoli, RNA patterns, RNA polymerase activity as well as the relative integrity of RNA synthesized *in vitro*. However, the possibility of partial extraction of RNA polymerase, and especially of activation of endogenous RNases as discussed in Section IV,B, cannot be ruled out at present. Further studies with nucleolar RNA, DNA, and proteins including RNA polymerase are under way in this laboratory with the aid of these procedures.

ACKNOWLEDGMENTS

The authors would like to thank all the colleagues who made invaluable contributions to the work described in this paper. The work on which this article was written was supported by grants from Ministry of Education, Science, and Culture, Japan, Prince Takamatsu Cancer Research Fund and from Naito Memorial Fund for Scientific Research.

REFERENCES

Chauveau, J., Moulé, Y., and Rouiller, C. (1956). *Exp. Cell Res.* **11**, 317.
Desjardins, R., Smetana, K., Steele, W. J., and Busch, H. (1963). *Cancer Res.* **23**, 1819.
Dounce, A. L. (1963). *Exp. Cell Res., Suppl.* **9**, 126.
Furusawa, M., Ikawa, Y., and Sugano, H. (1971). *Proc. Jpn. Acad.* **47**, 220.
Gribnau, A. A. M., Schoenmakers, J. G. G., and Bloemendal, H. (1969). *Arch. Biochem. Biophys.* **130**, 48.
Grummt, I. (1975). *Eur. J. Biochem.* **57**, 159.
Grummt, I., and Lindigkeit, R. (1973). *Eur. J. Biochem.* **36**, 244.
Grummt, I., Loening, U. E., and Slack, M. W. (1975). *Eur. J. Biochem.* **59**, 313.
Higashinakagawa, T., Muramatsu, M., and Sugano, H. (1972). *Exp. Cell Res.* **71**, 65.
Liau, M. C., and Perry, R. P. (1969). *J. Cell Biol.* **42**, 272.
Liau, M. C., Hnilica, L. S., and Hurlbert, R. B. (1965). *Proc. Natl. Acad. Sci. U.S.A.* **53**, 626.
McConkey, E. H., and Hopkins, J. W. (1964). *Proc. Natl. Acad. Sci. U.S.A.* **51**, 1197.
Maggio, R., Siekevitz, P., and Palade, G. E. (1963). *J. Cell Biol.* **18**, 293.
Monty, K. J., Litt, M., Kay, E. R., and Dounce, A. L. (1956). *J. Biophys. Biochem. Cytol.* **2**, 127.
Muramatsu, M. (1970). *Methods Cell Physiol.* **4**, 195–230.
Muramatsu, M. (1973). *Methods Cell Biol.* **7**, 23–51.
Muramatsu, M., and Busch, H. (1967). *Methods Cancer Res.* **2**, 303–359.
Muramatsu, M., Smetana, K., and Busch, H. (1963). *Cancer Res.* **23**, 510.
Muramatsu, M., Higashinakagawa, T., Ono, T., and Sugano, H. (1968). *Cancer Res.* **28**, 1126.
Muramatsu, M., Shimada, N., and Higashinakagawa, T. (1970). *J. Mol. Biol.* **53**, 91.
Muramatsu, M., Hayashi, Y., Onishi, T., Sakai, M., Takai, K., and Kashiyama, T. (1974). *Exp. Cell Res.* **88**, 345.
Nakamura, T., Rapp, F., and Busch, H. (1967). *Cancer Res.* **27**, 1084.
Onishi, T., and Muramatsu, M. (1976). *In* "Methods in Chromosomal Protein Research" Vol. 1 (G. Stein and J. Stein, eds.), Vol. 1. Academic Press, New York. In press.
Penman, S., Smith, I., and Holtzman, E. (1966). *Science* **154**, 786.
Rees, K. R., Rowland, G. F., and Varcoe, J. S. (1962). *Biochem. J.* **86**, 130.
Ro, T.-S., Muramatsu, M., and Busch, H. (1964). *Biochem. Biophys. Res. Commun.* **14**, 149.
Sakai, M., Hidaka, S., and Muramatsu, M. (1976). In preparation.
Scherrer, K., and Darnell, J. E. (1962). *Biochem. Biophys. Res. Commun.* **7**, 486.
Willems, M., Wagner, E., Laing, R., and Penman, S. (1968). *J. Mol. Biol.* **32**, 211.
Yoshikura, H., Hirokawa, Y., and Yamada, M. (1967). *Exp. Cell Res.* **20**, 225.

Chapter 14

Isolation of Plasma Membrane Vesicles from Animal Cells

DONALD F. H. WALLACH AND
RUPERT SCHMIDT–ULLRICH

Tufts-New England Medical Center, Department of Therapeutic Radiology,
Division of Radiobiology,
Boston, Massachusetts

I.   General Introduction

Plasma membrane fractionation constitutes a very large part of biomembrane research. Many aspects of the field have been reviewed extensively in recent years (e.g., Steck and Wallach, 1970; Steck, 1972; DePierre and

Karnovsky, 1973; Wallach and Lin, 1973; Wallach and Winzler, 1974), and a number of specific fractionation procedures are described in this series.

It is the aim of this article to deal with the fractionation and purification of plasma membranes as *sealed vesicles*. This approach, initiated by Wallach and Kamat (1964), has proved to be generally useful, but is now gaining increased attention because of its potential for the subfractionation of membrane "domains" and its utility in producing membrane fragments suitable for transport studies (cf. review by Hochstadt *et al.,* 1976).

We provide here basic concepts and general guidelines for those who wish to isolate plasma mebrane vesicles from cell types for which adequate fractionation procedures have not yet been developed. We will also give specific technical examples. However, we must first stress that, although the same fractionation *principles* may apply to a very wide variety of cell types, different cell categories, for example diverse lymphocyte species (Ferber *et al.,* 1972; Schmidt-Ullrich *et al.,* 1974), normal and neoplastically transformed cells (Wallach, 1975; Schmidt-Ullrich *et al.,* 1976a), may require significantly different fractionation conditions for the isolation of plasma membrane fractions that are representative and equivalent in yield and purity.

When membranes from *different* cell categories, or even *normal* and *neoplastic* variants of the *same* cell type, are to be compared, serious, often unrecognized obstacles can arise because the membranous organelles of diverse cell classes may behave differently in a given fractionation scheme. The greatest difficulties are encountered in plasma membrane fractionations for reasons detailed previously (Steck and Wallach, 1970; DePierre and Karnovsky, 1973; Wallach and Lin, 1973; Wallach and Winzler, 1974). First, most normal cells do not have a uniform surface topology. This is true even for "free"cells, such as leukocytes, but is most marked in solid tissues, where different areas of the cell surface become highly specialized. Second, changes in differentiation, such as occur in neoplastic transformation, are nearly always accompanied by some alteration in surface topology. This circumstance is particularly problematic *in vivo,* since the manner of plasma membrane fractionation then depends critically on cell interactions. However, the problem is not obviated *in vitro* since cultured normal and neoplastic cells commonly exhibit different and nonuniform surface topologies and surface topology can vary with culture conditions (Truding *et al.,* 1974; Pastan *et al.,* 1974). Third, difficulties can arise from changes in membrane composition, for example differences between cells in the complements of phospholipids, cholesterol, and glycolipids. Such chemical modifications tend to alter the physical properties of the membranes involved, producing different stabilities and varying sedimentation characteristics during differential and/or isopycnic ultracentrifugation.

The impact of these complications is apparent in experimentation on liver–hepatoma systems (Emmelot *et al.*, 1964, 1970; Lieberman *et al.*, 1967; Emmelot and Bos, 1969a,b; Bergelson, 1972; Moyer and Pitot, 1973). Most authors equate "liver plasma membrane" with the "bile-front" moiety of hepatocytes. This can be isolated in the form of rather large fragments using a method originally developed by Neville (1960), in which liver is homogenized in hypotonic (1 m*M*) NaHCO₃, pH 7.5. The isolation relies upon the stability of the various connections between apposed hepatocytes which together lead to the formation of bile canaliculi and "bile-fronts." However, such intercellular associations are abnormal to a greater or lesser extent in all hepatomas. Hence, application of a procedure designed to isolate a surface specialization which is defective or lacking in a tumor will not yield the same product as that obtained with a normal tissue. The different behavior of the membranes in bile-front isolates from normal liver and hepatomas is illustrated by the deviant distribution of marker enzymes in sucrose gradients. The protein and ATPase activities of rat hepatoma "bile-fronts" equilibrate at higher buoyant densities than are found with components of normal liver "bile-fronts" (Emmelot and Bos, 1966), and studies on a rat hepatoma (HW) (Wattiaux and Wattiaux-de Coninck 1968a,b) show a rather inconsistent distribution of microsomal membrane acid phosphatase. The fractionation difficulties are not restricted to microsomal membrane isolates. Thus, hepatoma mitochondria exhibit sedimentation characteristics different from those of normal mitochondria (Wattiaux and Wattiaux-de Coninck, 1968b). The same appears to be true for lysosomes and peroxisomes (Wattiaux and Wattiaux-de Coninck, 1968a,b). Smooth and rough "microsomal membranes" from hepatomas also fractionate differently from their normal equivalents (Moyer and Pitot, 1973; Moyer *et al.*, 1970).

Different plasma membrane fractionation characteristics have also been reported for normal and polyoma-transformed BHK21 fibroblasts (Graham, 1972) and for normal hamster lymphocytes and a neoplastic variant induced by simian virus 40 (Schmidt-Ullrich *et al.*, 1976a).

In conclusion:

1. One cannot necessarily use identical fractionation schemes for the isolation of a given membrane class from different cell types.

2. Intrinsic, functional membrane markers are not necessarily reliable for comparisons between different cell types. The composition and enzyme complement of diverse membranes can vary dramatically—for example, during differentiation and between the normal and neoplastic state. Clearly *multiple criteria* are essential and *complete balance sheets* must be maintained.

3. Due to differences in the membrane differentiation of different cell types, methods relying on membrane specializations cannot be considered reliable until proved to be so, in a specific case.

II.   Membrane Markers

To purify plasma membrane vesicles it is necessary to quantitatively monitor the fate of surface *and* cytoplasmic membranes following cell disruption. Electron microscopy is not satisfactory for this purpose, particularly when one deals with membranes lacking distinguishing surface characteristics, or when quantitation is desired. Most investigators therefore employ various membrane markers.

For *plasma membranes* such markers might be (a) membrane enzymes, (b) groups that can be labeled in intact cells from the extracellular space, and (c) membrane receptors that react exclusively with extracellular ligands. For *intracellular* membranes, only enzymes have proved to be generally useful. Markers for *both* surface and cytoplasmic membranes should be used concurrently to establish quantitative estimates of yield and purity.

An obvious principle, but one not always easily applied, is that marker procedures—indeed entire fractionations—may lead to very ambiguous results, unless heterogeneity in the cell population to be fractionated is minimized (cf. Wallach and Lin, 1973). Cell purification methods have been summarized by Wallach and Lin (1973) and newer specific approaches are treated in this series (e.g., Crissman *et al.,* 1975).

A.   Marker Enzymes

The use of diverse enzymes as markers to monitor the fate of surface and cytoplasmic membranes following cell disruption has been well reviewed recently (e.g., Steck, 1972; Wallach and Lin, 1973; Wallach and Winzler, 1974), and we will deal with this topic here only to introduce an element of caution and to give examples that we have found reliable in the fractionation of diverse lymphoid cells.

The vectorial orientation of certain membrane marker enzymes, particularly ATP phosphohydrolases, can cause unanticipated ambiguities, particularly if part or all of the membrane vesicles are sealed to molecules such as ATP. Whether the vesicles are sealed or not depends very much on environmental variables and/or conditions of cell disruption. Thus erythrocyte ghosts prepared by the method of Dodge *et al.* (1963) readily hydrolyze externally added ATP, whereas intact cells do not. The former are permeable to ATP, but intact cells are not. In contrast, ghosts prepared by the halothane method (Lin *et al.*, 1975) do not utilize added ATP, because their membranes are sealed to this metabolite. In a fractionation utilizing ATP-phosphohydrolases as exclusive markers, sealed ghosts might not be detected.

But the situation can be more complex, because certain cells, e.g., various neoplastic mouse cells (Acs *et al.,* 1954; Wallach and Ullrey, 1962a,b;

Rozengurt and Heppel, 1975), as well as some leukocytes (e.g., Luganova *et al.*, 1957), can hydrolyze extracellular ATP to ADP. The particular properties of these "ecto-ATPases" may provide unique plasma membrane markers, particularly for ATP-impermeant vesicles. It should be noted, however, that neoplastic cells may differ from their precursors in the presence or activity of "ecto-ATPases" (Rozengurt and Heppel, 1975). The use of ATPase activity can thus be quite ambiguous. This is true even if the assay is for Na^+-K^+-ATPase (ouabain sensitive), because of the possible presence of both sealed and unsealed vesicles and the competition for a common substrate of "endo" and "ecto" ATP-phosphohydrolases.

5'-Nucleotidase, one of the most frequently used plasma membrane marker enzymes, is not always reliable. First, its activity in some cells, e.g., certain lymphoid cells, is low (Schmidt-Ullrich *et al.*, 1974, 1976a). Second, in some cells, e.g., liver, the enzyme is not associated exclusively with the plasma membrane (Widnell, 1972).

Turning now to our specific experience in fractionating diverse lymphocytes, we have used the following marker enzymes to monitor *plasma membrane*: (a) alkaline *p*-nitrophenylphosphatase (EC 3.1.3.1), determined essentially as described by Emmelot and Bos (1966), but using 1 *M* diethanolamine buffer (pH 9.5) as suggested by Rick *et al.* (1972). (b) Mg^{2+}-stimulated Na^+, K^+-dependent ATPase (EC 3.6.1.3) as described by Wallach and Ullrey (1962b). To monitor *endoplasmic* reticulum, we used NADH-oxidoreductase (EC 1.6.4.3) estimated as described by Wallach and Kamat (1966) and glucose-6-phosphatase (EC 3.1.3.9) assayed as recommended by Swanson (1955). Possible contamination of the plasma membrane fraction by *mitochondria* was checked, using succinate dehydrogenase (EC 1.3.99.1) as marker for the inner mitochondrial membrane (Bosman *et al.*, 1968) and monoamine oxidase (EC 1.4.3.4) as marker for outer mitochondrial membrane. The latter enzyme was assayed as described by Wurtman and Axelrod (1963), but using twice the concentration of [^{14}C]tryptamine. Possible lysosomal contamination was monitored by β-glucuronidase (EC 3.2.1.31) activity determined as described by Gianetto and de Duve (1955) using *p*-nitrophenolglucoronic acid as substrate. A summary of our results is given in Table I.

B. Chemical Labeling

Many attempts have been made to chemically couple small marker molecules (fluorescent, radioactive) to the plasma membranes by chemical techniques, to thereby follow the fate of the plasma membrane following cell disruption. This strategy has been extensively reviewed (e.g., Wallach and Winzler, 1974; Carraway, 1975). We do not consider this approach generally useful at this time, because (a) the small molecules used generally permeate

TABLE

CHARACTERIZATION OF PLASMA MEMBRANE FRACTIONS ISOLATED FROM NORMAL

	Protein[a,b,d] (mg)	Membrane-^{125}I[a,b,e,f]		Na$^+$,K$^+$-ATPase[a,b,d,e,g]	
		Recovery	PF	Recovery	PF
Normal	0.9 (4.2%)	56.2	11.4	69.9 (65%)	16.7
GD248	2.09 (3.6%)	60.0	17.2	76.1 (62%)	15.8

[a] Recoveries are given as totals and as percent of the homogenate value (in parentheses).
[b] Results from three independent preparations.
[c] The plasma membrane vesicles exhibited no detectable activity of cytochrome oxidase, monoamine oxidase, or β-glucuronidase, indicating negligible (<0.1%) contamination by mitochondria or lysosomes, respectively. The recoveries of NADH-oxidoreductase and glucose-6-phosphatase in the plasma membrane vesicles were 5.8%, indicating elimination of >95% of endoplasmic reticulum.
[d] The values of three independent determinations varied not more than 15%.

into the cell interior even in the case of erythrocytes, and (b) the labeling conditions are lethal to most cell types other than erythrocytes.

C.   Enzymic Labeling

1.   LACTOPEROXIDASE-CATALYZED RADIOIODINATION

This is a most important technique for radiolabeling proteins exposed at the external plasma membrane surfaces of intact cells (Marchalonis, 1969; Phillips and Morrison, 1971). The technique has also been utilized for the vectorial membrane labeling of isolated mitochondria (Huber and Morrison, 1973; Astle and Cooper, 1974), endoplasmic reticulum (Kreibich et al., 1974), sarcoplasmic reticulum (Thorley-Dawson and Green, 1973), and chloroplasts (Arntzen et al., 1974).

In the presence of H_2O_2, low concentrations of iodide and a pH near neutrality lactoperoxidase catalyzes the iodination of accessible tyrosines (and possibly some histidines) of soluble or membrane-associated proteins. The reaction is believed to proceed primarily through an iodinated form of lactoperoxidase that transfers iodine atoms directly onto the protein to be labeled (Morrison et al., 1971), yielding primarily radiolabeled mono-iodotyrosines.

Iodination of —C= C— bonds of unsaturated lipids may also occur, but can be avoided by use of butylated hydroxytoluene (~ 1 μg/ml). H_2O_2 can

I

Alkaline phosphatase[a,b,d,e,h]		Phospholipid[b,d] (nmoles/ mg protein)	Cholesterol[b,d] (nmoles/ mg protein)	Cholesterol/ phospholipid (molar ratio)	DNA[b,d,i] (μg/mg protein)	RNA[b,d,i] (μg/mg protein)
Recovery	PF					
229.5 (58%)	12.2	812 ± 60	749 ± 74	0.92 ± 0.1	15.3	24.4
585.8 (55%)	18.2	780 ± 44	1090 ± 36	1.42 ± 0.15	12.7	20.5

[e] Purification factor (PF) = [specific activity (per mg protein) of fraction]/[specific activity of homogenate].

[f] Membrane-^{125}I refers to ^{125}I covalently bound to membranes after lactoperoxidase-catalyzed radioiodination of whole cells.

[g] Nanomoles of inorganic phosphorus liberated · mg protein $^{-1}$ · minutes^{-1}.

[h] Nanomoles p-nitrophenol liberated · mg protein^{-1} · minutes^{-1}.

[i] DNA and RNA correspond to 0.05% and 0.2%, respectively, of homogenate.

either be added in repeated microincrements (e.g., 0.2 μmole per 10^8 cells in two doses 10 minutes apart at 30°C) or generated continuously by glucose oxidase in the presence of glucose (Hubbard and Cohn, 1972). Under optimal conditions, lactoperoxidase-catalyzed radioiodination of intact erythrocytes causes no cell lysis, yields 97% of the incorporated label in membrane proteins and only 3% with hemoglobin.

A circumstance to be seriously considered in applying the lactoperoxidase approach is the tendency for the enzyme to iodinate soluble proteins (including itself) more readily than membrane proteins; in general tyrosyl residues of soluble proteins will be more accessible and the proteins more mobile. It is imperative, therefore, to exclude nonmembrane proteins from the iodination reaction. Serum proteins, e.g., from culture media, can interfere seriously, as can impurities in the lactoperoxidase preparations (and/or glucose oxidase) employed.

Related complications arise when the technique is employed with protein-secreting cells, e.g., lymphocytes. Thus Marchalonis et al. (1971) have used the lactoperoxidase technique with various lymphoid cells and argue that only the external surfaces of these cells' plasma membranes are labeled in the process. Their data depend heavily on radioautographic studies of labeled cells. The data indicate that 95% of the cells show grains, but each section exhibits only 20–50 grains, many of which are very large and cannot be localized to within more than a few hundred angstroms from the surface membrane. Moreover, lymphoid cells, particularly Ig-secreting lymphocytes, exhibit extremely rapid turnover rates of surface radiolabeled proteins, including IgG; these rates are much greater than those obtained using metabolic labeling (cf. review by Wallach and Schmidt-Ullrich, 1975). Moreover, in our own studies of diverse lymphoid cells (Schmidt-Ullrich *et al.*, 1974, 1976a), we invariably find that 20–30 % of the label taken up by whole, viable cells ends up in covalent association with proteins that are not sedimentable at up to $2 \times 10^7 g \cdot$ min, although the membrane-associated label concentrates in the plasma membrane fraction. Pinocytosis has been excluded. We therefore assume that the soluble iodinated proteins are entities that are loosely associated with the external surfaces of intact cells but separate from these surfaces with time and/or upon cell disruption.

Similar considerations may apply to the high-molecular weight glycoproteins (LETS protein) accessible to lactoperoxidase-catalyzed iodination in the case of normal cultured fibroblasts, but seemingly deficient in neoplastically transformed fibroblasts (cf. review by Wallach and Schmidt-Ullrich, 1975). Data by Yamada and Weston (1974) indicate that this protein is rather loosely associated with cell surfaces and can exchange between cells. Moreover, work by Hamilton and Nielsen-Hamilton (1976) shows that this protein is readily dissociated from transformed cells (but less easily from untransformed fibroblasts) by frequent changes of medium.

In using the lactoperoxidase procedure on untested systems, one should also be alert to the fact that iodination can deleteriously affect biomembranes. Thus the outer membrane of rat liver mitochondria breaks after incorporation of 0.2 nmole or more of iodine per milligram of protein (Huber *et al.*, 1975).

The specific procedure we have found reliable in the case of diverse lymphocytes (Schmidt-Ullrich *et al.*, 1974, 1976a) is as follows: About 10 % of the cells to be fractionated are radioiodinated with ^{125}I. For this we suspend about 5×10^7 cells in 1.0 ml of Dulbecco's phosphate-buffered saline (PBS), 10^{-6} M in NaI and chilled to $\sim 4°C$. We then add 0.01 ml of 0.1 % butylated hydroxytoluene solution in PBS to avoid lipid peroxidation. This is followed by 0.05 ml of lactoperoxidase (Boehringer/Mannheim, Germany; 1 mg/ml PBS) and 0.01 ml of Na ^{125}I (0.1 mCi). After warming to 25°C we start iodination, adding first 0.02 ml and then 2×0.01 ml of 0.003 % H_2O_2 at 60-second intervals. We terminate the iodination by addition of 9 ml of cold PBS, 1 % in fetal calf serum and 5 mM in NaI. The cells are then washed with PBS, 5 mM in NaI, and then mixed with unlabeled cells. The distribution of ^{125}I-labeled protein is followed by counting aliquots of intact cells and of all subcellular fractions after precipitation with iced 10 % trichloroacetic acid and two washes with cold 10 % trichloroacetic acid. The diluted samples contain $\sim 10^5$ cpm/10^7 cells.

Under these conditions the viability of the lymphocytes is not impaired. About 60 % of the ^{125}I-labeled membrane protein is recovered in our plasma membrane fraction (Table I).

2. GALACTOSE OXIDASE-CATALYZED RADIOLABELING OF MEMBRANE GLYCOPROTEINS AND GLYCOLIPIDS

Galactose oxidase oxidizes externally exposed hydroxymethyl residues at the 6-position of galactosyl- of N-acetyl-galactosaminyl residues of glyco-proteins and glycolipids in the plasma membranes of intact cells, producing aldehydic products, e.g.,

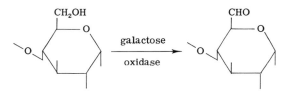

Addition of tritiated borohydride (NaB^3H_4) during (Steck and Dawson, 1974) or after (Gahmberg and Hakomori, 1973) enzyme action reduces and radiolabels the aldehydic products, e.g.,

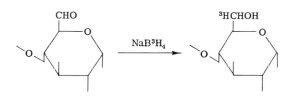

The galactose oxidase procedure is more selective than lactoperoxidase-catalyzed iodination as far as proteins are concerned, since labeling is restricted to *glyco*proteins. On the other hand glyco*lipids* are also labeled. Also, since galactosyl residues are often linked to N-acetylneuraminic acid, interfering with enzyme action, labeling can often be enhanced by pretreat-ment with neuraminidase (Gahmberg and Hakomori, 1973).

As pointed out by Steck and Dawson (1974) labeling via the galactose oxidase procedure involves several obstacles: (a) most enzyme preparations contain proteases; these can generally be inactivated by heating at $\sim 50°C$ for 30 minutes (b) the galactaldehyde generated by enzyme action can react with protein amino groups, which, after addition of NaB^3H_4 yield labeled, cross-linked products that are not dissociated by conventional procedures of membrane protein analysis (dodecyl sulfate polyacrylamide electrophor-esis; isoelectric focusing). This possibility can be reduced by addition of NaB^3H_4 coincident with galactose oxidase. NaB^3H_4 may also reduce various groups in membranes irrespective of galactose oxidase action. The extent of this process can be evaluated by measuring the degree of labeling with NaB^3H_4, without galactose oxidase.

D.   Receptors for Extrinsic Macromolecular Ligands

The use of receptors for external macromolecules as plasma membrane markers has major potential. However, one should be aware of the fact that the limited accessibility of ligand to receptor can also complicate evaluation of the total marker content of a cell dispersion or homogenate. For example, considering surface antigen or lectin receptors, we know that some of these lie at external surfaces. However, we do not always know whether some of these receptors do not also occur on intracellular membranes. Certainly some intracellular membranes do contain receptors for certain lectins, for example concanavalin A. At the same time it is possible that some receptors lie also on the internal surfaces of various membrane vesicles and are thus not accessible to macromolecular markers. In such cases one would first have to destroy membrane permeability in a way that leaves the ligand–receptor interaction unimpaired.

1.   HORMONE RECEPTORS

The plasma membranes of numerous cells bear hormone receptors that can potentially serve as plasma membrane markers. This potential has been recently realized in membrane fractionation by Chang *et al.* (1975), who monitored the fate of ^{125}I-labeled insulin that had been bound to intact rat fat cells prior to cell disruption. It was found that, upon fractionation of the subcellular elements in continuous sucrose gradients, the label associates preferentially with plasma membrane vesicles. Preferential association with plasma membrane was found also when the insulin was added to the cell homogenate rather than intact cells. The results suggest that the insulin receptors are localized exclusively on the plasma membrane and not also on intracellular membranes. The receptor hormone complex is rather stable at 4°C and can be dissociated by adding an excess of unlabeled insulin. Applications of this particular approach are limited because rather few cells appear to have insulin receptors (for example fat cells and lymphocytes) and these occur at very low surface density.

2.   LECTIN RECEPTORS

The use of lectin receptors as plasma membrane markers has also been explored by Chang *et al.* (1975). Their data (Table II) convincingly document that wheat agglutinin (WGA) can serve as a highly reliable marker for plasma membranes of rat fat cells, rat hepatocytes, and cultured human lymphocytes. The authors utilize [^{125}I]WGA in their studies. The lectin is reacted with the intact cells before they are disrupted and fractionated by differential centrifugation and sucrose-gradient ultracentrifugation. The data show that the lectin complex with plasma membrane is extremely tight. Exchange with added free lectin occurs at a very low rate indeed. Moreover, when the lectin

TABLE II

FRACTIONATION OF RAT FAT CELLS AND RAT LIVER AND HUMAN LYMPHOCYTES LABELED WITH ^{125}I-LABELED WHEAT GERM AGGLUTININ (WGA)[a]

Sample	Rat fat cells		Human lymphocytes		Rat liver	
	SA	TA	SA	TA	SA	TA
Homogenate	3.9×10^5	1.5×10^6	3.5×10^5	2.8×10^6	2.8×10^3	7.3×10^5
Mitochondria	7.6×10^4	1.6×10^5	1.3×10^5	2.3×10^5	2.0×10^2	1.0×10^4
Nuclei	3.0×10^4	2.5×10^4	5.0×10^4	8.0×10^4	4.3×10^3	1.0×10^5
Membrane fractions[b]	1.9×10^6 (MS)	7.7×10^5 (PM)	1.4×10^6 (MS)	1.2×10^6 (MS)	8.1×10^3 (HM) 8.5×10^3 (MS	2.5×10^5 (HM) 2.2×10^5 (MS)
Purification	5X		4X		3X	
Recovery	46%		43%		28% (MS) 34% (HM)	

[a] From Chang et al. (1975). SA and TA = specific and total activity. i.e., cpm of [^{125}I]WGA bound per milligram and total protein, respectively.
[b] PM; plasma membrane; MS, microsomes; HM, heavy membranes.

is equilibrated with cell homogenate the lectin reacts almost exclusively with what is identified as plasma membrane by other criteria.

On the other hand, the receptor–ligand complex can be easily dissociated using *N*-acetyl-*D*-glucosamine or ovomucoid. Important also for more subtle subfractionations is that the WGA–membrane complex does not dissociate upon exposure to dextran and therefore can be used as a marker in dextran density gradients. In contrast to WGA, concanavalin A is not a useful plasma membrane marker.

3.   VIRUS RECEPTORS

Plasma membranes bear surface components that specifically bind certain viruses. Such binding sites can be used to identify the membranes, monitor their fractionation, and study the distribution of the receptor sites on the membrane.

This approach has been recently used by Roesing *et al.* (1975), who have coupled Coxsackie virus V3 (^{32}P-labeled) to HeLa cells prior to cell fractionation and plasma membrane isolation. Upon ultracentrifugation of the membrane material in the sucrose step gradients (25–50%), they isolated four discrete bands. More than 70% of the radioactive virus was recovered in the two high density bands (ρ 1.16 and ρ 1.21, respectively). These bands also contained 40% of the Na^+-K^+-activated ATPase and 50% of the 5'-nucleotidase activity.

The data show that, since the virus membrane complex is very stable, the virus can be used as a sensitive plasma membrane marker. The experiments do not allow one to evaluate whether the virus particle can also be used as a density perturbant (see below), since the fractionation pattern of the virus-free membranes was not reported.

E.   Dodecyl Sulfate Polyacrylamide Gel Electrophoresis

High-resolution dodecyl sulfate polyacrylamide gel electrophoresis can be used to distinguish between diverse membrane categories in some cells. This is illustrated in Table III for the case of plasma membrane and endoplasmic reticulum vesicles in the microsomal membranes from thymocytes (Schmidt-Ullrich *et al.*, 1974).

Electropherograms of the microsomal membranes exhibit 10 dominant protein bands labeled 1–10 according to decreasing molecular weight (MW). Many components include several less intensely stained bands, designated by subnumerals, e.g., 3.1. Component 2 (apparent MW 155,000) consists of three closely spaced bands, designated 2.1, 2.2, and 2.3. After separation of endoplasmic reticulum from plasma membrane vesicles by isopycnic ultracentrifugation on dextran density gradients (see below), we recover all sodium dodecyl sulfate–polyacrylamide gel electrophoretic components in

TABLE III

DODECYL SULFATE–POLYACRYLAMIDE GEL ELECTROPHORESIS OF CALF THYMOCYTE
MEMBRANE FRACTIONS[a,b]

Peptide No.	Apparent molecular weight	Coomassie blue staining (% of total in gel)		
		Total microsomal membranes	Plasma membranes	Endoplasmic reticulum
1	280,000	2.9	4.5	2.9
1.1–1.3	210,000	4.6	5.5	3.2
2.1–2.3	155,000	5.7	8.0	5.0
3	120,000	3.9	5.0	3.1
3.1	108,000	2.2	3.2	1.6
4	79,000	10.6	15.0	6.9
4.1	66,000	NR[c]	NR	1.2
5	60,000	8.3	10.5	7.2
5.1	55,000	4.2	4.5	5.8
6	44,500	9.4	10.4	4.1
7	36,000	6.4	8.0	5.0
7.1	32,000	2.8	1.9	4.0
8	26,500	8.3	3.5	8.9
8.1	23,500	4.6	1.6	5.5
9	21,000	9.1	7.0	8.9
10	17,000	10.4	1.4	12.1
10.1	15,000	3.3	—	3.4

[a] From Schmidt-Ullrich et al. (1974).
[b] Soluble intravesicular proteins are eluted by osmotic shock prior to analysis.
[c] NR, not resolved.

one or another of the various gradient fractions. However, the proteins in the plasma membrane fraction consist principally of high molecular weight components (1–7), whereas components 7.1–10.1 predominate in endoplasmic reticulum. Moreover, the glycosylated proteins (1, 2.1–2.3, 3–5.1) are concentrated in plasma membrane vesicles.

These data suggest that one can develop reliable electrophoretic "fingerprints" for diverse biomembranes.

III. Cell Disruption

A. Disruption Medium

This disruption medium should be one in which the cells to be fractionated will remain viable for at least several hours at ice-bath temperatures. [Minimal criteria for viability would be normal exclusion of Trypan blue or normal fluorochromasia (Rotman and Papermaster, 1966)]. The medium should be buffered near pH 7.4, but at low temperatures, where metabolic activity is suppressed, high buffering capacity is not necessary. Tris buffers are suitable for many cells, provided the pH is adjusted for the temperature used in fractionation. However, certain cell types, for example many lymphoid cells, rapidly lose viability in Tris buffers; HEPES buffers or phosphate buffers (less than 10 mM) should be used in such cases.

Buffered, isosmotic, or slightly hyposmotic sucrose, is perhaps the most widely used disruption medium. This is largely because it is generally satisfactory for the separation of diverse subcellular organelles by differential centrifugation. However, some cell types, for example lymphoid cells, either aggregate rapidly in sucrose-containing media (Ferber et al., 1972), or even lose viability rapidly (Schmidt-Ullrich et al., 1974). In addition disruption media of low ionic strength, such as isosmotic sucrose, may lead to unusual associations of normally predominantly soluble proteins with plasma membrane fragments (Kant and Steck, 1973; Shin and Carraway, 1973).

A major argument for the use of low-ionic strength media is the tendency of small membrane vesicles, as well as other small subcellular, particulate fragments, to aggregate nonspecifically; this trend becomes small at low ionic strengths, where electrostatic repulsions between particles of like charge can be large. Dispersions of small particles, such as small membrane vesicles, tend to be unstable because the total free energy—owing to the large interfacial area between the dispersed phase (vesicles) and continuous phase (suspension medium)—can be decreased by aggregation of the vesicles. The interfacial energy will be positive and aggregation will occur when the forces of attraction between the vesicles exceed repulsive interactions, that is, when particle collisions become inelastic.

The number, c, of collisions per second between particles in a suspension is given by

$$c = 4kTN^2/3\eta \qquad (1)$$

where kT is the thermal energy unit (4.14×10^{-14} gm $\cdot$ cm^2 sec^{-2} at 25°C), N is the particle number per cm^3, and η is the viscosity (0.01 gm $\cdot$ cm^{-1} sec^{-1} for dilute aqueous solutions at 25°C. For aqueous suspensions containing 2×10^{11} particles, the collision rate is about 2×10^{11} encounters per second.

In terms of subcellular fractionations, the rate of dissociation of particle aggregates, once formed, is small, and the rate of decrease, due to inelastic collisions, in the number of vesicles can be approximated by

$$-dn/dt = 4kTN^2/3\eta \cdot e^{-(W_R/kT)} \qquad (2)$$

where W_R is the net energy barrier to aggregation. Then the time, $t_{1/2}$ (sec), for the vesicle number, N, to drop to half the initial value, N_0, is

$$t_{1/2} = (3\eta/4kTN_0)\, e^{(W_R/kT)} \tag{3}$$

for aqueous dispersions at $4°C$, $t_{1/2}$ becomes $(3.8 \times 10^{11}/N_0)\, e^{(W_R/kT)}$

W_R is due primarily to electrostatic repulsions between particles of like charge. Our calculations (Wallach et al., 1966), give ionic strengths yielding $t_{1/2}$ values at $25°C$ for typical plasma membrane vesicles, with an average radius of 700 Å, an area of about 18×10^3 Å^2 per net negative charge on their external surfaces, and particle concentrations of 2×10^{10} to 2×10^{14} per cm^3, corresponding to 0.01 to 100 mg of membrane protein per milliliter.

A thorough analysis of the effect on the aggregation rates of plasma membrane vesicles, and endoplasmic reticulum vesicles from Ehrlich carcinoma of pH, ionic strength, and ionic composition (Wallach et al., 1966), demonstrates the importance of the ionic variables on the "colloidal stability" of membranes vesicles. The data also suggest that very adequate fractionations can be achieved in salt-containing media.

An example drawn from our fractionations of thymocytes (Schmidt-Ullrich et al., 1974) is useful to illustrate this matter. These cells are essentially smooth spheres with maximal radii of $\sim 4 \times 10^4$ Å and are fractionated at a cell concentration not exceeding $5 \times 10^7/$cm^3. The surface area per cell is 2×10^8 Å^2, yielding a total plasma membrane area of 10^{18} Å^2. This is disrupted into vesicles of 700 Å average radius, i.e., 6.15×10^6 Å^2 per vesicle. If all cells are disrupted, this gives a yield of 1.6×10^{11} particles per cm^3. Assuming that twice that particle concentration arises from the shearing of intracellular membranes (probably an exaggerated value, we arrive at a

TABLE IV

COMPUTED EFFECTS OF IONIC STRENGTH ON
VESICLE AGGREGATION[a]

Ionic strength (M)	$t_{1/2}$ (min)[b]
0.152	3
0.122	63
0.102	234
0.054	425,000

[a] Recalculated from Wallach et al. (1966) for $4°C$ ($= 0.0157$ gm·cm^{-1} sec^{-1} and $kT = 386 \times 10^{-14}$ gm·cm^2 sec^{-2}), a particle concentration, N_0, of $5 \times 10^{11}/$cm^3, a particle radius of 700 Å, and an area per charge of 1.8×10^3Å^2.

[b] See Eq. (3).

particle concentration of $5 \times 10^{11}/cm^3$). The computed effect of varying ionic strength (monovalent ions) on such a suspension (Table IV) indicates that fractionation should be practical near physiological salt concentrations; this is the case (Schmidt-Ullrich et al., 1974, 1976a).

A very important variable in the disruption medium is the concentration of *divalent* cations, calcium and magnesium. This is because the stability of nuclei following cell disruption depends very markedly on the presence of divalent cations, and nuclear disruption almost always leads to poor yields of plasma membrane, as well as considerable cross contamination. We cannot provide any fixed rules concerning this variable since we find that, even in the lymphoid cell category, the nuclei of some cell types are adequately stabilized by 0.25 mM Mg^{2+}, where other lymphoid cells require ten times that magnesium concentration. We therefore recommend systematic exploration of this variable during the determination of optimal conditions for cell disruption and maintenance of nuclear integrity.

B.   Measurement of Nuclear Integrity

When testing for conditions that are optimal for cell disruption in a particular case one must decide how to evaluate nuclear integrity after cell disruption. This step is of the essence, although not usually employed with sufficient rigor. The simplest and most widely used approach is a microscopic one; cell counts and nuclear counts are performed before and after cell disruption. Ideally, after cell disruption no intact cells remain, but the nuclear count is unchanged. This procedure is generally adequate, however, where possible it may be profitable to radiolabel cellular DNA by a pulse of radioactive thymidine, followed by a chase with "cold" thymidine and to monitor the amount of label remaining in the supernatant after sedimentation of the nuclei and large granules, respectively. Under ideal conditions virtually all the label will be in the nuclear pellet and only traces in the large granule fraction (mitochondrial DNA). No label should appear in the high-speed sediment containing the plasma membrane vesicles.

In any event if either criterion indicates less than optimal maintenance of nuclear integrity, modifications of the concentration of divalent cation in the medium and/or conditions of mechanical disruption will be necessary.

Fractionations are generally carried out in the cold. One of the major reasons for the use of temperatures of $0°–4°C$ in cell disruption and cell fractionation is to inhibit the action of possible degradative enzymes that might be liberated and/or activated during cell disruption or subcellular fractionation, thus modifying native membrane structure. It should be noted, however, that many degradative enzymes are appreciably active in the cold and that chilling cannot be assumed *a priori* to prevent membrane degradation. On the other hand, the possible deleterious effects of degrading enzymes can again be minimized by operating at low cell concentrations.

C. Techniques for Cell Disruption

For purposes of isolating high yields of purified plasma membrane vesicles, it is necessary to use cell disruption techniques that spare other cellular organelles. For this reason the use of osmotic stress, pH extremes or detergents, with/without mechanical shear, are in general not desirable for eukaryotic cells.

The most commonly used approach to cell disruption involves mechanical (possibly including liquid) shearing of cells by a Potter–Elvehjem or Dounce homogenizer. These devices consist of precision-bore glass tubes with smooth or rough walls and motor- or hand-driven Teflon or glass pestles. Homogenizers can be obtained with very precise clearances between the pestle and the tube wall. The Dounce homogenizer, generally considered more "gentle" has a spherical head and is usually operated manually.

Homogenizers are generally used in a rather empirical fashion, depending on the notion of the investigator as to how a plasma membrane should fragment. For example, those who wish to maintain the membrane in large pieces, use homogenizers with large clearances and apply shear in few strokes with a gentle vertical motion. However, there is evidence, (cf. review by Wallach and Lin, 1973) that even very gentle "homogenization" of rat liver ruptures at least 15% of lysosomes and strips at least 10% of the mitochondria of their outer membranes. Another problem is that even under the best conditions the amount of shear applied may vary considerably from cell to cell, leading to a rather wide size range within the population of membrane fragments. Finally, the technique becomes more and more difficult to apply as the nuclear radius approaches that of the cell. This is a common problem, particularly in lymphoid cells (Fig. 1), and very often produces a situation where it is impossible to rupture a significant proportion of cells without at the same time damaging an undesirable proportion of nuclei.

As pointed out previously (Wallach and Lin, 1973), high shear approaches such as sonication and use of the French press also tend to be nonselective in their disruption action, affecting intracellular membranes to the same extent as plasma membranes. Moreover, sonication can produce high local levels of temperature, as well as free radicals, both of which can produce molecular damage to both plasma membranes and intracellular membranes. The high pressures employed in operating a French press are also not innocuous and may lead to the denaturation of both soluble and membrane proteins.

A highly satisfactory general technique of cell disruption, *gas-bubble*

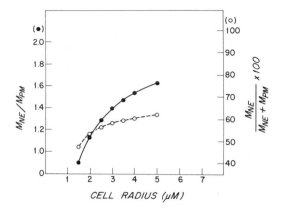

FIG. 1.   Variation of the proportions of membrane mass contributed by the nuclear envelope (M_{NE}) and plasma membrane (M_{PM}) in lymphoid cells of diverse radii. The nuclear envelope is treated as a double membrane. ●———●, Ratio; ○----○, %.

nucleation, or commonly referred to as "nitrogen cavitation," was first introduced by Wallach *et al.* (1960), using a technique first used to rupture bacteria and then developed further by Hunter and Commerford (1961) to produce intact nuclei from rat hepatocytes.

In this approach, cell suspensions are equilibrated with 30–80 atmospheres of an inert gas (usually nitrogen) and then returned with minimum liquid shear to atmospheric pressure. It was originally thought that cell rupture occurred due to intracellular cavitation of nitrogen bubbles. However, at the low pressures used, diffusion of the gas of the cell predominates upon return to atmospheric pressure. It appears most likely that cell disruption occurs when small, spinning bubbles nucleate at the plasma membrane and at sheets of endoplasmic reticulum, effectively producing cell rupture by liquid shear. It is conceivable also that counter diffusion of gases across the membranes could lead to supersaturation of membranes and formation of microbubbles within the membrane.

The technique of gas bubble nucleation provides for the quantitative disruption of diverse cell types, at controlled temperatures, without local heating, in inert atmshpere and under conditions where all cells are subjected to the same amount of shear. Also, as reviewed by Wallach and Lin (1973), the technique is sparing of all intracellular organelles other than endoplasmic reticulum, which also vesiculates. In addition, the method yields vesicles of rather uniform size. Finally, the technique produces vesicles that are sealed to small molecules and can be used for the study of the transport mechanisms involved in the uptake of a number of cellular nutrients (see, for example, Hochstadt *et al.*, 1976).

We stress that in the application of the gas nucleation technique release of cell suspensions from the "bomb" should be slow, that is 10–20 ml per minute. Rapid release causes nonselective disruption of cells and subcellular organelles at the valve orifices by flow-shear, which is not the desired disruption principle. The cells should emerge *intact* from the pressure vessel and rupture only upon return to atmospheric pressure outside the pressure vessel. In some cases, heavy foaming cannot be avoided without addition of a few microliters of a high alcohol, for example decanol, to the receiving flask prior to release of the cell suspension from the pressure vessel. This strategem is rarely necessary, but it does not interfere with subsequent fractionation processes. The gas nucleation technique is most effective when used with cell suspensions, but can also be employed with small tissue fragments.

Some cell types require very special cell disruption techniques. An example from our experience is the adipocyte, whose large size and fat content create rather special problems. We (Avruch and Wallach, 1971) have found that the plasma membranes of adipocytes can be effectively vesiculated by forcing these cells (diameter 50–100 mm) through stainless steel photoetched screens with apertures of about 200 mm.

In all cases the shear condition for a particular cell type has to be adjusted to achieve the maximum cell rupture without significant nuclear damage. The optimal conditions are particularly easy to determine in cases where the gas nucleation technique can be applied. Here, the cells, suspended in a medium compatible with their viability are introduced into the appropriate "bomb" (for example, the apparatus manufactured by Artisan Metal Products, Waltham, Massachusetts), equilibrated with a low pressure of nitrogen (for example, 40 atmospheres for 20–30 minutes) and a few drops released gently onto a microscope slide to measure the number of undisrupted cells (exclusion of trypan blue) and the number of intact nuclei (phase contrast; Giemsa staining). This process is repeated with increments of 5 atmospheres of gas pressure until optimal disruption has been obtained.

IV.   Centrifugal Fractionation of Membrane Vesicles

After cell disruption, various organelles, as well as vesicular fragments of plasma membrane and endoplasmic reticulum, can be partially separated by differential centrifugation. This portion of the fractionation process is rather simple and has been adequately described elsewhere (e.g., Steck, 1972; Wallach and Winzler, 1974). In general, nuclei will sediment below 1.8×10^4 $g \cdot min$; mitochondria, lysosomes, and peroxisomes between 2×10^4 $g \cdot min$ and 4×10^5 $g \cdot min$. Small membrane vesicles generally will not pellet below

$5 \times 10^6\,g \cdot \text{min}$, but in simple differential centrifugation they may be trapped artifactually among larger particles.

Effective separation and purification of plasma membrane vesicles requires the use of sophisticated density gradient techniques. In the following we present some basic principles involved in the use of these methods.

A.   Membrane Fragments as Sealed Vesicles

The membrane vesicles that sediment as part of the "microsomal fraction" upon conventional differential centrifugation, using conventional fractionation media, are minute, negatively charged sacs that have pinched off primarily from plasma membrane and endoplasmic reticulum in the process of the fragmentation of these structures during cell disruption. Contained in the aqueous spaces within these vesicles are soluble proteins trapped during the process of vesiculation. In the case of plasma membrane vesicles, whose pinching off process encloses small volumes of the cytoplasmic space (Fig. 2), the entrapped soluble proteins tend to be those of the soluble cytosol. In the case of endoplasmic reticulum vesicles, which pinch off in an opposite direction (their cytoplasmic faces become the external faces of the isolated vesicles) soluble contents of the vesicles tend to be cisternal material from endoplasmic reticulum. Rather high protein concentrations might be expected to occur within endoplasmic reticulum vesicles isolated from protein-secreting cells.

The retention of soluble proteins within the isolated vesicles depends very markedly upon the composition of the fractionation medium. This can be

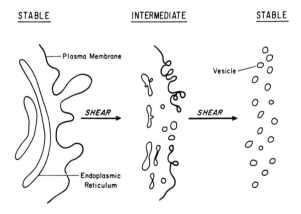

FIG. 2.   Presumed effect of liquid and/or mechanical shear on plasma membrane and endoplasmic reticulum shearing causes a "pinching off" process in both types of membranes, leading to the formation of relatively stable vesicles. From Wallach and Lin (1973) with permission of *Biochim. Biophys. Acta.*

adjusted to produce the release of the trapped proteins by, for example, hyposmotic shock, a technique that can be used to coincidentally introduce "reporter" molecules, cofactors, and substrates into the vesicles (Wallach *et al.*, 1965).

We (Wallach *et al.*, 1965; unpublished) have shown that when microsomal vesicles are exposed to hyposmotic solutions of low ionic strength, they become transiently permeable to macromolecules. By inclusion of a suitable enzyme in such "lysing solution," this can be introduced into the microsomal vesicles, providing an indicator for the permeability properties of the vesicles as well as a probe for the intravesicular environment. We have used horseradish peroxidase (MW 44,000) because of its high solubility even at low ionic strengths, its inertness toward components of the membrane structure, its small, permeant substrate, the very high sensitivity of peroxidase assay techniques and availability of the enzyme in high purity.

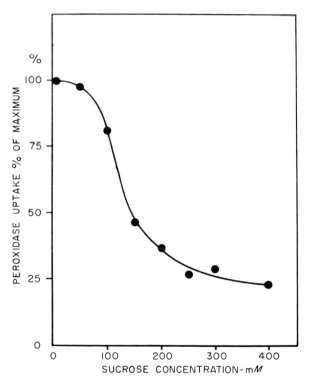

FIG. 3.   Effect of sucrose on the introduction of horseradish peroxidase into intraluminal space of microsomal membrane vesicles from Ehrlich ascites carcinoma. Microsomal membranes were exposed for 1 minute at 20°C to peroxidase-containing sucrose (pH 7.4) solution before adjustment to 0.15 *M* NaCl, 1.0 m*M* MgCl$_2$, washing in this medium, and assay. From Wallach *et al.* (1965).

We have found that upon exposure of microsomal membrane vesicles to transient osmotic shock, the enzyme can be "shocked into the vesicles" to a degree, depending upon its concentration in the "lysing solution" and the following variables:

1. Timing. The enzyme can be "shocked" into the intravesicular state ("uptake") only if present in the medium at the *instant* osmotic shock is initiated.

2. pH. Enzyme entry is slight at pH 6.6 and below and increases with rising pH.

3. Osmotic activity of the medium. Entry is inhibited by sucrose or glucose in proportion to their concentration, up to 0.25 M (Fig. 3). These findings indicate that the walls of the microsomal membrane vesicles are only poorly permeable to sucrose.

4. Ionic strength. Enzyme uptake is inhibited by increasing ionic strength (univalent electrolytes) between 2.5 mM and 75 mM (Fig. 4).

5. Ionic composition. Uptake of enzyme is almost completely inhibited by calcium (or magnesium) concentrations well below 1 mM (Fig. 5).

All in all, the data indicate that microsomal membrane vesicles behave as reasonable osmometers and that their permeability properties are such as to justify their use in transport studies (see, for example, Quinlan and Hochstadt, 1975).

The behavior of the vesicles in gradients of polysucrose or polyglucose, even after osmotic shock, further supports this contention: when equilibrated in density gradients of dextran (MW 50,000 to 150,000), the vesicles behave in a fashion indicating that their walls are not permeable to these polymers.

B.   Osmotic Behavior of Sealed Vesicles during Density Gradient Ultracentrifugation

SIMPLE OSMOTIC EFFECTS

Microsomal membrane fragments, whether of surface or intracellular origin, are vesicles 1000–2000 Å in diameter, bounded by charged, semipermeable lipoprotein walls about 80 Å thick. All these vesicles change in volume and density with variations of their ionic and osmotic milieu. Particles of diverse origins respond in different ways to environmental alterations, however, and this permits their separation by ultracentrifugation in suitable density gradients (Wallach and Kamat, 1966; Wallach, 1967; Stech *et al.*, 1970; Steck and Wallach, 1970).

Detailed treatments of the behavior of sealed vesicles in media of different osmotic activities and ionic properties have been presented by Wallach (1967) and Steck *et al.*, (1970).

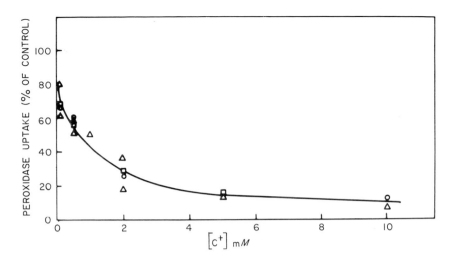

FIG. 4. Effect of univalent electrolyte, C^+, on introduction of horseradish peroxidase into intraluminal space of microsomal vesicles (from Ehrlich ascites carcinoma). □, Choline Cl; △, NaCl; ○, KCl. Exposure for 1 minute at pH 7.4 before addition of "sealing solution". From Wallach *et al.* (1965).

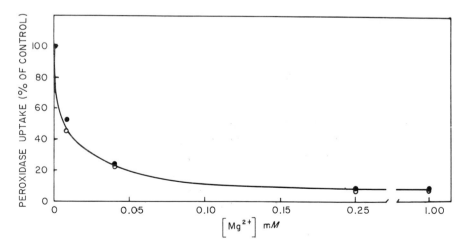

Fig. 5. Effect of divalent cations, Mg^{2+}, on introduction of peroxidase into microsomal membrane vesicles. ●, $CaCl_2$; ○, $MgCl_2$. Procedure as in Figs. 3 and 4. From Wallach *et al.* (1965).

Considering a vesicle as a sphere, its total volume, V_T, will consist of three compartments: (a) V_w is the volume of the anhydrous membrane shell, with internal radius r_1 and external radius r_2. (b) V_h is the volume of the water of hydration associated with the membrane shell; this can be considered as composed of layers, of thickness t, one on the inside and one on the outside surface of the wall. (c) V_f is the volume of fluid within the vesicle. Then

$$V_T = V_w + V_h + V_f \tag{4}$$

when $t \ll r_1$, approximate expressions for V_T, V_w, V_h, and V_f are:

$$V_T = 4/3\pi \, (r_2 + t)^3 \tag{5}$$

$$V_w = 4/3\pi \, (r_2^3 - r_1^3) \tag{6}$$

$$V_h = 4\pi t(r_2^2 + r_1^2) \tag{7}$$

$$V_f = 4/3\pi \, (r_1 - t)^3 \tag{8}$$

While the total mass of anhydrous membrane material is constant in a given particle, V_w, V_f, V_h, and therefore V_T, can vary with environmental conditions.

The vesicle volume depends upon the balance between cohesive focus maintaining the integrity of the vesicle wall, and opposing electrostatic processes, namely, Gibbs–Donnan effects and coulombic repulsions. Both produce expansion of V_f, and thus V_T, at low ionic strength, at pH levels away from the effective "isoelectric point," and in the absence of polyvalent counterions. In the present model, however, expansion of V_T cannot occur without some alteration of the properties of the wall. Therefore, if the wall volume V_w stays constant, the wall thickness $r_2 - r_1$ must diminish, but if the wall thickness remains constant, its structural elements must become less tightly packed. Alterations of the ionic environment can thus also influence V_h although this is $< 10\%$ of V_T in the case of vesicles with radii $\geqq 500$ Å. Osmotic mechanisms can act in the same or opposite direction as electrostatic effects, depending on whether the concentration of the impermeant solutes is greater within or without the vesicles. At very small particle radii and under ionic conditions generating high net surface charge, coulombic repulsions can impair membrane integrity, leading to porous vesicles, or small membrane sheets.

Contraction of volume may proceed (a) without alteration in shape, in which case there must be a change in wall thickness and/or wall density, or (b) through folding of the wall, in which instance the thickness and density of the wall may remain unaltered. In the latter situation, Eq. (4) no longer applies directly but can be used to calculate volumes in terms of equivalent spheres or shells.

The mass of a given vesicle is the sum of the masses of the three compartments. Accordingly, the density, ρ_T, of the vesicle can be derived from the equation

$$\rho_T V_T = \rho_w V_w + \rho_h V_h + \rho_f V_f \qquad (9)$$

and will equal

$$\rho_T = (\rho_w V_w + \rho_h V_h + \rho_f V_f)/V_T \qquad (10)$$

where ρ_w, ρ_h, ρ_f, and ρ_T are the densities of the anhydrous wall, water of hydration, internal fluid, and whole vesicle, respectively.

Substituting from Eqs. (4)–(8):

$$\rho_T = \left[\rho_w(r_2^3 - r_1^3) + \rho_h 3t(r_2^2 + r_1^2) + \rho_t(r_1 - t)^3\right]/(r_2 + t)^3 \qquad (11)$$

By binomial expansion:

$$(r_2 + t)^{-3} \approx r_2^3(1 - 3t/r_2) \quad \text{and} \quad (r_1 - t)^3 \approx (r_1^3 - 3r_1^2 t) \qquad (12)$$

Therefore:

$$\rho_T = \left[\rho_w\left[1 - (r_1/r_2)^3\right] + \rho_h 3t\frac{(r_2^2 + r_1^2)}{r_2^3} + \rho_f\frac{(r_1^3 - 3r_1^2 t)}{r_2^3}\right](1 - 3t/r_2) \qquad (13)$$

In very small vesicles ρ_T will approach the density of the hydrated wall, (then second- and third-order terms in t can no longer be neglected). In very large sacs, ρ_T approaches the density ρ_f of the fluid within (Fig. 6). Quite clearly, environmental variables that affect V_f and V_h can influence vesicle density.

Particle densities can be measured by means of ultracentrifugal equilibration in density gradients. The values of ρ_T obtained for given values of ρ_w,

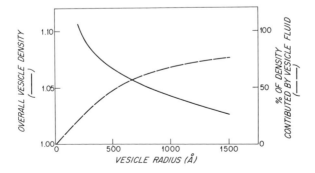

FIG. 6. Variation of ideal vesicle density, ρ_T, with vesicle radius according to Eq. (13) ρ_T on left ordinate. Proportion of ρ_T contributed by density of fluid in vesicle is given on right ordinate.

ρ_h, and ρ_f, however, depend upon the gradient system used. This is illustrated by the following three cases (Fig. 7).

a. Density gradients composed of permeant solutes: When gradients of such substances (e.g., glycerol) are used, Eq. (13) simplifies, and the observed densities will reflect only on the densities and volumes of the wall and the hydration layers. Alterations of membrane hydration by different experimental variables can be detected in such gradients, as well as by replacement of H_2O with D_2O (de Duve et al., 1959).

b. Density gradients composed of nonpermeant small solutes. The *sucrose* gradients most widely used in membrane fractionations fall into this category. In such systems, gradients of density also yield large osmotic gradients. These engender contraction of V_f as a vesicle sediments. Therefore the densities obtained in such gradients approach those found in case (a). Since V_f will be contracted, alterations of ionic environment will not produce

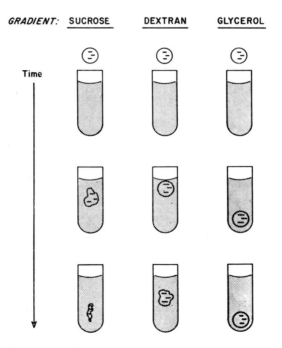

FIG. 7.   Schematization of the behavior of an internally charged membrane vesicle in three gradient systems. The vesicle wall is permeable to glycerol, but not to sucrose or dextran. In glycerol the vesicle equilibrates at the density of the hydrated membrane wall. In sucrose the vesicle is collapsed by the increasing osmotic activity of increasing sucrose concentrations. It does not collapse completely owing to electrostatic effects. In dextran the osmotic contributions of the gradient are slight. From Wallach and Lin (1973) with permission of *Biochim. Biophys. Acta.*

large changes in ρ_T. Variations in wall hydration may, however, effect significant changes in ρ_T.

c. Density gradients composed of uncharged nonpermeant macromolecules. Typical examples of this situation are gradients of polysucrose (Ficoll; cf. Wallach and Kamat, 1964, 1966) or of polyglucose (dextran; e.g., Steck et al., 1970). In such gradients osmotic effects due to the gradient solute are small (Fig. 7) (Steck et al., 1970), and then Eq. (13) becomes a reasonable approximation for vesicle behavior. In general ρ_f will be much smaller than ρ_w, and ρ_T will depend markedly on V_f. Also, alterations of V_f resulting from electrostatic and osmotic mechanisms will produce substantial changes in ρ_T. Alterations in V_h will generate less dramatic effects.

In our original explorations (e.g. Wallach and Kamat, 1964), we employed polysucrose (Ficoll) gradients but have turned to dextran (polyglucose) (e.g., Steck et al., 1970; Schmidt-Ullrich et al., 1974, 1976a) because the latter can be easily obtained free of ionic and toxic contaminants. In addition polyglucose can be commercially obtained in batches of narrow molecular-weight ranges that allow sharper separations in density gradients.

2. DONNAN EFFECTS

The density, ρ_T, of a membrane vesicle can also be described by the equation (Steck et al., 1970)

$$\rho_T = (Z + Q \cdot \rho_w)/(Z/\rho_{med} + Q) \tag{14}$$

where ρ_{med} is the density of the gradient medium, Q is the effective charge density (meg/gm membrane) on the inner membrane surface and $Z = (S^2 + 4mS)^{1/2}$, where S is the molal concentration of impermeant solute, and m the molal concentration of permeant univalent electrolyte.

The above model does not consider several factors that might produce deviations from ideal behavior: (1) Charges fixed to the membrane may produce anomalous distributions and activities of their Donnan counterions at the osmotic boundary. (2) The effective charge, Q, probably differs from the net fixed charge usually assumed in Donnan equilibria. (3) Forces other than osmotic effects are neglected. Electrostatic repulsions among fixed charges of the same sign would contribute to the expansive potential and, in general, mimic Donnan effects. In addition, mechanical (elastic) forces may dominate as vesicles expand or collapse fully. Small vesicles could become maximally inflated before reaching buoyant equilibrium. For example, a vesicle of internal radius ~ 500 Å, wall thickness $= 75$ Å, and wall density $= 1.15$ cannot achieve a density below 1.058; a similar vesicle of radius $= 1000$ Å can equilibrate below $\rho = 1.032$. Thus, high-resolution fractionation of homologous vesicles according to size may be possible using isopycnic centrifugation. (4) Charges fixed to the outer surface of the membrane have been neglected. These could alter equilibrium density by affecting the Donnan-osmotic equilibrium, the electrostatic potential, and the hydration of the membrane. (5) The formulation assumes that no soluble solutes were held within the vesicle. While vesicles derived from plasma membrane and endoplasmic reticulum can be so prepared, native organelles may contain soluble macromolecules, e.g., charged proteins, that may make a significant contribution to the Donnan effect. Then the determinants of ρ_T must include the masses and densities of nondiffusible, hydrated solids, and Q becomes the total effective intravesicular fixed charge. Our hypothesis (Steck *et al.*, 1970) indicates that vesicles with identical membranes might be fractionated *according to unique soluble proteins they may contain.*

Our hypothesis leads to the following predictions:

1. The size of a vesicle does not generally influence its equilibrium density. Thus, a dispersion of vesicles of common origin should equilibrate at a single density, independent of membrane mass.

2. When ρ_w is determined independently, Q can be derived from measurements of ρ_T and Z at buoyant equilibrium.

3. The fixed charge concentration in a vesicle will be a function of the medium and not the vesicle. Thus, all vesicles in osmotic equilibrium with a given medium will have the same fixed charge concentration, Z, irrespective of the size of their membrane or their charge.

4. pH and divalent cations can influence ρ_T by titrating fixed charges within a vesicle. This prediction is supported by the data of Wallach and Kamat (1964), Wallach *et al.* (1966), and Steck *et al.* (1970).

5. Closely related vesicles can be separated ultracentrifugally when they differ in their complement of titrable membrane proteins; i.e. ρ_T will vary widely with changes of Q at low values of Q, i.e. titration of charges fixed at internal membrane surfaces, altering Z, can produce major changes in ρ_T.

C.   A Practical Example: Isolation of Plasma Membrane Vesicles from Lymphoid Cells

The fractionation scheme we have found most useful for high-yield isolation of plasma membrane vesicles from normal and leukemic GD 248 hamster lymphocytes is as follows (Schmidt-Ullrich *et al.*, 1976a)

Cell rupture and subsequent steps are carried out at 4°C. Cells are disrupted by "nitrogen cavitation." For this, GD 248 cells are resuspended in 0.065 M NaCl, 0.01 M HEPES, pH 7.4, 0.002 M $MgCl_2$, 0.075 M KCl to a concentration of 2 to 5×10^7/ml. The disruption buffer for normal lymphocytes is identical except that 0.00025 M $MgCl_2$ is used. Up to 100 ml of cell suspension are equilibrated with 30 atm of N_2 for 20 minutes, using gentle stirring. After release from the bomb the GD 248 homogenates are made 0.002 M

in EDTA (trisodium salt), using a 0.1 M stock solution. For normal cells the final EDTA concentration is 0.001 M.

Nuclei are sedimented at 1.8 × 10^4 $g \cdot$ min. The "large granule fraction," containing mitochondria and lysosomes, as well as some plasma membrane fragments, is collected at 4 × 10^5 $g \cdot$ min. To recover surface membrane material from this pellet, the large granule fraction is subfractionated on a Dextran-150 gradient with density steps of 1.05, 1.10, 1.12, and 1.16. The gradient buffer was 0.001 M HEPES, pH 7.5, 0.0001 M MgCl$_2$. The "large granule pellet" material from ~ 10^9 cells, resuspended in the EDTA-containing disruption buffers used for the normal and transformed cells, was distributed equally into the four dextran gradient steps, overlaid with buffer, and then centrifuged to equilibrium at 5 × 10^7 $g \cdot$ min (Spinco rotor SW 50.1; 5-ml tubes).

The band at density 1.05, containing plasma membrane markers, is combined with the microsomal fraction collected at 1.2 × 10^7 $g \cdot$ min from the 4 × 10^5 $g \cdot$ min supernatant. The resulting combination is washed first with 0.01 M HEPES, pH 7.5, and then with 0.001 M HEPES, pH 7.5 (1.2 × 10^7 $g \cdot$ min). These washes were for the purpose of removing soluble cytoplasmic proteins trapped in the membrane vesicles. The membranes are then separated into plasma membrane and endoplasmic reticulum using gradients of dextran-150, buffered with 0.001 M HEPES, pH 8.2 and 0.0001 M in MgCl$_2$ for GD 248 cells or 0.0005 M in MgCl$_2$ for normal lymphocytes. Each gradient consists of a 3.0 ml of dextran-150 cushion (density 1.09 ± 0.005) overlaid by a 1.0 ml continuous gradient with a density range 1.0–1.09. The membrane material to be fractionated, first dialyzed against 400 ml of 0.001 M HEPES, pH 8.2, 0.0001 M MgCl$_2$ (0.0005 MgCl$_2$ for normal cells) is loaded in 1-ml portions onto the top of the gradient. After ultracentrifugation at 10^8 $g \cdot$ min two closely located plasma membrane bands (that can sometimes fuse into one) are found in the continuous gradient. The endoplasmic reticulum pellets through the dextran cushion. The overall properties of the plasma membrane vesicles obtained by the above fractionation scheme are listed in Table I.

D. Density Perturbation Methods

Quite frequently membrane vesicles of various types can be isolated and/or fractionated selectively according to their contents. This approach has been applied very extensively in the purification of rat liver lysosomes in the form of "tritosomes" (Baudhuin et al., 1965). In these studies animals are fed about 80 mg of Triton WR1339 per 100 gm body weight, intravenously or intraperitoneally, 2 to 4 days before sacrifice. During this period the relatively nontoxic detergent concentrates in the lysosomes of hepatocytes. To isolate these "tritosomes," the livers are homogenized by conventional techniques, nuclei and "debris" are removed by low speed centrifugation, and lysosomes are pelleted together with mitochondria and peroxisomes ar approximately 340,000 $g \cdot$ min. These three organelles can then be separated in sucrose density gradients, because the densities of the detergent-laden lysosomes lie well below those of the other organelles.

The density perturbation principle has also been utilized by Wetzel and Korn (1969) and Heine and Schnaitman (1971) for *Acanthamoeba* and L cells, respectively, and has been recently reviewed in this series by Chara-lampous and Gonatas (1975). In essence the approach consists of exposing cells to low-density polystyrene latex particles which are engulfed by phago-cytosis. Once the cells are disrupted the particles with their surrounding membranes can be simply harvested by virtue of the low density of the polystyrene. In princple, this approach is very similar to that employed by Dowben *et al.* (1967) for the isolation of the eccrine membranes of mammary epithelial cells, by virtue of the low density of the milk that they enclose.

A nonbiological density perturbation strategem might be of use in the purification of membrane vesicles containing specific phosphatases. For example, accumulation of insoluble lead phosphates within microsomal vesicles during hydrolysis of glucose-6-phosphate in the presence of Pb^{2+} raises the density of the vesicles containing the enzyme (Leskes *et al.*, 1971).

A more general approach, termed "affinity density-perturbation" has been developed recently (Wallach *et al.*, 1972; Wallach, 1974). In this technique membrane fragments bearing a specific receptor, for example, an antigen or a lectin-binding protein, are separated centrifugally from vesicles lacking such receptors, following combination of the receptor with its specific ligand —coupled to a particle of very high density, the density perturbant. The vesicles may also be subfractionated, according to the number of receptors per vesicle.

In practice, isolated membrane vesicles are allowed to react with a specific density perturbant, and the mixture is centrifuged thereafter in density gra-dient until isopycnic equilibrium is achieved. The unreacted membrane vesicles will band at a low density and the unreacted density perturbant will go to a very high density. Vesicles that have reacted with the density pertur-bant will localize at intermediate densities. For localization and quantifi-cation, the vesicles and perturbants are labeled with different radioisotopes. In many instances, the membrane–ligand complex can be reversed by addition of competitive reagents.

Originally (Wallach *et al.*, 1972; Wallach, 1974) used a high-density bacteriophage as density perturbant. However, a far more satisfactory and practical procedure utilizing high-density latex spheres of precise size has been recently introduced (Lim *et al.*, 1975). The latex spheres are synthesized by aqueous-emulsion copolymerization of methacrylate derivatives containing hydroxl and carboxyl groups (Molday et al., 1975). Spheres can be produced of homogeneous density (1.22) and in homogeneous categories of size, ranging from 300 Å to 3400 Å. Ligands can be coupled directly to the acrylate spheres, using a variety of mild conditions, for example, carbodimide, cyanogen bromide, and glutaraldehyde.

We have found that the densities of the spheres can be increased considerably (for example, to about 1.35) by first coating them with bovine serum albumin, following this by covalent coupling of the ligand, using glutaraldehyde. This procedure also gives these spheres a more "biological" surface and tends to reduce nonspecific adsorption. Finally, it also allows much higher ligand loading.

The latex sphere system was first tested by Lim et al. (1975) using human erythrocyte membranes that had been coated with rabbit antihuman-erythrocyte immunoglobulin. The latex spheres were first coupled with goat immunoglobulin directed against rabbit IgG and then allowed to react with erythrocyte ghosts or vesicles derived therefrom. It was found that the density of the membranes in sucrose gradients increased in proportion to the number of beads per unit of cell surface area. The density increase with increasing concentrations of beads per unit surface area describes a saturation curve, leveling at the density of the beads in the sucrose (Lim et al., 1975). The observed membrane density, ρ_c, the number of spheres bound per membrane vesicle, the densities and volumes of the membranes, beads, and fluid inside of the vesicles are related by the formula

$$\rho_c = (\rho_m V_m + n\rho_b V_b + \rho_s V_s)/(V_m + nV_b + V_s) \tag{15}$$

where n is the number of beads bound per vesicle; ρ_m, ρ_b, ρ_s and V_m, V_b, and V_s are the densities (ρ) and volumes (V), respectively, of membranes (m), beads (b), and intravesicular fluids (s).

We have utilized the methacrylate bead approach for the subfractionation of the microsomal membrane vesicles obtained upon fractionation of lymphocytes of various types (Schmidt-Ullrich et al., 1976b). In these experiments we first coated 600 Å spheres with bovine serum albumin. After treatment with 1% (w/w) glutaraldehyde, we coupled ^{125}I-labeled concanavalin A to the beads using a modification of the glutaraldehyde method of Yahara and Edelman (1975). We have found that the spheres should not be loaded with more than 30 tetrameric concanavalin A molecules per sphere (average) since, at high loading levels, vesicles that have reacted with a large number of spheres tend to rapidly agglutinate. The reacted beads are purified by isopycnic centrifugation in CsCl step gradients (20, 30, 40%, CsCl in H_2O, w/w) and then dialyzed against Dulbecco's phosphate-buffered saline, pH 7.4, lacking Ca^{2+} and Mg^{2+}, but 20 mM in glycine (to inactivate excess glutaraldehyde). One volume of beads is dialyzed against 500 volumes of buffer at 4°C for 16 hours. The beads are again banded on a CsCl gradient and dialyzed extensively against phosphate-buffered saline before use.

The beads are allowed to react with ^{131}I-labeled microsomal membrane vesicles isolated from rabbit thymocytes and previously freed of cytoplasmic proteins (Schmidt-Ullrich et al., 1974), in 0.001 M HEPES, pH 7.4. This reaction is carried out under slow continuous stirring for 16 hours at 4°C with gentle stirring. Thereafter, the reaction mixture is fractionated in a CsCl density step gradient (20, 25, 30, 35, 40% in H_2O, w/w).

To block formation of membrane-perturbant complexes we included 0.1 M α-methyl-D-glucoside in the reaction mixture and gradients.

As shown in Fig. 8, whereas unreacted microsomal membrane vesicles settle on the 20% CsCl step, vesicles that have reacted with the affinity density-perturbant distribute in several bands of much higher densities. The perturbant itself equilibrates at very high density. Upon treatment of the membrane–perturbant complexes with 0.1 M α-methyl-D-glucoside, the centrifugation patterns show almost complete dissociation of the complexes. With a ligand such as Con A one can easily overload the system so that all vesicles band at a single, high density. Fractionation of vesicle categories then depends markedly on the ratio of beads to vesicles. This problem may be less severe with other ligands (with lower receptor density on the vesicle surfaces).

Although we have yet to determine the characteristics of the membrane fractions separated by this procedure, the experiments clearly demonstrate that affinity density perturbation can subfractionate a membrane vesicle population that appears to be homogeneous in a density gradient system without the perturbant.

The density of proteins (about 1.36) is sufficiently high so that, when specific proteins are allowed to react with membrane vesicles at high concentrations these proteins themselves can act as density perturbants. This fact has been utilized by Zachowski and Paraf (1974) to subfractionate right side out (RO) and inside out (IO) plasma membrane vesicles from cultured plasmacytoma cells (MOPC173). For this they utilized Con A copolymerized with bovine serum albumin (Avrameas and Trnynck, 1969). This copolymer reacts only with RO vesicles, because it is on the external surface of the plasma membrane that the carbohydrate residues which react with Con A are located. The aggregated, Con A-reacted vesicles can then be isolated by low speed differential centrifugation.

In a related experiment, Heine and Roizman (1973) increased the density of plasma membrane vesicles isolated from HEp-2 cells infected with herpes simplex virus, by reacting the membranes with antivirus antibody. They find

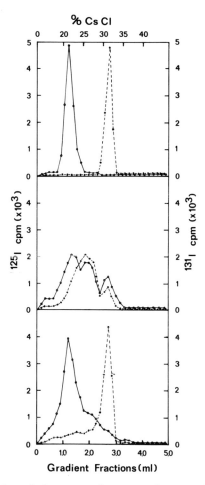

FIG. 8. Subfractionation of thymocyte plasma membrane vesicles by affinity density-perturbation. Top panel: distribution of ^{131}I-labeled membranes (———) and [^{125}I]concanavalin A beads (– – –) in separate density gradients. Middle panel: Distribution of membrane label (———) and density-perturbant label (– – – – –) after isopycnic equilibration of membrane-perturbant complexes. Bottom panel: Distribution of membrane (———) and density-perturbant (– – – – –) when formation of membrane-perturbant complexes is inhibited by 0.1 M α-methyl-D-glucoside (single gradient).

that vesicles from infected and unifected cells, labeled metabolically with [3H]- and [14C]amino acids, respectively, concentrate at the same density in the absence of antibody (in sucrose gradients). Reaction of antibody, in contrast, allows quantitative separation of the two membrane species.

V.   Aqueous Two-Phase or Liquid–Interface Partition

A classical procedure for the separation of small and large molecules utilizes their differential partition between two immiscible phases. Until recently this approach has been limited to systems where at least one of the phases was an organic solvent, a condition unsuitable for the fractionation, in a native state, of most macromolecules and not applicable to the separation of supramolecular assemblies, such as cells, viruses, and membranes.

It has been known for many years, however (Beijerinck, 1896), that, when aqueous solutions of certain polymers are mixed, two liquid phases may ultimately separate. Using this information Albertsson (1960) developed techniques for the partition of macromolecules and supramolecular complexes in polymer two-phase systems. Since then the technique has been extensively developed (Albertsson, 1970, 1971). The use of the technique for the separation of cells has been recently reviewed in this series by Walter (1975).

The theory of phase separation in polymer mixtures has been reviewed by Albertsson (1960, 1971) and by Konigsveld (1968). In principle, when aqueous solutions of two different polymers are mixed, these form two separate phases when the attractive forces between like molecules exceed those between unlike species. This occurs only above certain "critical" polymer concentrations (Albertsson, 1960, 1971). When phase-separation does occur, each phase contains predominantly one polymer class, but the activities of water and most small molecules included in the system are identical in the two phases. However, some salts, phosphates in particular (Johansson, 1970; Reitherman et al., 1973), do not partition equally, leading to an electrical and zeta potential between the two phases.

When a mixture of particles with diverse surface properties is included in such a polymer system and the two phases are well mixed, the components can often be resolved by repeated phase separations. Particles of equivalent surface charge and/or equivalent surface tension may be separated if they differ in other surface properties, respectively.

In theory, the partition coefficient, K, of a particle relates to its surface properties as follows:

$$K = C_a / C_b = e^{-\lambda A/RT} \qquad (16)$$

where C_a and C_b represent the concentrations in phases a and b respectively; A is the particle surface area; λ refers to factors such as the chemical nature of the macromolecule (membrane); R is the gas constant; and T is the absolute temperature. The partition coefficients of proteins range from 0.1 to 10, and very large particles tend to collect in one phase only. When the partition coefficients of the components of a particle mixture differ by < 10, full separation of these components requires repeated sequential partitions, i.e., countercurrent extraction.

The role of A is clearly very important, and, for spherical particles, K will change approximately as the square of particle radius. λ may be more critical still. Thus, if λ has a small positive value, K can still be large when A is is large, but if λ is changed (e.g., by titration of an ionizable group) to a small negative value, K will become very small (Albertsson, 1970).

As originally pointed out by Albertsson (1960), large particles tend to collect at the interface between the liquid layers. He suggested that this occurs during mixing of the phases, when the interfacial area of the mechanically separated phase droplets is very extensive. When mixing is stopped and the droplets coalesce, the particles remain at the interface even though the interfacial area diminishes. This is the basis for liquid–interface partition.

Interestingly, the collection of particles at the interface does *not* follow a linear adsorption isotherm (where the ratio of the amount of material adsorbed at the interface to the concentration of the material in the bulk phase is constant), but obeys the relation:

$$G = (C_a V_a)/N_a \qquad (17)$$

where G is a constant, C_a is the concentration of material in phase a, V_a is the volume of phase a, and N_a is the amount of material at the interface. This means that for a given concentration C_a, the amount of material N_a at the interface will vary with the volume V_a of phase a.

The polymer two-phase approach was first applied to the separation of plasma membranes by Brunette and Till (1971), and their method has been utilized by several workers (e.g., Stone *et al.*, 1974; Wickus *et al.*, 1974) for comparisons of plasma membranes from normal and neoplastic cells. In this approach cells were disrupted by "Dounce homogenization" in a hypotonic medium, 10^{-3} M in $ZnCl_2$. After removal of "debris" by low-speed centrifugation, the membranes were purified by two centrifugal partitions into the interface formed by a two-phase polymer mixture of dextran and polyethylene glycol. The membranes purified in this manner consist of large sheets and probably collect at the interface because of their size rather than according to any molecular criterion specific for plasma membranes.

To our knowledge only Steck (1974) has exploited the real potential of phase separation for membrane fractionation; he did this for the purpose of separating normally oriented (right-side out; RO) and inside-out (IO) vesicles derived from erythrocyte plasma membranes.

Steck (1974) employed the same polymers used by Brunette and Till (1971): dextran and polyethylene glycol. Ten percent aqueous solutions (w/v) of dextran T500 (Pharmacia) and ethylene glycol 6000 were combined at $\sim 5°C$

with 0.1–0.4 ml of packed erythrocyte vesicles and appropriate volumes of 0.1 M NaCl, 0.05 M Tris-borate, pH 8.0, to give final concentrations of 4% (w/v) dextran T500, 3.2% (w/v) polyethylene glycol 6000, 0.001 M NaCl, and 0.0005 M Tris-borate, pH 8.0. After blending, phase separation was allowed to proceed for at least 10 minutes on ice, and was completed by centrifuging at 2500 $g \cdot$ min at 5°C. The RO vesicles tend to collect in the upper phase, the IO vesicles in the lower. The appropriate phases were therefore collected and separated from the polymer mixture by centrifugation at 450,000 $g \cdot$ min, after first diluting out the polymer. IO vesicles favor the bottom phase by a factor of 20–100. A single partition thus yields highly purified RO vesicles in the upper phase (yield 32%). RO vesicles in contrast favor the top phase by a factor of 5, leaving some of the lower phase contaminated with RO vesicles. The purification factor of IO vesicles is not improved by further partitions, perhaps owing to their instability, but nearly 100% purity could be achieved for RO vesicles.

Steck (1974) points out that satisfactory separation of vesicles occurs only under narrowly defined conditions. Thus the polymer and membrane concentrations should be at the levels listed and partition must be carried out in the cold. Ionic composition is also critical. Thus, high ionic strengths lead to accumulation of all vesicles at the interface, halide ions drive all vesicles into the bottom phase and phosphate buffers tend to push all vesicles into the top phase.

Potential progress in the use of partitioning approaches to membrane fractionations may come through application of the "affinity partitioning" principle (Flanagan and Barondes, 1975). This would involve use of *polymer-ligands* that are specific for unique membrane receptors and exhibit well defined partitioning characteristics. Derivatized dextrans represent possible candidates.

Shanbhag and Axelsson (1975) have introduced a modified aqueous two-phase system useful for the separation of macromolecules according to hydrophobicity. This involves use of fatty acid esters of polyethylene glycol, rather than polyethylene glycol itself. The usefulness of this strategem to membrane fractionation remains to be established.

VI. Electrophoretic Techniques

A third method with possible potential for the purification of plasma membrane vesicles involves separation according to their surface charge density. A variety of approaches have been tried to achieve this goal. Indeed Davenport (1964) was able to achieve reasonable separations of mitochondria and "microsomal" membranes from rat liver, using field strengths of 3.2 V/cm for 16 to 20 hours in the cold, in a special electrophoretic apparatus employing linear density gradients. We (D. F. H. Wallach and M. P. Perez-Esandi, unpublished) were able to achieve similar separations of mitochondria and microsomal fragments, in the case of Ehrlich ascites carcinoma cells, but could not separate the "microsomal" membrane fraction of these cells into its component entities. Sellinger and Borens (1969) attempted to separate membrane fragments from brain cortex by zonal density gradient electrophoresis, but also failed to obtain adequate separation of membrane subfractions by this method.

This hitherto unsatisfactory situation has changed recently with the introduction of continuous, preparative, free-flow electrophoresis for the fractionation of subcellular particles of rat liver (Heidrich et al., 1970; Stahn et al., 1970; Henning and Heidrich, 1974). This method has also been successfully applied for the separation of normally oriented and inside out membrane vesicles derived from human erythrocytes (Heidrich and Leutner, 1974; Hannig and Heidrich, 1974). The electrophoresis apparatus employed in these studies consists of a cooled, perpendicularly posed chamber (dimensions 550 × 100 × 1 mm) through which buffer moves continuously downward in *laminar* flow. An electric field is applied perpendicularly to the direction of buffer flow. Field strengths of up to 100 volts per centimeter are used. Since all membrane particles to be separated bear a net negative surface charge, these are pumped into the cathodal side of the chamber near the top. The particles separate electrophoretically during their flow with the buffer and are collected in multiple small fractions at the bottom of the apparatus as shown in Fig. 9 (Hannig and Heidrich, 1974). Considerable care has to be applied in choosing the appropriate buffer. This is particularly important to avoid severe particle aggregation during electrophoresis. Buffers of very low ionic strength are not suitable because of adverse temperature elevation, whereas very high ionic strengths are undesirable because they promote particle aggregation.

As demonstrated in Fig. 9 (Hannig and Heidrich, 1974), all subcellular organelles and membranes of liver cells other than lysosomes and mitochondria tend to overlap in their elution profiles. Only lysosomes, which have a high electrophoretic mobility, and mitochondria and peroxisomes, which have comparatively low electrophoretic mobilities, can be separated without significant overlap. On the other hand, at present lysosomal preparations are still contaminated with plasma membrane components. There is also some contamination of the mitochondrial fraction with peroxisomes and the endoplasmic reticulum. However, if the subcellular organelles are prepurified by appropriate centrifugal methods, highly purified mitochondrial membranes (Heidrich et al., 1970) or lysosomes (Stahn et al., 1970; Henning and Heidrich,

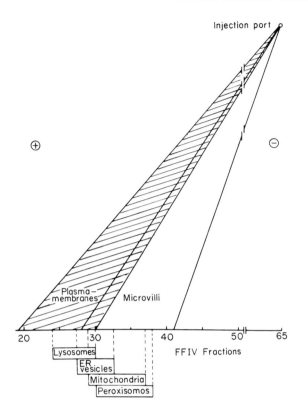

FIG. 9. Schematic representation of the electrophoretic mobility of subcellular organelles and membrane vesicles from rat liver cells; continuous preparative free-flow electrophoresis in an FF4 apparatus. From Hannig and Heidrich (1974) with permission of the authors and Academic Press.

1974) can be isolated by free-flow electrophoresis. In addition, Hannig and Heidrich (1974) were able to purify fractionated lysosomes with a high cholesterol and sphingomyelin content ("tritosomes"). These are the lysosomal particles obtained when rats are fed with a diet of Triton WR-1339, the detergent being taken up by some liver lysosomes preferentially.

The separation of normally oriented and inverted plasma membrane vesicles from human erythrocytes (Hannig and Heidrich, 1974; Heidrich and Leutner, 1974) indicates the potency of this electrophoretic method, since these particles cannot be separated in normal *sucrose* gradients. On the other hand very adequate separation of these particles can be rather simply attained by the use of dextran density gradients (Steck, 1974). As far as more complex membrane mixtures are concerned, further improvement in separation can be expected on theoretical grounds and may be feasible at the practical level when the complexities of free-flow electrophoresis are sufficiently understood and can be adequately controlled (Hannig *et al.*, 1975).

VII.   Concluding Remarks

Even in the living state, the surfaces of animal cells commonly pinch off vesicular membrane fragments. Disruption of cellular integrity extends this process and can often yield the entire plasma membrane in the form of vesicular sacs.

Accepting the proclivity of membranes to vesiculate, we have designed a fractionation strategy to isolate and purify plasma membranes in vesicular form. This strategy involves disruption of pure cell populations in a manner that produces a relatively uniform vesicle population, separation of the vesicles from large particles and soluble material by differential centrifugation, and purification of the vesicles according to their osmotic and Donnan-osmotic properties. The last step employs isopycnic ultracentrifugation in inert polymer-density gradients of low osmotic activity and of defined ionic compositions.

We have presented general principles and overall guidelines of this fractionation strategy, together with indications for appropriate marker techniques. We have also introduced several newly developed fractionation tactics that may prove useful in the isolation and purification of membrane vesicle populations with specific biological or biochemical characteristics.

Finally, we provided examples of specific applications. However, we have avoided presentation of a generalized fractionation method, since the approaches described appear generally applicable *in principle*, but, *in practice*, must be tailored to the specific cell systems under scrutiny.

ACKNOWLEDGMENTS

Supported by awards from the American Cancer Society (DFHW), from the Max-Planck-Gesellschaft zur Förderung der Wissenschaften (RSU), and from the National Cancer Institute, USPHS (CB 4400).

REFERENCES

Acs, G., Ostrowski, W., and Straub, F. B. (1954). *Acta Physiol. Acad. Sci. Hung.* **6**, 261.
Albertsson P.-Å. (1960). "Partition of Cell Particles and Macromolecules," 1st ed. Wiley, New York.
Albertsson, P.-Å. (1970). *Adv. Protein Chem.* **24**, 309.
Albertsson, P.-Å. (1971). "Partition of Cell Particles and Macromolecules," 2nd ed. Wiley (Interscience), New York.

Arntzen, C. J., Armond, P. A., Zeltinger, C. S., Vernotte, C., and Briantais, J. M. (1974). *Biochim. Biophys. Acta* **347**, 329.

Astle, L., and Cooper, C. (1974). *Biochemistry* **13**, 154.

Avrameas, S., and Trnynck, T. (1969). *Immunochemistry* **6**, 53.

Avruch, J., and Wallach, D. F. H. (1971). *Biochim. Biophys. Acta* **233**, 334.

Baudhuin, P., Beaufay, H., and de Duve, C. (1965). *J. Cell Biol.* **26**, 219.

Beijerinck, N. W. (1896). *Zentralbl. Bakteriol. Parasitenkd.* **2**, 627 and 698.

Bergelson, L. D. (1972). *Prog. Chem. Fats Other Lipids* **13**, 1.

Bosman, H. B., Hagopian, A., and Eylar, E. H. (1968). *Arch. Biochem. Biophys.* **128**, 51.

Brunette, D. M., and Till, J. E. (1971). *J. Memb. Biol.* **5**, 215.

Carraway, K. L. (1975). *Biochim. Biophys. Acta* **415**, 379.

Chang, K. -J., Bennett, J., and Cuatrecasas, P. (1975). *J. Biol. Chem.* **250**, 488.

Charalampous, F. C., and Gonatas, N. K. (1975). *Methods Cell Biol.* **9**, 259.

Crissman, H. A., Mullaney, P. F., and Steinkamp, J. A. (1975). *Methods Cell Biol.* **9**, 180.

Davenport, J. B. (1964). *Biochim. Biophys. Acta* **88**, 177.

de Duve, C., Berthet, J., and Beaufay, H. (1959). *Prog. Biophys. Biophys. Chem.* **9**, 325.

De Pierre, J. W., and Karnovsky, M. L. (1973). *J. Cell Biol.* **56**, 275.

Dodge, J. T., Mitchell, C., and Hanahan, D. J. (1963). *Arch. Biochem. Biophys.* **110**, 119.

Dowben, R. M., Brunner, J. R., and Philpott, D. L. (1967). *Biochim. Biophys. Acta* **135**, 1.

Emmelot, P., and Bos, C. J. (1966). *Biochim. Biophys. Acta* **266**, 494.

Emmelot, P., and Bos, C. J. (1969a). *Int. J. Cancer* **4**, 705.

Emmelot, P., and Bos, C. J. (1969b). *Int. J. Cancer* **4**, 723.

Emmelot, P., Bos, C. J., Benedetti, E. L., and Rümke, P. (1964). *Biochim. Biophys. Acta* **90**, 126.

Emmelot, P., Feltkamp, C. A., and Vaz Dias, H. (1970). *Biochim. Biophys. Acta* **211**, 43.

Ferber, E., Imm, W., Resch, K., and Wallach, D. F. H. (1972). *Biochim. Biophys. Acta* **266**, 494.

Flanagan, S. D., and Barondes, S. H. (1975). *J. Biol. Chem.* **250**, 1484.

Gahmberg, C. G., and Hakamori, S. -I. (1973). *J. Biol. Chem.* **248**, 4311.

Gianetto, R., and de Duve, C. (1955). *Biochem. J.* **59**, 433.

Graham, J. M. (1972). *Biochem. J.* **130**, 1113.

Hamilton, J., and Nielsen-Hamilton, M. (1976). To be published.

Hannig, K., and Heidrich H. -G. (1974). *In* "Methods in Enzymology" (S. Fleischer, L. Oacker, and R. Estabrook, eds.), Vol. 31, part A, p. 746. Academic Press, New York.

Hannig, K., Wirth, H., Meyer, B. -H., and Zeiller, K. (1975). *Hoppe-Seyler's Z. Physiol. Chem.* **356**, 1209

Heidrich, H. -G., and Leutner, G. (1974). *Eur. J. Biochem.* **41**, 37.

Heidrich, H. -G., Stahn, R., and Hannig, K. (1970). *J. Cell Biol.* **46**, 137.

Heine, J. W., and Roizman, B. (1973). *J. Virol.* **11**, 810.

Heine, J. W., and Schnaitman, C. A. (1971). *J. Cell Biol.* **47**, 703.

Henning, R., and Heidrich, H. -G. (1974). *Biochim. Biophys. Acta* **345**, 326.

Hochstadt, J., Quinlan, D. C., Rader, R. L., Li, C. C., and Dowd, D. (1976). *Methods Membr. Biol.* **5**, 117.

Hubbard, A. L., and Cohn, Z. A. (1972). *J. Cell Biol.* **55**, 390.

Huber, C. T., and Morrison, M. (1973). *Biochemistry* **12**, 4274.

Huber, C. T., Edwards, H. H., and Morrison, M. (1975). *Arch. Biochem. Biophys.* **168**, 463.

Hunter, M., and Commerford, L. (1961). *Biochim. Biophys. Acta* **47**, 580.

Johansson, G. (1970). *Biochim. Biophys. Acta* **22d**, 387.

Kamat, V. B., and Wallach, D. F. H. (1965). *Science* **148**, 1343.

Kant, J. A., and Steck, T. L. (1973). *J. Biol. Chem.* **248**, 8457.

Konigsveld, R. (1968). *Adv. Colloid Interface Sci.* **2**, 2.

Kreibich, G., Hubbard, A. L., and Sabatini, D. (1974). *J. Cell Biol.* **60**, 616.

Leskes, A., Siekevitz, P., and Palade, G. E. (1971). *J. Cell Biol.* **49**, 264.

Lieberman, I., Lansing, I. A., and Lynch, W. E. (1967). *J. Biol. Chem.* **242**, 736.

Lim, R. W., Molday, R. S., Huang, H. V., and Yen, S. -P. S. (1975). *Biochim. Biophys. Acta* **394**, 377.

Lin, P. -S., Wallach, D. F. H., Mikkelsen, R. B., and Schmidt-Ullrich, R. (1975). *Biochim. Biophys. Acta* **40**, 73.

Luganova, I. S., Seits, I. F., and Teodorovich, V. I. (1957). *Dokl. Akad. Nauk SSSR* **113**, 149.

Marchalonis, J. J. (1969). *Biochem. J.* **113**, 299.

Marchalonis, J. J., Cone, R. E., and Santer, V. (1971). *Biochem. J.* **124**, 921.

Molday, R. S., Dreyer, W. J., Rambaum, A., and Yen, S. P. S. (1975). *J. Cell Biol.* **64**, 75.

Morrison, M., Bayse, G. S., and Webster, R. G. (1971). *Immunochemistry* **8**, 289.

Moyer, G. H., and Pitot, H. C. (1973). *Cancer Res.* **33**, 1316.

Moyer, G. H., Murray, R. K., Khairallah, L. H., Suss, R., and Pitot, H. C. (1970). *Lab. Invest.* **23**, 108.

Neville, D. M. (1960). *J. Biophys. Biochem. Cytol.* **8**, 413.

Pastan, I., Anderson, W. B., Carchanan, R. A., Withingham, M. C., Russell, T. R., and Johnson, G. S. (1974). *In* "Control of Proliferation in Animal Cells" (B. Clarkson and R. Baserga, eds.), p. 563. Cold Spring Harbor Lab., Cold Spring Harbor, New York.

Phillips, D. R., and Morrison, M. (1971). *Biochemistry* **19**, 1766.

Quinlan, D. C., and Hochstadt, J. (1975). *Proc. Natl. Acad. Sci. U.S.A.* **71**, 5000.

Reitherman, R., Flanagan, S. D., and Barondes, S. H. (1973). *Biochim. Biophys. Acta* **297**, 193.

Rick, W., Fritsch, W. -P., and Szasz, G. (1972). *Dtsch. Med. Wochenschr.* **97**, 1828.

Roesing, T. G., Roselli, P. A., and Crowell, R. L., (1975). *J. Virol.* **15**, 654.

Rotman, B., and Papermaster, B. (1966). *Proc. Natl. Acad. Sci. U.S.A.* **55**, 134.

Rozengurt, E., and Heppel, I. (1975). *Biochem. Biophys. Res. Commun.* **67**, 1581.

Schmidt-Ullrich, R., Ferber, E., Knüfermann, H., and Wallach, D. F. H. (1974). *Biochim. Biophys. Acta* **332**, 175.

Schmidt-Ullrich, R., Wallach, D. F. H., and Davis, F. D. G. (1976a). *J. Natl. Cancer Inst.* In press.

Schmidt-Ullrich, R., Bieri, V., and Wallach, D. F. H. (1976b). In preparation.

Sellinger, O. Z., and Borens, R. M. (1969). *Biochim. Biophys. Acta* **174**, 176.

Shanbhag, V. P., and Axelsson, C. -G. (1975). *Eur. J. Biochem.* **60**, 17.

Shin, B. C., and Carraway, K. L. (1973). *J. Biol. Chem.* **248**, 1436.

Stahn, R., Maier, K. -P., and Hannig, K. (1970). *J. Cell Biol.* **46**, 576.

Steck, T. L. (1972). *In* "Membrane Molecular Biology" (C. F. Fox and A. Keith, eds.), p. 76. Sinauer, Assoc., Stamford, Connecticut.

Steck, T. L. (1974). *Methods Membr. Biol.* **2**, 245.

Steck, T. L., and Dawson, G. (1974). *J. Biol. Chem.* **249**, 2135.

Steck, T. L., and Wallach, D. F. H. (1970). *Methods Cancer Res.* **5**, 92.

Steck, T. L., Straus, J. H., and Wallach, D. F. H. (1970). *Biochim. Biophys. Acta* **203**, 385.

Stone, K. R., Smith, R. E., and Joklik, W. K. (1974). *Virology* **58**, 86.

Swanson, M. A. (1955). *In* "Methods in Enzymology" (S. P. Colowick and N. O. Kaplan, eds.), Vol. 2, p. 541. Academic Press, New York.

Thorley-Dawson, D. A., and Green, N. M. (1973). *Eur. J. Biochem.* **40**, 403.

Truding, R., Shelanski, M. L., Daniels, M. P., and Morell, P. ((1974). *J. Biol. Chem.* **249**, 3973.

Wallach, D. F. H. (1967). *In* "Specificity of Cell Surfaces" (B. Davis and L. Warren, eds.), p. 129. Prentice-Hall, Englewood Cliffs, New Jersey.

Wallach, D. F. H. (1974). *In* "Methods in Enzymology" (W. B. Jakoby and M. Wilchek, eds.), Vol. 34, p. 171. Academic Press, New York.

Wallach, D. F. H., and Kamat, V. B. (1964). *Proc. Natl. Acad. Sci. U.S.A.* **52**, 721.
Wallach, D. F. H., and Kamat, V. B. (1966). *In* "Methods in Enzymology" (E. F. Neufeld and V. Ginsburg, eds.), Vol. 8, p. 164. Academic Press. New York.
Wallach, D. F. H., and Lin, P. -S. (1973). *Biochim. Biophys. Acta* **300**, 211.
Wallach, D. F. H., and Schmidt-Ullrich, R. (1975). *In* "Membrane Molecular Biology of Neoplastic Cells" (D. F. H. Wallach, ed.), Chapter 2. Elsevier, Amsterdam.
Wallach, D. F. H., and Ullrey, D. (1962a). *Cancer Res.* **22**, 228.
Wallach, D. F. H., and Ullrey, D. (1962b). *Biochim. Biophys. Acta* **64**, 526.
Wallach, D. F. H., and Winzler, R. (1974). *In* "Evolving Strategies and Tactics in Membrane Research", Chapter 1. Springer-Verlag, Berlin and New York.
Wallach, D. F. H., Soderberg, J., and Bricker, L. (1960). *Cancer Res.* **20**, 397
Wallach, D. F. H., Kamat, V. B., and Murphy, F. L. (1965). "Symposium on Invasiveness of Tumor Cells, Paris." Int. Union Against Cancer.
Wallach, D. F. H., Kamat, V. B., and Gail, M. H. (1966). *J. Cell Biol.* **30**, 601.
Wallach, D. F. H., Kranz, B., Ferber, E., and Fischer, H. (1972). *FEBS Lett.* **21**, 29.
Walter, H. (1975). *Methods Cell Biol.* **9**, 25.
Wattiaux, R., and Wattiaux-de Conink, S. (1968a). *Eur. J. Cancer* **4**, 193.
Wattiaux, R., and Wattiaux-de Conink, S. (1968b). *Eur. J. Cancer* **4**, 201.
Wetzel, M. G., and Korn, E. D. (1969). *J. Cell Biol.* **43**, 90.
Wickus, G. G., and Robbins, P. W. (1973). *Nature (London) New Biol.* **245**, 65.
Wickus, G. G., Branton, P. E., and Robbins, P. W. (1974). *In* "Control of Proliferation in Animal Cells" (B. Clarkson and R. Baserga, eds.), p. 541. Cold Spring Harbor Lab., Cold Spring Harbor, New York.
Widnell, C. (1972). *J. Cell Biol.* **52**, 542.
Wurtman, R. J., and Axelrod, J. (1963). *Biochem. Pharmacol.* **12**, 1439.
Yahara, I., and Edelman, G. (1975). *Proc. Natl. Acad. Sci. U.S.A.* **72**, 1579.
Yamada, K. M., and Weston, J. A. (1974). *Proc. Natl. Acad. Sci. U.S.A.* **71**, 3492.
Zachowski, A., and Paraf, A. (1974). *Biochem. Biophys. Res. Commun.* **57**, 787.

Chapter 15

Rapid Isolation of Nuclear Envelopes from Rat Liver

R. R. KAY

Imperial Cancer Research Fund,
London, England

AND

I. R. JOHNSTON

Biochemistry Department,
University College,
London, England

I. Introduction

The nuclear envelope isolation procedure to be described here has the virtues of being rapid, straightforward, and readily applicable to large-scale envelope preparation (Kay *et al.*, 1972). The isolated envelopes show ultrastructural preservation, including the major features of the nuclear pore, as good as or better than that obtained using other techniques, and they also possess a considerable range of enzymic activities. On these and other considerations (Kay and Johnston, 1977), nuclear envelopes so isolated should provide suitable material for investigation of envelope-associated proteins, DNA, and RNA and for further subfractionation studies, such as the recently reported isolation of elements of the nuclear pore (Aaronson and Blobel, 1975).

In common with most other nuclear envelope isolation procedures (e.g., Kashnig and Kasper, 1969; Franke *et al.*, 1970; Zbarsky, 1972; Berezney, 1974; and for a review of various procedures, see Kay and Johnston, 1977), this one involves the initial isolation of nuclei from the tissue. A slight modification of the method of Widnell and Tata (1964) has been routinely used, it involves homogenization of the liver and sedimentation of the nuclei from the homogenate at low speed, followed by centrifugation at high speed

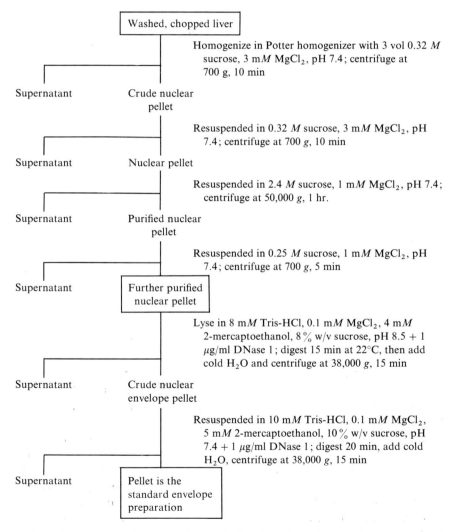

CHART 1.   Summary of steps involved in the isolation of nuclei and nuclear envelopes. The supernatants to the left are discarded.

through dense sucrose. Envelopes have also been successfully isolated from nuclei prepared by two other procedures (Blobell and Potter, 1966; Lawson *et al.*, 1969), and so it is likely that in fact many others would also be satisfactory.

Isolated nuclei are lysed by suspension in buffer at pH 8.5 in the presence of low concentrations of $MgCl_2$ and DNase 1. After a short period of digestion, the resulting nuclear ghosts are spun down and then resuspended and digested again at pH 7.4. This double DNase 1 digestion is employed because a prolonged single digestion leads to precipitation of digested chromatin along with the envelopes (Kay *et al.*, 1972). The envelopes pelleted from this second digestion constitute the standard preparation; if required they can be further purified by isopycnic centrifugation on sucrose density gradients. The steps involved in envelope isolation are outlined in Chart 1 and described in detail below for about 100 gm of rat liver.

II. Isolation of Rat Liver Nuclei

The method is based on that of Widnell and Tata (1964).

The following solutions (all ice-cold) are used: (1) 0.32 M sucrose, 3 mM $MgCl_2$, brought to pH 7.4 with $NaHCO_3$ (approx. 1 liter); (2) 2.4 M sucrose, 1 mM $MgCl_2$, brought to pH 7.4 with $NaHCO_3$ (approx. 500 ml); (3) 0.25 M sucrose, 1 mM $MgCl_2$, brought to pH 7.4 with $NaHCO_3$ (approx. 200 ml). All procedures are carried out on ice or at 4°C.

The livers from rats (about 12 albino Wistars of 180 gm body weight), killed by cervical dislocation, are rapidly removed and chopped with scissors in a 50 ml beaker and then transferred to a larger beaker, which has been preweighed, containing about 200 ml of 0.32 M sucrose, 3 mM $MgCl_2$, pH 7.4. When all the livers have been combined in this beaker, the beaker is reweighed to obtain the total liver weight, and the livers are then drained in a strainer and washed through with more 0.32 M sucrose, 3 mM $MgCl_2$, pH 7.4, to remove as much blood as possible. Fresh 0.32 M sucrose, 3 mM $MgCl_2$, pH 7.4, is added to the liver in a ratio of 3 :1 (v/w) and the mixture homogenized in portions by 25 up and down strokes at 1200 rpm in a glass/Teflon Potter homogenizer having a clearance of 0.25 mm. The homogenate is filtered through two layers of nylon bolting cloth, and the nuclei are pelleted from it by centrifugation in a swing-out rotor at 700 g for 10 minutes. The supernatant and the loose, uppermost layer of the pellet (consisting mainly of mitochondria) are aspirated off. The pellet is then resuspended by gentle stirring with a pestle and vortex mixing to the original volume in more 0.32 M sucrose, 3 mM $MgCl_2$ pH 7.4, and the nuclei are again pelleted at

700 g for 10 minutes. After the supernatant has been aspirated off, the crude nuclear pellet is resuspended, as above, in 2.4 M sucrose, 1 mM MgCl$_2$, pH 7.4, so that the material remaining from 1 gm of original liver has 4.1 ml of dense sucrose added to it. This mixture is centrifuged at 50,000 g for 60 minutes in an angle rotor (the material from 100 gm liver approximately fills a Spinco 30 rotor). At the end of this period the nuclei form a pellet at the bottom of the tube, with a tail up one side, while red cells, mitochondria, and cell debris float as a red plug on top of the sucrose. This plug is removed with a spatula, the sucrose is drained off, and any remnants of the floating material are wiped away with tissue. Finally, the nuclei are resuspended in 50–100 ml of 0.25 M sucrose, 1 mM MgCl$_2$, pH 7.4, and, after centrifugation at 700 g for 5 minutes produce a grayish pellet of highly purified nuclei.

III.   Isolation of Nuclear Envelopes from Purified Nuclei

The following solutions (at 22°C, except where otherwise stated) are used: (1) 0.1 mM MgCl$_2$; (2) 10 mM Tris-HCl, 0.1 mM MgCl$_2$, 5 mM 2-mercaptoethanol, 10% w/v sucrose, pH 8.5 (the mercaptoethanol is added just before use); (3) 10 mM Tris-HCl, 0.1 mM MgCl$_2$, 5 mM 2-mercaptoethanol, 10% w/v sucrose, pH 7.4 (the mercaptoethanol is added just before use); (4) 100 μg/ml DNase 1 (Sigma, electrophoretically pure, dissolved freshly on the day and kept on ice).

The final nuclear pellet is drained, and the walls of the centrifuge tube are wiped dry to reduce the amount of MgCl$_2$ carried into the lysis step, since excessive MgCl$_2$ protects the nuclei from efficient lysis. Nuclei are next resuspended in 0.1 mM MgCl$_2$ so that the nuclei from about 4 gm of liver are contained in 1 ml. DNase 1 is added immediately to 5 μg/ml, followed by 4 volumes of 10 mM Tris-HCl, 0.1 mM MgCl$_2$, 5 mM 2-mercaptoethanol, and 10% w/v sucrose pH 8.5. After vortex mixing, the digestion is continued for 15 minutes at 22°C, during which time the nuclei, which appear initially as dense spheres with distinct nucleoli (Fig. 1A), lyse and give rise to much less distinct nuclear "ghosts" (Fig. 1B). Digestion is terminated by the addition of an equal volume ice-cold distilled water, followed by centrifugation of the lysate at 38,000 g for 15 minutes in an angle rotor. The resulting pellet consists of nuclear "ghosts" and large fragments of envelope together with about 10% of the total nuclear DNA. The pellet is drained and resuspended in 5 volumes of 10 mM Tris-HCl, 0.1 mM MgCl$_2$, 5 mM 2-mercaptoethanol, 10% w/v sucrose pH 7.4, and DNase 1 is added to 1 μg/ml. Digestion is continued for 20 minutes at 22°C before being terminated by the addition of ice-cold water and centrifugation, as before. The resulting pellet comprises

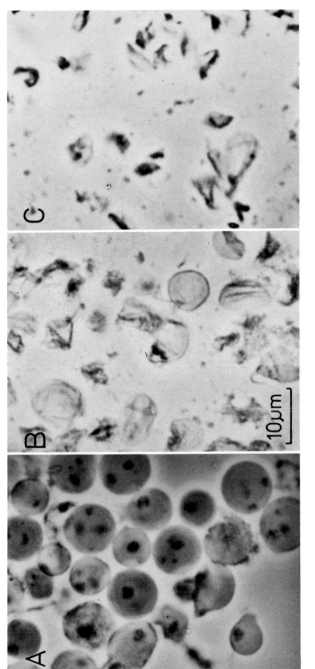

FIG. 1. Phase contrast light micrographs of nuclei and nuclear envelopes. (A) Highly purified nuclei, suspended in 0.25 M sucrose, 1 mM MgCl$_2$, pH 7.4. Nucleoli are clearly visible as darker regions in the nuclei. (B) Nuclei after lysis and during their first DNase 1 digestion at pH 8.5. Most of the nuclear contents have been released, and the nucleoli have broken down, leaving nuclear "ghosts" and large membrane fragments. (C) The standard nuclear envelope preparation, resuspended in 0.25 M sucrose, 1 mM MgCl$_2$, pH 7.4. In comparison with (B) the envelopes are more refractile, and many "ghosts" appear to have collapsed on themselves. (A) and (B) are taken from Kay et al. (1972). All at × 690.

the standard nuclear envelope preparation. After resuspension, phase contrast microscopy shows it to consist of large, sometimes refractile, membrane fragments (Fig. 1C) among which an intact nucleus can only very rarely be distinguished. This basic preparation can be obtained in 80 minutes from the isolated nuclei, so that the entire isolation procedure takes about 4.5 hours to perform, from the time of killing the rats. It has been used routinely for studies of envelope-associated proteins (see the next section) and for further subfractionation (Aaronson and Blobel, 1975).

Some further purification of the envelope can be obtained by centrifugation to equilibrium in a sucrose density gradient. This process is quite time-consuming and leads to a loss of phospholipids from the envelope (see Table I) and has not usually been used, but it is useful to demonstrate the association of a particular component with the envelope, or, if envelope-associated DNA is of interest, to reduce contamination by intact nuclei (Kay et al., 1971; Kay and Johnston, 1974). For this step the basic nuclear envelope pellet is resuspended in a few milliliters of sucrose of density 1.26 gm/ml made up in 10 mM Tris-HCl, 10 mM NaCl, 1 mM EDTA, 5 mM 2-mercaptoethanol, pH 7.4. This envelope suspension, in a tube to fit the Spinco SW 25.1 swing-out rotor is overlayered with sucrose covering the density range 1.24 to 1.16 gm/ml (either as a continuous density gradient or in steps of 0.02 gm/ml) made up in the same buffer. The gradient is centrifuged at 50,000 g for 14 hours at 4°C, after which the envelopes can be seen floating (as a double band on a continuous gradient) in sucrose of density 1.17 to 1.19 gm/ml (Kay et al., 1971). These bands are aspirated off and diluted several-fold with buffer, and the envelopes are finally pelleted at 50,000 g for 60 minutes.

IV.   Characterization of the Basic Nuclear Envelope Preparation

In thin section, under the electron microscope (Fig. 2A and B) the basic envelope preparation consists mainly of large sheets of double membrane, connected in places at nuclear pores. The outer nuclear membrane is studded with ribosomes, while amorphous material, probably chromatin, is attached to the inner one. Both in thin section and after negative staining (Fig. 2C and D) the annular and central grannules of the nuclear pore complex are readily visible, and the complex itself appears to be quite well preserved. The preparation also contains some stretches of single membrane and also ribosome-studded vesicles, often with the ribosomes on the inside. Presumably these structures are produced by the separation, in places, of the inner and outer nuclear membranes. Recognizable nuclei, nucleoli, or mitochondria

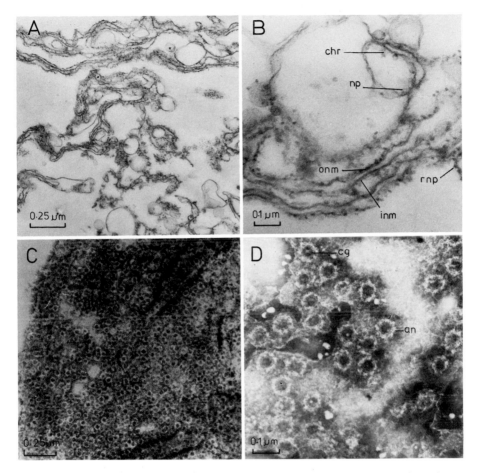

Fig. 2.   Electron micrographs of isolated nuclear envelopes. (A) and (B) Ultrathin sections;
(C) and (D) air-dried envelopes negatively stained. (A) Survey field. (B) Higher power view, in
which nuclear pores (np) can be distinguished connecting in places the outer and inner nuclear
membranes (onm, inm). The outer membrane is distinguished by the presence of ribosomes (rnp),
and the amorphous material attached to the inner one is probably chromatin (chr). (C) A large
membrane fragment, showing many nuclear pores, which are shown at higher power in (D),
where the pore annulus (an) and an occasional central granule (cg) can be seen. Taken from
Kay *et al.* (1972). (A) × 18,150; (B,D) × 33,000; (C) × 8250.

are extremely rare; the last result is in agreement with the absence of detectable succinate dehydrogenase in the preparation, indicating the presence of
less than 1 % mitochondrial protein.

A chemical analysis of nuclei, nuclear envelopes, and nuclear envelopes

TABLE I

CHEMICAL COMPOSITION OF ISOLATED NUCLEI, NUCLEAR ENVELOPES ISOLATED BY
THE STANDARD PROCEDURE, AND NUCLEAR ENVELOPES FURTHER PURIFIED BY
DENSITY GRADIENT CENTRIFUGATION[a]

Fraction	DNA	RNA	Phospholipid
Nuclei	277 ± 25	47 ± 6	57 ± 1
Nuclear envelopes (standard procedure)	60 ± 16	55 ± 4	407 ± 84
Nuclear envelopes (from sucrose density gradient)	56 ± 10	58 ± 11	157 ± 28

[a] Results are expressed as micrograms per milligram of protein with standard errors. Taken in part from Kay et al. (1972).

further purified by isopycnic centrifugation is given in Table I. The recovery of protein in the envelope fraction is $7.4 \pm 0.8\%$ of that in the original nuclear fraction, together with 50–60% of the phospholipid, about 9% of the RNA and $1.3 \pm 0.2\%$ of the DNA. After isopycnic centrifugation in sucrose density gradients, made up in 10 mM Tris-HCl, 10 mM NaCl, 1 mM EDTA, 5 mM 2-mercaptoethanol, pH 7.4, 80–90% of the loaded DNA is recovered in the membrane zone, together with a similar percentage of the protein and RNA but much less of the phospholipids, indicating their preferential loss in this step.

The presence of DNA in the envelope preparation is naturally of some interest, and as a result of the controversy about the involvement of the envelope in DNA replication (for reviews, see Kay and Johnston, 1973; Franke and Scheer, 1974; Kasper, 1974), the possibility that this DNA may be a contaminant has to be considered. It is currently hard to envisage an experiment to rule out definitely this possibility, but the following observations argue strongly against it: (1) Reconstruction experiments indicate only 5–7% of envelope-associated DNA results from the reassociation of released chromatin with the envelopes during isolation (Kay et al., 1972). (2) There is a high recovery of DNA with envelopes centrifuged in sucrose density gradients as described above. (3) Envelope-associated DNA replicates preferentially late in S phase in regenerating rat liver (Kay et al., 1971). (4) Finally, envelope-associated DNA from rat and mouse liver nuclei is enriched in repetitive DNA (Kay and Johnston, 1974). In combination these results show that the DNA in the nuclear envelope fraction is a distinctive fraction of the total nuclear DNA and is closely associated with the isolated envelopes, probably as a result of a previous in vivo attachment.

Proteins typical both of the endoplasmic reticulum (glucose-6-phosphatase, NADH and NADPH cytochrome c reductases, and cytochrome P-450)

TABLE II

SOME ENZYMIC ACTIVITIES OF NUCLEAR ENVELOPES ISOLATED BY THE STANDARD PROCEDURE WITH SOME CORRESPONDING VALUES FOR NUCLEI AND MICROSOMES[a,b]

Fraction	Glucose-6-phosphatase	NADH-cytochrome c reductase	NADPH-cytochrome c reductase	Cytochrome P-450	DNA polymerase	RNA polymerase	Poly (A) polymerase
Nuclei	0.22 ± 0.03	1.08 ± 0.21	0.93 ± 0.09	ND	683 ± 50	302 ± 142	114 ± 32
Nuclear envelopes	1.60 ± 0.21	2.67 ± 0.45	0.51 ± 0.04	0.08	90 ± 10	46 ± 48	56 ± 20
Microsomes	1.76 ± 0.34	7.59 ± 1.5	0.34 ± 0.05	0.80	—	—	—

[a] Values without standard errors are from fewer than three determinations; ND = not detected. Taken in part from Kay et al. (1972) and from unpublished work.

[b] Units: glucose-6-phosphatase: micromoles of glucose 6-phosphate hydrolyzed in 10 minutes per milligram of protein; NADH and NADPH cytochrome c reductases: micromoles of cytochrome c reduced in 10 minutes per milligram of protein; cytochrome P-450: nanomoles per milligram of protein; DNA polymerase: picomoles of [^{3}H]TMP incorporated in 10 minutes per milligram of protein; RNA polymerase: picomoles of [^{3}H]AMP incorporated in 10 minutes per milligram of protein, in an assay that contained in a final volume of 100 μl: 5 μmoles of Tris-HCl pH 8.0, 0.15 μmole of MgCl$_2$, 12.5 μmoles of sucrose, 0.05 μmole of [^{3}H]ATP and 0.05 μmole each of CTP, GTP, and UTP; poly(A) polymerase: picomoles of [^{3}H]ATP incorporated in 10 minutes per milligram of protein, in the same assay as for RNA polymerase, but lacking UTP, GTP, and CTP and with the addition of 5 μg of poly(A) per assay.

and of the nucleus (DNA polymerase, RNA polymerase, and poly (A) polymerase) are found in the isolated envelopes (Table II). Of these proteins, the envelope fraction is the major nuclear location for glucose-6-phosphatase and possibly for cytochrome P-450, but the recoveries of the other microsomal activities are only 10–20 % of that present in the original nuclear pellet. Only about 1–4 % of the total nuclear DNA polymerase, RNA polymerase, and poly (A) polymerase is recovered in the envelope fraction. Interestingly, and in contrast to DNA polymerase, 1 M NaCl extracts very little of the envelope-associated poly (A) polymerase, opening the possibility that a small fraction of the total nuclear activity of this enzyme may be tightly bound to the envelope (R. R. Kay, unpublished work).

In general and with the exception of poly (A) and RNA polymerases, which have not been extensively investigated by other workers, and glucose-6-phosphatase, which is lost or inactivated in some preparations, the proteins listed in Table II have been found at comparable or lower specific activities in envelopes isolated by many other procedures (see Franke, 1974; Kasper, 1974). Thus, from a biochemical point of view these results indicate good preservation of the envelopes during isolation.

Overall, both morphological and biochemical criteria show that nuclear envelopes isolated as described are very well preserved by current standards. Contamination by mitochondria, intact nuclei, and released chromatin can be assessed and is negligible or small. One obvious lacuna in this characterization is the absence of a good means for measuring contamination by endoplasmic reticulum. However, neither electron microscopy, where the morphology of the double nuclear membrane is distinctive, nor isopycnic centrifugation, in which the two membrane types have different buoyant densities, reveal any substantial contamination from this source. Further, because of the high purity of the original nuclear preparation, such contamination does not seem very likely.

ACKNOWLEDGMENTS

We wish to thank the Medical Research Council for financial support.

REFERENCES

Aaronson, R. P., and Blobel, G. (1975). *Proc. Natl. Acad. Sci. U.S.A.* **72**, 1007–1011.
Berezney, R. (1974). *In* "Methods in Cell Biology" (D. M. Prescott, ed.), Vol. 8, pp. 205–228. Academic Press, New York.
Blobel, G., and Potter, V. R. (1966). *Science* **154**, 1662–1665.
Franke, W. W. (1974). *Philos. Trans. R. Soc. London, Ser. B* **268**, 67–93.
Franke, W. W., and Scheer, U. (1974). *In* "The Cell Nucleus" (H. Busch, ed.), Vol. 1, pp. 219–347. Academic Press, New York.
Franke, W. W., Deumling, B., Ermen, B., Jarasch, E., and Kleinig, H. (1970). *J. Cell Biol.* **46**, 379–395.

Kashnig, D. M., and Kasper, C. B. (1969). *J. Biol. Chem.* **244**, 3786–3792.

Kasper, C. B. (1974). *In* "The Cell Nucleus" (H. Busch, ed.), Vol. 1, pp. 349–384. Academic Press, New York.

Kay, R. R., and Johnston, I. R. (1973). *Sub-Cell. Biochem.* **2**, 127–166.

Kay, R. R., and Johnston, I. R. (1974). *Experientia* **30**, 472–473.

Kay, R. R., and Johnston, I. R. (1977). *Tech. Biochem. Biophys. Morphol.* **3** (in press).

Kay, R. R., Haines, M. E., and Johnston, I. R. (1971). *FEBS Lett.* **16**, 233–236.

Kay, R. R., Fraser, D., and Johnston, I. R. (1972). *Eur. J. Biochem.* **30**, 145–154.

Lawson, D., Wilson, P. W., Barker, D. C., and Kodicek, E. (1969). *Biochem. J.* **115**, 263–268.

Widnell, C. C., and Tata, J. R. (1964). *Biochem. J.* **92**, 313–317.

Zbarsky, I. B. (1972). *In* "Methods in Cell Physiology" (D. M. Prescott, ed.), Vol. 5, pp. 167–198. Academic Press, New York.

Chapter 16

Isolation of Plasma Membranes from Cultured Muscle Cells[1]

STEVEN D. SCHIMMEL

Department of Biochemistry,
University of South Florida College of Medicine,
Tampa, Florida

CLAUDIA KENT

Department of Biochemistry,
Purdue University,
West Lafayette, Indiana

AND

P. ROY VAGELOS

Merck, Sharp & Dohme Research Laboratories
Rahway, New Jersey

[1] This work was carried out at the Department of Biological Chemistry, Washington University School of Medicine, St. Louis, Missouri, and was supported in part by NSF Grant 6B-38676X and NIH Grants 2-R01-HL-10406 and CA-12964. S.D.S. and C.K. were recipients of American Cancer Society Postdoctoral Fellowships PF-689 and PF-787, respectively.

I. Introduction

The development of skeletal muscle, both *in vivo* and *in vitro*, proceeds via the fusion of mononucleated myogenic cells to form multinucleated myotubes which elaborate muscle-specific proteins and become contractile (for review, see Fischman, 1973; Hauschka, 1972). We have postulated that alterations in the surface of these cells must play a critical role in this developmental process. In order to define these myogenic surface alterations, we have undertaken a study of the plasma membranes of embryonic chick muscle cells developing in monolayer culture. The initial step in this study has been to establish a plasma membrane isolation procedure, described below, which provides plasma membrane fractions of reproducible purity and yield from cells at various stages of myogenesis.

The various techniques available for the isolation and characterization of plasma membranes from a variety of tissues and cells have been extensively reviewed (Warren *et al.*, 1968; de Duve, 1971; Hinton, 1972; DePierre and Karnovsky, 1973; Wallach and Lin, 1973). The isolation of surface membranes of skeletal muscle tissue has been described by several workers (Kono and Colowick, 1961; McCollester, 1962; Rosenthal *et al.*, 1965; Kidwai *et al.*, 1973). Each of these muscle membrane preparations was derived from whole tissue and, therefore, contained elements of connective tissue and basement membrane. The plasma membrane fractions described below were isolated from cultures of myogenic cells which were devoid of basement membrane and which contained only small numbers of contaminating fibroblasts. We have used this procedure to obtain plasma membranes from myogenic cells prior to fusion, during nearly synchronous fusion, and following fusion. It is anticipated that the availability of this procedure will aid in the elucidation of the molecular events that occur at the cell surface during myogenesis and membrane fusion. Most of these results have been published elsewhere (Schimmel *et al.*, 1973; Kent *et al.*, 1974).

II. Cell Culture Conditions

Chick muscle cells were cultured by trypsinization of breast tissue dissected from 12-day embryos (Bischoff and Holtzer, 1968). The cells were preplated for 10 minutes at 37°C to remove some of the fibroblastic cells that adhere to the dish. The nonadherent cells were then plated on 150-mm collagen-coated dishes at 3.5 to 5×10^6 cells/dish in 20 ml of the appropriate medium (see below). All cultures were incubated at 37°C in a humidified atmosphere of 5% CO_2–95% air, and 10 ml of the culture medium was replaced daily. All culture media contained; 83.4% Eagle's minimum essential medium, 8.3% horse serum [Grand Island Biological Co. (GIBCO), Grand Island, New York], 8.3% chick-embryo extract (high speed supernatant of quickly frozen 50% extract of 11-day embryos in EBSS[2]) and 1% antibiotic–antimycotic solution (GIBCO). Cultures grown continuously for 66 hours in this culture medium (high-calcium medium—1860 μM Ca^{2+}) contained broad multinucleated myotubes, unfused myoblasts, and a small number of fibroblasts.

To obtain cultures of unfused myogenic cells, 5 μg of 5-bromodeoxyuridine (BrdU) per milliliter was added to the culture medium (Coleman et al., 1969). Myogenic cells grown in this medium assumed a flat, stellate, fibroblastlike morphology rather than their usual rounded, bipolar appearance. In contrast to their appearance when grown in the presence of BrdU, cells grown in medium containing 5–10% of the usual concentration of calcium (low-calcium medium—160 μM Ca^{2+}) retained the bipolar morphology of myogenic cells but were unable to fuse (Shainberg et al., 1969). This low-calcium medium contained 83.4% Eagle's minimum essential medium, calcium and magnesium free, 8.3% each of horse serum and embryo extract, 680 μM $MgSO_4$, 200 μM EGTA, and 1% antibiotic–antimycotic solution. The restoration of a high calcium concentration (1960 μM) to the medium by the addition of 1.8 μmoles of $CaCl_2$ per milliliter initiated a rapid, synchronous burst of fusion such that multinucleated myotubes were prominent within 4 hours (Kent et al., 1974). Thus, by adjusting the calcium concentration in the medium, we obtained cultures consisting primarily of fusion-competent mononucleated cells which, upon the addition of calcium, fused rapidly to form multinucleated myotubes.

Cultures of myotubes were obtained by adding 10^{-5} M 5-fluorodeoxyuridine (FdU), an inhibitor of DNA synthesis, to high-calcium medium 24–48 hours after establishment of the cultures. After an additional 24–48 hours, most proliferating mononucleated cells had detached from the dish and nondividing myoblasts had fused to form myotubes (Coleman and Coleman, 1968). Fibroblasts were obtained by trypsinization of subcutaneous connective tissue from the groin region of 12-day chick embryos. These cells were plated at 0.25×10^6 cell per milliliter of high-calcium medium, grown for 3 days, subcultured at the same density, and harvested after 2 days.

[2] Abbreviations: BrdU, 5-bromodeoxyuridine; CMF, calcium- and magnesium-free Earle's balanced salt solution; EBSS, Earle's balanced salt solution; EGTA, ethyleneglycolbis(β-aminoethyl ether)-N,N-tetraacetic acid; FdU, 5-fluorodeoxyuridine; TEA, triethanolamine.

III.   Analytical Procedures

A.   Assay of Enzymic Markers

Ouabain-sensitive Na^+, K^+-ATPase and 5'-nucleotidase were assayed as previously described (Schimmel *et al.*, 1973). The activity of the ATPase was calculated from the difference in inorganic phosphate liberated in the presence and in the absence of 1 mM ouabain. In this assay, the procedure for the determination of inorganic phosphate (Ames, 1966) was modified by shortening the time of incubation with the ascorbate-molybdate reagent to reduce the background due to hydrolysis of ATP in the acidic solution. Phosphodiesterase I was assayed by measuring the release of *p*-nitrophenol from *p*-nitrophenyl 5'-thymidylate essentially as described by Touster *et al.* (1970), except that the absorbance at 400 nm was determined directly after the reaction was terminated by the addition of 0.4 *M* NaOH. Leucyl-β-naphthylamidase (leucine aminopeptidase) activity was measured by the determination of the amount of β-naphthylamine released from leucyl β-naphthylamide (Hubscher *et al.*, 1965). The microsomal marker, NADPH (TPNH)-dependent cytochrome *c* reductase (Phillips and Langdon, 1962), and the mitochondrial marker, succinate-dependent cytochrome *c* reductase, were assayed spectrophotometrically as described by Sottocasa *et al.* (1967).

B.   External Labeling of Plasma Membranes with $[^{125}I]\alpha$-Bungarotoxin

Pure α-bungarotoxin was a gift from Dr. Regis B. Kelly. It was iodinated as described by Berg *et al.* (1972) by allowing 0.5 mg of toxin to react with 5 mCi of carrier-free $Na^{125}I$ (Industrial Nuclear Co., St. Louis, Missouri). Cultures to be labeled were washed twice with EBSS and incubated for 10 minutes at 37°C in 10 ml of culture medium containing 30 pmoles (8.5×10^6 cpm) of labeled toxin. The cultures were then washed 4 times with EBSS containing 2 mg of bovine serum albumin per milliliter, then twice with EBSS before they were harvested. The addition of 1 mM carbamylcholine to the incubation mixture resulted in a 90–95% decrease in binding of labeled toxin. Samples were counted in a Packard Model 3001 Auto-Gamma Spectrometer.

C.   Chemical Analysis

Protein was determined by the method of Lowry *et al.* (1951) with bovine serum albumin as standard. The distribution of DNA or RNA in subcellular fractions was determined in preparations from cultures grown for 3 days in the presence of 3 μM [*methyl*-^{3}H]thymidine (0.5 μCi/ml; New England Nuclear Corp., Boston, Massachusetts) or 3 μM [5-^{3}H]uridine (0.5 μCi/ml; Schwarz/Mann, Orangeburg, New York). RNA was determined from the 280 nm/260 nm absorbance ratio of plasma membrane samples dissolved in sodium dodecyl sulfate.

Lipids were extracted by the method of Bligh and Dyer (1959). The total lipid content was determined by the method of Marzo *et al.* (1971), except that the reaction volume was reduced to 0.5 ml and the final absorbance was measured at 295 nm to increase the sensitivity of the assay considerably. The extinction coefficient for this assay was determined from the dry weight of lipid extract obtained from particulate homogenates. Total cholesterol was determined by the *o*-phthalaldehyde method (Rudel and Morris, 1973). Total phosphate was assayed by the method of Ames (1966). Glycerides were determined using the reagents and procedure supplied in the Boehringer test kit for neutral fats and glycerol (catalog No. 15904, Boehringer/Mannheim, New York) after separation of the neutral lipids by thin-layer chromatography (Kent *et al.*, 1974). The final reaction volume was reduced to 0.3 ml to increase sensitivity.

IV.   Isolation of Plasma Membranes

Monolayer cultures of muscle cells or fibroblasts were washed 4 times with cold EBSS or CMF. The cells from each dish were scraped into 10 ml of EBSS or CMF by mopping with a small sheet of perforated cellophane (Microbiological Associates, Bethesda, Maryland) and collected by centrifugation for 5 minutes at 1100 g. (All centrifugal forces refer to r_{av}.) This and all subsequent steps were done at 0°–4°C. The cell pellet was suspended in 15 ml of 0.25 M sucrose–1 mM triethanolamine · HCl (pH 7.4) (sucrose-TEA) per gram of cells (wet weight). The cells were broken in a Dounce homogenizer (Kontes Glass Co., Vineland, New Jersey) with the tight (B) pestle. Fifteen up-and-down strokes were sufficient for complete breakage, as judged by phase-contrast microscopy. The homogenate was centrifuged for 10 minutes at 1700 g; the supernatant was removed and centrifuged for 60 minutes at either 27,000 or 33,000 g. A slightly better yield of plasma membranes was obtained with 33,000 g, and that centrifugal force is now used routinely. The resulting pellet was suspended in 1.0 ml of sucrose-TEA and layered over a sucrose discontinuous gradient of the following composition: 0.5 ml of 55 % (w/w) sucrose, then 2.5 ml each of 40 %, 32 %, 27 %, and 20 % sucrose.

After centrifugation in an SW 41 rotor at 41,000 rpm (206,000 g) for 1.5 hours, the bands of turbid material at the interfaces in the gradient were removed and designated as follows: 8.3–20%, I; 20–27%, II; 27–32%, III; 32–40%, IV; and 40–55%, V. Each fraction was diluted to about 9 ml with 1 mM TEA · HCl (pH 7.4), and the particulate material was collected by centrifugation for 60 minutes at 105,000 g. The pellets were resuspended in 1 mM TEA · HCl for analysis. Samples were also taken at each step of the preparation before the gradient step, diluted with TEA · HCl, and centrifuged for 60 minutes at 105,000 g, so that only particulate protein was assayed for enzyme activity. At least 30 minutes before dilution and centrifugation, beef-pancreas deoxyribonuclease I (Sigma Chemical Co., St. Louis, Missouri), at a final concentration of 50 μg/ml, was added to the samples of homogenate and 1700 g pellet to facilitate later resuspension for assay.

As shown below (Section V,A), both fractions I and II were enriched in plasma membrane material. In order to obtain a single fraction equivalent to the combined fractions "I + II," we have subsequently modified the sucrose density gradient to consist of: 0.5 ml of 55%, 3.5 ml of 32%, 4.0 ml of 27%, and 2.5 ml of 13% (w/w) sucrose. The gradient is spun for approximately 16 hours, and the material collected at the 13–27% interface is taken as the plasma membrane fraction.

V.   Distribution, Purification, and Recovery of Membrane Markers

A.   Purification of Membrane Fractions Enriched in Na⁺, K⁺-ATPase

Three-day-old muscle cultures grown in high-calcium medium were labeled with [^{125}I]α-bungarotoxin, harvested, and then fractionated by differential and sucrose density gradient centrifugations as described above (Section IV). The distributions and recoveries of particulate protein, ^{125}I, and the various marker activities are shown in Table I. All recoveries were close to 100% at each step of the purification. After the initial low-speed centrifugation (1700 g), essentially all the nuclei (99% of the total DNA) and about 80% of the mitochondrial marker were in the pellet. In addition, 63% of the 5'-nucleotidase and 48% of the Na⁺,K⁺-ATPase activities were found in the pellet fraction. These results were not altered by more vigorous homogenization.

The distributions on the sucrose gradient of particles containing the two most widely used plasma membrane markers were quite different (Fig. 1). The specific acitvity of the Na⁺,K⁺-ATPase was highest in the least dense bands (I and II) at the top of the gradient, while that of the 5'-nucleotidase was maximal in fraction IV near the bottom of the gradient. The distribution

TABLE I

DISTRIBUTION AND ACTIVITIES OF PROTEIN AND SUBCELLULAR MARKERS IN THE FRACTIONATION OF CULTURED MUSCLE CELLS[a]

Fraction	Protein	[125I]α-Bungarotoxin[b]	Na+,K+-ATPase[c]	5'-Nucleotidase[c]	Phosphodiesterase I[c]	Leucyl-β-naphthylamidase[c]	NADPH-cytochrome c reductase[c]	Succinate-cytochrome c reductase[c]
Crude particulate fraction	(100)	13.1 (100)	17.8 (100)	5.11 (100)	6.08 (100)	3.09 (100)	9.00 (100)	14.3 (100)
	8.99[d]	118,000[e]	161[f]	<6.0[f]	54.6[f]	27.8[f]	80.8[f]	129[f]
1700 g Supernatant	(32.8)	28.1(70.5)	30.2(55.5)	7.25(46.3)	9.59(51.9)	4.88(51.8)	13.5 (49.5)	9.43(21.5)
1700 g Pellet	(80.6)	5.1(31.3)	10.6(47.7)	2.54(62.8)	4.07(53.3)	1.74(45.3)	5.74(52.0)	17.3 (97.6)
28,000 g Supernatant	(6.0)	11.0 (5.1)	9.0 (3.1)	1.80 (2.1)	3.13 (3.1)	2.10 (4.1)	2.80 (1.9)	0.76 (0.3)
28,000 g Pellet	(21.5)	33.0(54.6)	38.8(46.6)	8.73(36.8)	11.3 (40.0)	5.24(36.3)	15.4 (36.8)	6.88(10.3)
I + II	(4.3)	93.6(29.8)	94.1(22.8)	4.35 (4.3)	17.9 (12.7)	8.08(12.9)	8.99 (4.4)	0.58 (0.2)
Total gradient	(20.3)	(52.0)	(45.3)	(32.4)	(36.8)	(42.1)	(27.8)	(19.2)

[a] Muscle cultures were grown for 66 hours in high-calcium medium and fractionated as described (Section IV). Numbers in parentheses are the percent recovery of the initial protein or activity in the crude particulate fraction.

[b] Specific radioactivity expressed as counts per minute per microgram of protein.

[c] Enzyme-specific activities expressed as nmoles min^{-1} mg of protein^{-1}.

[d] Total milligrams of protein in the crude particulate fraction.

[e] Total counts per minute.

[f] Total activities, expressed as nmoles min^{-1}.

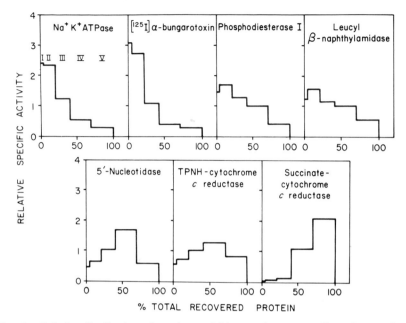

FIG. 1. Relative distribution of marker activities on the sucrose discontinuous density gradient. Muscle cultures grown for 66 hours in high-calcium medium were labeled with [¹²⁵I]α-bungarotoxin, homogenized in sucrose-TEA, and fractionated. Fractions were collected at each interface as follows: I, 8.3–20 % sucrose; II, 20–27 %; III, 27–32 %; IV, 32–40 %; and V, 40–55 %. This is the same experiment as in Table I. Relative specific activity was calculated as the percent in each fraction of the total activity recovered from the gradient divided by the percent in that fraction of the total protein recovered from the gradient.

of the other two putative plasma membrane markers, leucyl-β-naphthyl-amidase and phosphodiesterase I, were essentially identical with each other and intermediate between those of the Na^+,K^+-ATPase and the 5′-nucleo-tidase. This ambiguity was resolved by determining the distribution of the [¹²⁵I]α-bungarotoxin label, which was nearly identical to that of the Na^+,K^+-ATPase. Since the toxin binds to acetylcholine receptors at the muscle cell surface (Berg *et al.*, 1972; Vogel *et al.*, 1972; Hartzell and Fambrough, 1973), the codistribution of the toxin and the Na^+,K^+-ATPase indicated that the latter is a reliable marker for the plasma membranes of these cells.

The overall purification of the Na^+,K^+-ATPase in fractions I and II combined was 5.3-fold with respect to particulate protein (7.4-fold with respect to total cell protein), and the yield of enzymic activity was 23 % (Table I). In this same preparation, there was a 7.2-fold increase in the specific radioactivity of bound [¹²⁵I]α-bungarotoxin with a 30 % yield of radioactivity. The purified plasma membrane fractions (I plus II) contained 4.4 % of the initial microsomal marker activity, NADPH-dependent cytochrome c reductase, 0.2% of the mitochondrial succinate-dependent cytochrome c reductase, 0.02% of the cellular DNA, and 2.7% (183 μg/mg protein) of the cellular RNA.

B. Membrane Preparations from Mononucleated Cells and Myotubes

An important criterion for the suitability of the plasma membrane isolation procedure to our needs was the demonstration of its reproducibility with respect to cells at all stages of development. To examine this, cells were grown under various conditions that enriched the cultures for cells at specific developmental stages (see Section II). The distributions of the various marker activities were essentially identical in membrane preparations from all cultures, indicating the general applicability of the isolation procedure. The purification and yield of Na^+,K^+-ATPase activity in gradient fractions I plus II were very similar for preparations from the different cell types (Table II). When FdU- and BrdU-grown cells were labeled with $[^{125}I]\alpha$-bungarotoxin before harvest, the distribution of radioactivity on the gradient closely followed that of the Na^+,K^+-ATPase activity. (For BrdU-grown cells, specific binding of the toxin was less than 7% of that found with cells grown in high calcium medium without BrdU.)

VI. Chemical Composition of Plasma Membranes

The lipid compositions of the plasma membrane and crude particulate fractions were determined for cells grown in high-calcium medium, in low-calcium medium to block fusion, in low-calcium medium followed by high-calcium medium to synchronize fusion, and in FdU-containing medium to obtain myotubes. Table III lists the total lipid, phospholipid, cholesterol, and neutral glyceride contents of the crude particulate and plasma membrane fractions. The plasma membranes contained high levels of total lipid and phospholipid per milligram of protein compared to the crude particulate fractions. In both the plasma membrane and crude particulate fractions the phospholipids accounted for 50–60% of the total lipids. In agreement with findings for other cell types, the ratio of cholesterol to phospholipid was higher in the plasma membranes than in the homogenates. There was somewhat less triglyceride plus diglyceride in the plasma membranes than in the homogenates, and no monoglyceride was detected. The compositions of

TABLE II

Specific Activities, Purification, and Yield of Na^+,K^+-ATPase Activity[a] in Plasma Membrane Preparations from Cells at Various Stages of Development

	Muscle BrdU	Muscle low Ca^{2+}	Muscle high Ca^{2+}	Muscle FdU	Fibroblasts
Crude particulate fraction[b]	12.7	16.4	17.8	16.6	6.86
Plasma membrane (I + II)	63.5	107	94.1	110	34.8
Purification (fold)	5.0	6.5	5.3	6.6	5.2
Yield[c]	23.3	23.9	22.8	25.4	17.6

[a] Enzyme activity is expressed as nmoles min^{-1} mg of $protein^{-1}$.

[b] Crude particulate fraction refers to that portion of the total cell homogenate found in the pellet after centrifugation at 105,000 g for 1 hour.

[c] Total activity recovered in fractions I + II as percent of total activity in the crude particulate fraction.

the phospholipid classes and fatty acids were also determined (Kent *et al.*, 1974), and there was no reproducible variation in the content of any lipid class with stage of development.

VII. Comments

Of prime importance in the establishment of a subcellular fractionation procedure is the availability of suitable markers for the various cell organelles. These have generally been confined to the determination of membrane-associated components, particularly the assay of enzymic activities which have been ascribed to a particular organelle. As the number of cell types from which subcellular fractions have been obtained grows, it is becoming increasingly clear that specific enzymic markers cannot be assigned universally to specific organelles. In an extensive review of enzymic markers employed in plasma membrane isolation studies, Solyom and Trams (1972) concluded that none of the commonly used marker enzymes characterizes plasma membranes universally. The use, therefore, of the most frequently cited plasma membrane marker enzyme activities, 5'-nucleotidase and Na^+,K^+-ATPase, without positive verification that they are indeed associated with the plasma membrane of the cells under study, is a practice to be avoided. There are several reports of the differential fractionation of these two putative markers (Evans, 1969, 1970; Lutz and Frimmer, 1970). Widnell (1972) and Wallach and Ullrey (1962) have provided histochemical evidence for the

TABLE III

LIPID CONTENTS OF PLASMA MEMBRANE AND CRUDE PARTICULATE FRACTIONS[a,b]

	Total lipid per 1 mg protein (mg)	Phospholipid per 1 mg protein (μmoles)	Cholesterol per 1 μmole phospholipid (μmoles)	Glyceride[c] per 1 μmole phospholipid (μmoles)
Crude particulate fraction				
High-Ca^{2+}	0.48	0.41	0.41	0.12
FdU	0.79	0.50	0.46	0.11
Low-Ca^{2+}	0.59	0.40	0.41	0.06
Low-Ca^{2+}, then high-Ca^{2+} added				
1.5 Hours	0.67	0.50	0.46	0.12
4 Hours	0.83	0.51	0.39	0.08
Plasma membranes				
High-Ca^{2+}	2.3	1.67	0.61	0.07
FdU	2.8	2.01	0.55	0.04
Low-Ca^{2+}	2.3	1.77	0.58	0.04
Low-Ca^{2+}, then high-Ca^{2+} added				
1.5 Hours	2.6	1.79	0.57	0.05
4 Hours	2.6	1.82	0.57	0.08

[a] From Kent et al. (1974).
[b] High-calcium and low-calcium cultures were grown for 66 hours in medium containing 1860 μM or 160 μM calcium, respectively. High-calcium (1960 μM) was added to low-calcium cultures 1.5 or 4 hours before harvest. To some of the high-calcium cultures, FdU (10 μM) was added 40 hours after plating, and the cultures were harvested 88 hours after plating. Plasma membrane and crude particulate fractions were isolated, and protein and lipid contents were determined as described in Section III. The numbers listed are the averages of two or more determinations.
[c] Triglyceride plus diglyceride.

presence of 5'-nucleotidase activity in organelles other than the plasma membrane. In our initial efforts to establish a procedure for the isolation of plasma membranes from cultured muscle cells, we assayed all fractions for both the Na$^+$,K$^+$-ATPase and the 5'-nucleotidase and found that their distributions differed markedly. The assay of two additional enzymic activities, phosphodiesterase I and leucyl-β-naphthylamidase, which have been reported to be plasma membrane markers, resulted in still a third distribution. It was only through the use of an external surface labeling reagent, viz. radioactive α-bungarotoxin, that it was possible to unambiguously label the cell surface and identify the Na$^+$,K$^+$-ATPase as a bona fide marker for muscle cell plasma membranes. It is not possible at this time to assign subcellular localizations to the other putative plasma membrane marker activities. The 5'-nucleotidase is probably not microsomal, since conditions were found where this marker had a distribution markedly different from that of the microsomal marker (Schimmel *et al.*, 1973). It may be that the plasma membrane of these cells is heterogeneous with respect to its density and distribution of enzymic activities, as has been found for liver (Evans, 1969, 1970). It is interesting to note that in subcellular fractions from noncultured embryonic chick muscle there was a 5'-nucleotidase activity which had the same distribution as the Na^+,K^+-ATPase (C. Kent, unpublished observations).

The unambiguous labeling of the intact cell surface with α-bungarotoxin is illustrative of the potential usefulness of a variety of nonpenetrating reagents in the evaluation of membrane isolation procedures. The external labeling of cells may be used either to follow the distribution and purification of plasma membrane or to detect plasma membrane contaminants in other subcellular fractions. A growing number of external surface-labeling techniques is available for this purpose (Chang *et al.*, 1975; Dutton and Singer, 1975; Gahmberg and Hakomori, 1973; Hubbard and Cohn, 1972; Itaya *et al.*, 1975; Phillips and Morrison, 1971; Rifkin *et al.*, 1972). Critical reviews of many of these techniques may be found in Carraway (1975), DePierre and Karnovsky (1973), and Wallach (1972). Several labeling methods should be applicable to essentially any cell type. On the basis of our experience with cultured muscle cells, we strongly recommend the use of one or more of these external labeling procedures to verify results obtained with enzymic markers.

References

Ames, B. N. (1966). *In* "Methods in Enzymology" (S. P. Colowick and N. O. Kaplan, eds.), Vol. 8, pp. 115–118. Academic Press, New York.

Berg, D. K., Kelly, R. B., Sargent, P. B., Williamson, P., and Hall, Z. W. (1972). *Proc. Natl. Acad. Sci. U.S.A.* **69**, 147.

Bischoff, R., and Holtzer, H. (1968). *J. Cell Biol.* **36**, 111.

Bligh, E. G., and Dyer, W. J. (1959). *Can. J. Biochem. Physiol.* **37**, 911.

Carraway, K. L. (1975). *Biochim. Biophys. Acta* **415**, 379.

Chang, K. J., Bennet, V., and Cuatrecasas, P. (1975). *J. Biol. Chem.* **250**, 488.

Coleman, J. R., and Coleman, A. W. (1968). *J. Cell. Physiol.* **72**, Suppl. 1, 19.

Coleman, J. R., Coleman, A. W., and Hartline, E. J. H. (1969). *Dev. Biol.* **19**, 527.

de Duve, C. (1971). *J. Cell Biol.* **50**, 20d.

DePierre, J. W., and Karnovsky, M. L. (1973). *J. Cell Biol.* **56**, 275.

Dutton, A., and Singer, S. J. (1975). *Proc. Natl. Acad. Sci. U.S.A.* **73**, 2568.

Evans, W. H. (1969). *FEBS Lett.* **3**, 237.

Evans, W. H. (1970). *Biochem. J.* **116**, 833.

Fischman, D. A. (1973). *In* "The Structure and Function of Muscle" (G. H. Bourne, ed.), 2nd ed., Vol. 1, pp. 75–148. Academic Press, New York.

Gahmberg, C. G., and Hakomori, S. (1973) *J. Biol. Chem.* **248**, 4311.

Hartzell, H. C., and Fambrough, D. (1973). *Dev. Biol.* **30**, 153.

Hauschka, S. D. (1972). *In* "Growth, Nutrition, and Metabolism of Cells in Culture" (G. H. Rothblat and V. J. Cristofalo, eds.), Vol. 2, pp. 67–130. Academic Press, New York.

Hinton, R. H. (1972). *In* "Subcellular Components, Preparation and Fractionation" (G. D. Birnie, ed.), 3rd ed., pp. 119–156. Univ. Park Press, Baltimore, Maryland.

Hubbard, A. L., and Cohn, Z. (1972). *J. Cell Biol.* **55**, 390.

Hubscher, G., West, G. R., and Brindley, D. N. (1965). *Biochem. J.* **97**, 629.

Itaya, K., Gahmberg, C. G., and Hakomori, S. I. (1975). *Biochem. Biophys. Res. Commun.* **64**, 1028.

Kent, C., Schimmel, S. D., and Vagelos, P. R. (1974). *Biochim. Biophys. Acta* **360**, 312.

Kidwai, A. M., Radcliffe, M. A., Lee, E. Y., and Daniel, E. E. (1973). *Biochim. Biophys. Acta* **298**, 593.

Kono, T., and Colowick, S. P. (1961). *Arch. Biochem. Biophys.* **93**, 520.

Lowry, O. H., Rosebrough, N. J., Farr, A. L., and Randall, R. J. (1951). *J. Biol. Chem.* **193**, 265.

Lutz, F., and Frimmer, M. (1970). *Hoppe-Seyler's Z. Physiol. Chem.* **351**, 1429.

McCollester, D. L. (1962). *Biochim. Biophys. Acta* **57**, 427.

Marzo, A., Ghirardi, P., Sardini, D., and Meroni, G. (1971). *Clin. Chem.* **17**, 145.

Phillips, A. H., and Langdon, R. G. (1962). *J. Biol. Chem.* **237**, 2652.

Phillips, D. R., and Morrison, M. (1971). *Biochemistry* **10**, 1766.

Rifkin, D. B., Compans, R. W., and Reich, E. (1972). *J. Biol. Chem.* **247**, 6432.

Rosenthal, S. L., Edelman, P. M., and Schwartz, I. L. (1965). *Biochim. Biophys. Acta* **109**, 512.

Rudel, L. L., and Morris, M. D. (1973). *J. Lipid Res.* **14**, 364.

Schimmel, S. D., Kent, C., Bischoff, R., and Vagelos, P. R. (1973). *Proc. Natl. Acad. Sci. U.S.A.* **70**, 3195.

Shainberg, A., Yagil, G., and Yaffe, D. (1969). *Exp. Cell Res.* **58**, 163.

Solyom, A., and Trams, E. G. (1972). *Enzyme* **13**, 329.

Sottocasa, G. L., Kuylenstierna, B., Ernster, L., and Bergstrand, H. (1967). *J. Cell Biol.* **32**, 415.

Touster, O., Aronson, N. N., Jr., Dulaney, J. T., and Hendrickson, H. (1970). *J. Cell Biol.* **47**, 604.

Vogel, Z., Sytkowski, A. J., and Nirenberg, M. W. (1972). *Proc. Natl. Acad. Sci. U.S.A.* **69**, 3180.

Wallach, D. F. H. (1972). *Biochim. Biophys. Acta* **265**, 61.

Wallach, D. F. H., and Lin, P. S. (1973). *Biochim. Biophys. Acta* **300**, 211.

Wallach, D. F. H., and Ullrey, D. (1962). *Cancer Res.* **22**, 228.

Warren, L., Glick, M. C., and Nass, M. K. (1968). *In* "The Specificity of Cell Surfaces" (B. D. Davis and L. Warren, eds.), pp. 109–127. Prentice-Hall, Englewood Cliffs, New Jersey.

Widnell, C. C. (1972). *J. Cell Biol.* **52**, 542.

Chapter 17

Preparation of Plasma Membranes from Amoebae

J. E. THOMPSON

Department of Biology,
University of Waterloo,
Waterloo, Ontario, Canada

I. Introduction

Acanthamoeba castellanii is a unicellular protozoan that exists in either of two states of differentiation—a vegetative stage (amoeba) and a resting stage (cyst) (Neff, 1957; Neff *et al.*, 1964). The amoeba possesses a full complement of subcellular organelles and, unlike some of the larger amoebae, is surrounded by a naked plasma membrane (Bowers and Korn, 1968). Perhaps the most specialized property of the *Acanthamoeba* plasma membrane is a capacity to carry out endocytosis (Weisman and Korn, 1967; Korn and Weisman, 1967; Goodall and Thompson, 1971). In fact the amoebae appear to be incapable of actively transporting solutes (Bowers and Olszewski, 1972) and thus to be solely dependent upon phagocytosis and pinocytosis as means

303

for accumulating nutrients. A major consequence of this mode of cellular transport is the expenditure of large amounts of plasma membrane that must be replaced. Accordingly, the nutritional dependence of the organism upon endocytosis affords a unique opportunity for examining the dynamics of plasma membrane involvement in ingestion as well as the mechanisms of membrane turnover or exchange at the cell surface.

Procedures for isolating plasma membrane from *Acanthamoeba* are now available, and this has made at least partial characterization of the membrane possible. These procedures as well as techniques for culturing the amoebae are described in this chapter. In addition, selected properties of membrane preparations obtained by the different isolation procedures are compared.

II.   Culture Methods

Amoeboid cells of *Acanthamoeba castellanii* (strain Neff) grow axenically in the culture medium described in Table I. We prepare the medium in up to 10-liter batches. All the components are weighed out fresh except for vitamin B_{12}, which is stored frozen as a stock solution at a concentration of 0.1 mg/ml. Once all the components are in solution the mixture is adjusted to pH 7.0 with concentrated sodium hydroxide and filtered through Whatman No. 4 filter paper. It is then dispensed into 2-liter and 250-ml Erlenmeyer flasks— 1000 ml into the 2-liter flasks and 100 ml into the 250-ml flasks. The flasks are stoppered with Dispo plugs and autoclaved for 15 minutes. The inside surfaces of the flasks are previously coated with Siliclad to prevent the amoebae from adhering to the glass. The autoclaved medium can be stored

TABLE I

GROWTH MEDIUM FOR *Acanthamoeba castellanii*

Component	Final concentration
Yeast extract (Difco)	0.75%
Proteose peptone (Difco)	0.75%
Glucose	1.5%
KH_2PO_4	2 mM
$MgSO_4$	1 mM
$CaCl_2$	0.05 mM
Ferric citrate	0.1 mM
Thiamine hydrochloride	1.0 mg/liter
Biotin	0.2 mg/liter
Vitamin B_{12}	1.0 μg/liter

for up to 7 days at room temperature with no noticeable effect on growth.

The cells are cultured at 29°C in the dark on a gyratory shaker rotating at 100 rpm. Stock cultures are maintained in the 250-ml Erlenmeyer flasks containing 100 ml of medium. The stocks are transferred every 7 days by inoculating a fresh flask with 0.1 ml of culture. Cultures to be used for membrane isolation are grown in the 2-liter Erlenmeyer flasks containing 1000 ml of medium. These cultures are inoculated with 2 ml of a 7-day-old stock culture and allowed to grow for 48 hours or to a desired cell concentration.

The amoebae grow exponentially under these conditions with a mean generation time of about 7 hours to a population density ranging from 2 to 3.5×10^6 cells/ml (Fig. 1). Stationary-phase cultures have been monitored for the presence of cysts for up to 14 days after the culture was started by examining wet mounts with a phase-contrast microscope. By 7 days less than 1 % encystment had occurred. Even after 14 days fewer than 20 % of the organisms had encysted.

III. Plasma Membrane Isolation

A. Fractionation

A flow chart summarizing a fractionation procedure that yields highly purified amoeba plasma membranes is illustrated in Fig. 2. This procedure was first described by Schultz and Thompson (1969). It can be completed within less than 5 hours and produces 2–3 mg of plasma membrane protein from 2 to 3×10^9 cells.

All solutions used in the isolation procedure are precooled to 4°C, and all centrifugation steps are carried out at 4°C. Amoebae are cultured to population densities ranging from 2 to 5×10^5 cells/ml. For a single membrane preparation 2 to 3×10^9 cells are harvested from the growth medium by centrifugation at 800 g for 10 minutes and washed by resuspension in 5 mM NaHCO$_3$, pH 7.5, and recentrifugation at 800 g for 10 minutes. After total wet weight of the cell sediment is determined, the amoebae are resuspended in sufficient homogenizing medium (1 mM NaHCO$_3$, pH 7.5) to make a 20 % w/v suspension. The suspension is homogenized with 15 strokes of a Potter–Elvehjem homogenizer (Tri-R Instruments Inc.) rotating at 1700 rpm with a clearance of 0.13–0.18 mm. The homogenate is diluted with 4 volumes of homogenizing buffer, strained through 6 layers of cheesecloth, and centrifuged at 1500 g for 10 minutes. This yields a pellet consisting largely of nuclei, cell debris, and fragments of plasma membrane. The supernatant is

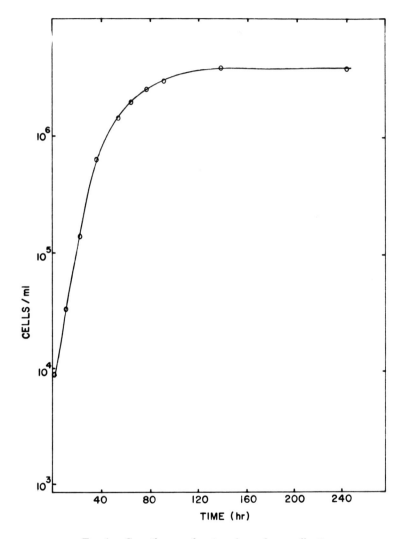

FIG. 1.   Growth curve for *Acanthamoeba castellanii*.

discarded, and the pellet is resuspended in a volume of homogenizing medium equivalent to 2.25 ml per original gram of packed cells. This new suspension is then centrifuged at the slightly lower speed of 1220 *g* for 10 minutes in order to partially separate the components of the original pellet. This second centrifugation produces a well packed pellet comprising nuclei and cell debris which is covered by a layer of loosely packed, fluffy material. The supernatant is removed and the upper, loosely packed portion of the pellet,

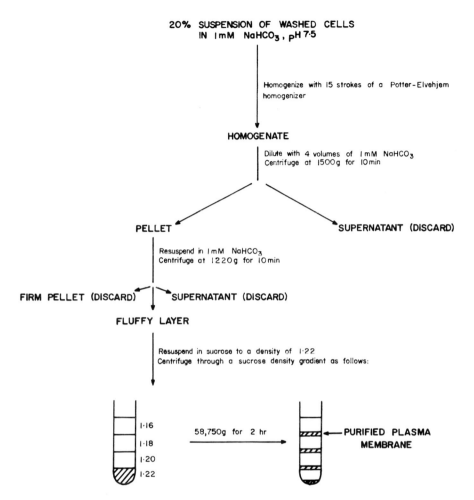

FIG. 2.   Flow chart summarizing a procedure for isolating plasma membranes from *Acanthamoeba castellanii.*

which is largely plasma membrane, is resuspended in a volume of homogenizing medium equivalent to 1.5 ml per original gram of packed cells. This suspension is then centrifuged again at 1220 *g* for 10 minutes. These procedures of resuspension and centrifugation are repeated until a homogeneous sediment of fluffy material, showing no visible trace of the original hard-packed bottom portion, is obtained. The partially purified plasma membranes are then resuspended in 10 ml of homogenizing medium.

A final purification is achieved by flotation centrifugation through a discontinuous sucrose gradient in a Spinco SW 25.1 rotor. The bottom layer of the gradient is prepared by mixing 3.1 ml of membrane suspension with 5.8 ml of sucrose, density 1.34, to give a final density of 1.22. The gradient is then completed by layering the following solutions into the tubes: 7 ml of sucrose, density 1.20; 7 ml of sucrose, density 1.18; 7 ml of sucrose, density 1.16. (All sucrose solutions are made up in 1 mM NaHCO$_3$, pH 7.5.) After centrifugation for 2 hours at 58,750 g a layer is present at each interface, and there is a pellet at the bottom of the tube. The layer at the uppermost interface is the purified plasma membrane fraction. It is removed from the gradient with a syringe, diluted with approximately 5 volumes of 1 mM NaHCO$_3$, pH 7.5, and centrifuged at 122,000 g for 1 hour. The resulting pellet is resuspended in 2–5 ml of the same medium and either stored at $-20°C$ or used directly for analysis.

The technique can be used to prepare plasma membranes from either exponential- or stationary-phase amoebae.

B.   Electron Microscopy

Isolated preparations of plasma membrane have been examined extensively by transmission electron microscopy. For this purpose freshly prepared membranes are washed twice by resuspension in 10 ml of 1 mM NaHCO$_3$, pH 7.5, and centrifugation at 122,000 g for 1 hour. The membrane pellet is fixed in glutaraldehyde (3 % in 0.05 M phosphate, pH 7.0) for 1 hour and washed over a period of 8 hours with 3 changes of the same phosphate buffer. The pellet is then broken up into small portions about 1 mm^2 and postfixed in osmium tetroxide (1 % in 0.05 M phosphate, pH 7.0) for 2 hours. The fixed specimens are washed in the same phosphate buffer, dehydrated in acetone and embedded in Vestopal (Kellenberger et al., 1958). Thin sections are poststained for 30 seconds with 1 % uranyl acetate and for 10 seconds with 0.1 % lead citrate in 0.1 N sodium hydroxide.

The isolated fraction typically contains vesicles and some strands of membrane, and in high-resolution micrographs the trilamellar feature of the membranes is clearly visible (Fig. 3). No recognizable contamination by mitochondria has been detected, although in routine scanning occasional ribosome-studded vesicles were seen.

C.   Enzymic Properties of the Membrane Preparation

1.   ASSAY PROCEDURES

As a further parameter by which to judge the purity of the isolated membranes the fraction was assayed for the following enzymes according to the methods indicated: succinate dehydrogenase (Pennington, 1961); glucose-

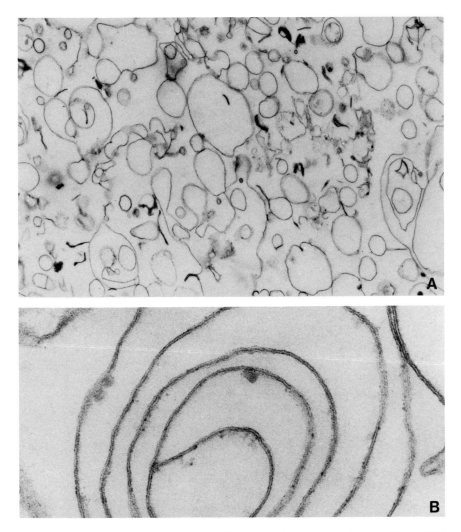

FIG. 3.   Transmission electron micrographs of isolated plasma membranes. (A) × 18,150
(B) × 89,750.

6-phosphatase (Hübscher and West, 1965); rotenone-insensitive NADH–
cytochrome c reductase (Sottocasa et $al.$, 1967); alkaline phosphatase
(Bosmann et $al.$, 1968); 5′-nucleotidase (Michell and Hawthorne, 1965);
Mg^{2+}-ATPase (Klein, 1962).
   Phosphate levels were determined by the method of King (1932) and protein

by the method of Lowry *et al.* (1951) using bovine serum albumin as a standard.

2. ENZYMIC CRITERIA FOR PURITY

Succinate dehydrogenase is a well recognized marker enzyme for mitochondrial inner membrane, and rotenone-insensitive NADH–cytochrome *c* reductase is known to be present on the membranes of endoplasmic reticulum and also on mitochondrial outer membrane. Both of these enzymes were either undetectable in the purified plasma membrane fraction or present at very low levels (Table II). When present, their specific activities were always less than 1 % of those for mitochondrial fractions in the case of succinate dehydrogenase, and for microsomal fractions in the case of NADH–cytochrome *c* reductase. Glucose-6-phosphatase is also considered to be a marker enzyme for endoplasmic reticulum, and it too was for the most part undetectable in the purified membrane fraction (Table II). In the event the enzymes were detectable they always showed enrichments of less than 1 relative to homogenate on a specific activity basis, indicating that the purification procedure serves to separate them away from the plasma membranes. Thus contamination by mitochondrial and endoplasmic reticulum membranes appears to be minimal.

Three other enzymes—alkaline phosphatase, 5′-nucleotidase, and Mg^{2+}-ATPase—showed substantial enrichments in the purified membrane

TABLE II

ENZYMIC CRITERIA OF PLASMA MEMBRANE PURITY[a]

Enzyme	Expt.	Homogenate	Plasma membrane	Enrichment[b]
Succinate	A	0.052	ND[d]	—
dehydrogenase[c]	B	0.137	0.024	0.18
	C	0.092	ND	—
Glucose-6-	A	38.5	ND	—
phosphatase[c]	B	40.6	ND	—
	C	30.6	17.3	0.56
NADH–cytochrome *c*	A	0.20	ND	—
reductase	B	0.25	0.02	0.08
(rotenone-insensitive)	C	0.13	0.04	0.31

[a] Succinate dehydrogenase activities are expressed as micromoles of reduced iodonitrotetrazolium chloride per milligram of protein per hour, glucose-6-phosphatase as micrograms of P per milligram of protein per hour, and NADH–cytochrome *c* reductase as micromoles of reduced cytochrome *c* per milligram of protein per hour.

[b] Enrichment, the ratio of the specific activity in the plasma membrane fraction to that of the corresponding homogenate.

[c] From Schultz and Thompson (1969).

[d] ND, not detectable.

fraction, indicating that they are purified by the isolation procedure and are probably associated with the plasmalemma (Table III). The enrichments for 5'-nucleotidase and alkaline phosophatase were similar ranging from about 10- to 24-fold. Corresponding values for Mg^{2+}–ATPase were lower, but this doubtless reflects the presence of the same enzyme activity in other cell membranes.

IV. Alternative Isolation Procedures

An alternative procedure for isolating plasma membranes from *Acanthamoeba castellanii* has been reported by Ulsamer *et al.* (1971), and for comparative purposes this procedure is also described. Details of the isolation are summarized in Fig. 4. The procedure is similar to the one described in Section III, A in that it again involves an initial purification of the membranes by low speed sedimentation from a homogenate followed by a final purification on a discontinuous sucrose gradient.

About 1×10^{10} amoebae are sedimented from the growth medium by centrifugation at 500 g for 5 minutes, washed in 0.01 M Tris chloride, pH 7.4, and resuspended in the same buffer at a population density of 2×10^7

TABLE III

ENZYMES ENRICHED IN THE PURIFIED PLASMA MEMBRANE FRACTION[a]

Enzyme	Expt.	Homogenate	Plasma membrane	Enrichment[b]
Alkaline phosphatase[c]	A	5.0	55.1	10.8
	B	3.1	73.4	23.6
	C	5.4	74.1	13.7
5'-Nucleotidase[d]	A	1.82	24.6	13.5
	B	1.66	27.8	16.7
	C	1.84	35.9	19.5
Mg^{2+}-ATPase[d]	A	7.7	19.2	2.5
	B	3.9	36.8	9.4
	C	3.8	28.7	7.5

[a] Alkaline phosphatase activities are expressed as micromoles of *p*-nitrophenol per milligram of protein per hour, and 5'-nucleotidase and ATPase activities as micromoles of P per milligram of protein per hour.
[b] Enrichment, the ratio of the specific activity in the plasma membrane fraction to that of the corresponding homogenate.
[c] From Wilkins and Thompson (1974).
[d] From Schultz and Thompson (1969).

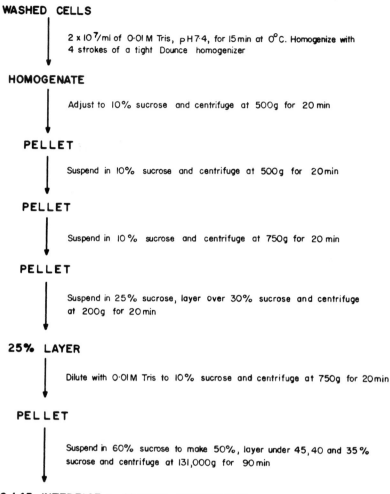

WASHED CELLS

2×10^7/ml of 0·01 M Tris, pH 7·4, for 15 min at $0°$C. Homogenize with 4 strokes of a tight Dounce homogenizer

HOMOGENATE

Adjust to 10% sucrose and centrifuge at 500g for 20 min

PELLET

Suspend in 10% sucrose and centrifuge at 500g for 20 min

PELLET

Suspend in 10% sucrose and centrifuge at 750g for 20 min

PELLET

Suspend in 25% sucrose, layer over 30% sucrose and centrifuge at 200g for 20 min

25% LAYER

Dilute with 0·01M Tris to 10% sucrose and centrifuge at 750g for 20min

PELLET

Suspend in 60% sucrose to make 50%, layer under 45, 40 and 35% sucrose and centrifuge at 131,000g for 90 min

40/45 INTERFACE = PLASMA MEMBRANES

Fig. 4.   Flow chart summarizing an alternative procedure (Ulsamer *et al.*, 1971) for isolating plasma membranes from *Acanthamoeba castellanii*.

cells/ml. After 15 minutes, by which time the cells have swollen, the suspension is homogenized with four strokes of a 50-ml Dounce homogenizer. The homogenate is adjusted to 10% sucrose by adding the appropriate amount of 60% sucrose solution, and centrifuged at 500 *g* for 20 minutes. (All sucrose solutions are made up in homogenizing buffer, and their concentrations are expressed as grams per 100 ml of solution.) The pellet is resuspended in half the original volume of 10% sucrose, centrifuged at 500 *g* for 20 minutes, resuspended again, and centrifuged at 750 g for 20 minutes. These low speed centrifugations are designed to remove mitochondria, microsomes, and ribosomes. The resulting pellet is resuspended in a volume of 25% sucrose equivalent to half the original homogenate volume. Up to 100 ml of this suspension are layered over 40 ml of 30% sucrose and centrifuged at 200 g for 20 minutes in order to separate fragments of plasma membrane from nuclei and intact cells. The 25% layer is retained, diluted to a sucrose concentration of 10% by adding homogenizing buffer, and centrifuged at 750 g for 20 minutes. This yields a pellet of partially purified plasma membranes.

The membranes are further purified by flotation density gradient centrifugation in a Spinco SW 27 rotor. The membrane pellet is resuspended in enough 60% sucrose to render the final sucrose concentration 50%, and then diluted with 50% sucrose to a volume equivalent to 1 ml per 10^8 cells used in the fractionation. To prepare the gradients, 16 ml of this suspension are placed in the bottom of each centrifuge tube, and the following sucrose solutions are layered on top: 45% (8 ml), 40% (7 ml), and 35% (6 ml). The material that collects at the 40/45 interface after centrifugation at 131,000 g for 90 minutes is purified plasma membrane.

Enzymic analyses of these membranes have revealed that they are biochemically equivalent to those isolated by the procedure described in Section III, A. The specific activities of alkaline phosphatase, 5'-nucleotidase, Mg^{2+}–ATPase, succinate dehydrogenase, and acid phosphatase are closely similar for the two preparations (Table IV). Acid phosphatase, although

TABLE IV

COMPARISON OF PLASMA MEMBRANES PREPARED BY TWO DIFFERENT PROCEDURES[a]

| | | Plasma membrane | |
	Homogenate	Procedure described in Section III,A	Procedure described in Section IV
Alkaline phosphatase	23	325	284
5'-Nucleotidase	2	31	23
Mg^{2+}–ATPase	11	37	28
Succinate dehydrogenase	0.4	0	0
Acid phosphatase	12	6	2
Phospholipid (mg/mg protein)	—	0.37	0.43
Sterol (mole/mole phospholipid)	—	0.96	0.98

[a] From Ulsamer et al. (1971). Enzyme activities are expressed as micromoles product per milligram of protein per hour.

not a membrane-bound enzyme, is present in phagosomes, and its low levels in the membrane fractions suggest that contamination by phagosomes is minimal. The sterol:phospholipid ratio of the amoeba plasma membrane is approximately 1, and the phospholipid:protein ratio by weight is about 0.43 (Table IV). A detailed account of the membrane lipid composition can be found in the article by Ulsamer et al. (1971).

Chlapowski and Band (1971) have also isolated plasma membrane from Acanthamoeba. However, the purity of the fraction was judged by morphological criteria only, and thus a direct comparison with membranes prepared by the procedures described herein is not possible. Nevertheless, there is no reason to assume that the procedure of Chlapowski and Band does not also yield highly purified plasma membranes.

V.   Plasma Membrane Enzymes

It would seem reasonable to interpret the high enrichments of alkaline phosphatase and 5′-nucleotidase in preparations of purified plasmalemma as indicating that they serve as markers for this membrane. Both enzymes are also present in preparations of phagosomal membranes. Moreover their specific activities in preparations of phagosomal and plasma membraes are of comparable magnitudes (Ulsamer et al., 1971). Since phagosomal membranes are derived from the plasmalemma during endocytosis, this observation supports the proposal that alkaline phosphatase and 5′-nucleotidase are associated with the plasma membrane. However, there may only be one enzyme, a nonspecific alkaline phosphatase, accounting for both activities. It seems likely that the 5′-nucleotidase activity is attributable to alkaline phosphatase operating suboptimally since adenosine 5′-monophosphate, p-nitrophenylphosphate, and β-glycerophosphate are all hydrolyzed maximally at alkaline pH, but adenosine 5′-monophosphate is the least effective substrate of the three, even at neutral pH (Ulsamer et al., 1971).

Of relevance to the proposal that the amoeba plasma membrane possesses a nonspecific alkaline phosphatase is a recent study in which cytochemical tests on whole cells failed to demonstrate its presence on the cell surface (Bowers and Korn, 1973). The enzyme was instead identified as being present on the contractile vacuole membrane. These observations do not absolutely preclude the presence of an alkaline phosphatase on the plasmalemma because upward of 75% of the total enzyme activity is eliminated by the reagents required for cytochemistry. It is thus conceivable that a plasma membrane phosphatase could be selectively and completely inactivated by this technique.

Nevertheless, in the absence of positive evidence to the contrary, there is a distinct possibility that alkaline phosphatase is associated exclusively with membrane of the water expulsion vacuole. Granted this assumption, the high enrichments of 5′-nucleotidase and alkaline phosphatase activities in preparations of purified plasma membrane must at least partially reflect the fact that the contractile vacuole becomes associated with the plasmalemma during discharge (Vickerman, 1962; Pal, 1972; Daniels, 1972). Since *Acanthamoeba castellanii* eliminates a volume of water approximately equal to its weight in 15–30 minutes (Pal, 1972), a substantial proportion of the cells in a population would be engaged in water expulsion at any one time. The vacuolar membrane associated with the cell surface during discharge would in essence be part of the plasmalemma, albeit transiently, and would isolate with it during fractionation, thus accounting for the presence of alkaline phosphatase. It has, in fact, been observed that the levels of alkaline phosphatase associated with the cell surface decline as the need for water expulsion decreases (Wilkins and Thompson, 1974). Since the contractile vacuole is involved in osmoregulation, it is possible that enzymes associated with its membrane are involved in ion transport. However, it is not yet clear what role alkaline phosphatase plays in the activities of this membrane.

More recently it has been established that certain enzymes of phospholipid metabolism are associated with isolated *Acanthamoeba* plasma membrane. These include phospholipase A, lysophospholipase, acyl-CoA hydrolase and palmitoyl-CoA synthetase (Victoria and Korn, 1975). These enzymes are presumably involved in phospholipid turnover, and in this context their presence on the plasma membrane may be related to the nutritional dependence of the amoebae upon endocytosis. It has been proposed (Lucy, 1970) that membrane fusion, which is a feature of endocytosis, may be contingent upon a localized biochemical modification of phospholipids at the point of fusion.

The enrichment of Mg^{2+}–ATPase activity in plasma membrane preparations indicates that this enzyme is also on the amoeba surface. However, it appears to be insensitive to Na^+ and K^+ and thus, apart from the prospect that it may be involved in endocytosis, the significance of its presence on the amoeba plasmalemma is not clear.

VI.   Macromolecular Composition of Amoeba Plasma Membrane

The macromolecular composition of amoeba plasma membrane has proved to be singularly uncomplicated in comparison with the surface membranes of mammalian and bacterial cells (Korn and Wright, 1973; Wilkins and Thompson, 1974). Electrophoresis of solubilized membrane has revealed that a single polypeptide comprises 40–50 % of the total membrane protein. There are other proteins in the membrane including enzymes, but their relative proportions are minute by comparison. The only other major components are two phosphonoglycans which stain when the gels are treated with periodic acid–Schiff reagent.

A spectrophotometric scan of a Coomassie blue-stained gel for plasma membrane is illustrated in Fig. 5A. Solubilization of the membranes in sodium dodecyl sulfate and polyacrylamide gel electrophoresis were carried out using the technique described by Fairbanks et al. (1971). The protein banding pattern features a major band at a relative mobility (R_f) of about 0.8 which corresponds to a polypeptide with a molecular weight of approximately 15,000 and accounts for 40–50 % of the total protein. Three additional protein bands of high enough intensity to register in spectrophotometric scans are also apparent at R_f values ranging from 0.5 to 0.7. These bands are more discernible in our scans than in those of Korn and Wright (1973) presumably because we applied 50–60 μg of protein to the gels whereas they applied only 3–12 μg. It has been our experience that when 10 μg of protein or less are applied, little else other than the major protein band running at an R_f of 0.8 can be seen. In addition there are 6–8 minor bands of lower intensity corresponding to proteins ranging in molecular weight from 35,000 to about 70,000. These bands were not of sufficient intensity to register distinctly in spectrophotometric scans, but were clearly visible by eye.

The scan in Fig. 5A is for membrane which was not lipid-extracted prior to solubilization. However, removal of the lipid by treatment with chloroform–methanol (2:1) does not alter the protein banding pattern.

Korn and Wright (1973) have reported that actin is sometimes present in plasma membrane fractions isolated by the procedure of Ulsamer et al. (1971), although we do not find it in significant amounts in membranes prepared by the procedure described in Section III, A. Actin is readily detected by electrophoresis of the isolated membranes because it shows up as a major peak in spectrophotometric scans comparable in magnitude to that for the 15,000 molecular weight protein (Korn and Wright, 1973). From electron micrographs it appears to be present as filaments within the plasma membrane vesicles (Pollard and Korn, 1973), suggesting that it is not part of the membrane per se. It is not known why actin is only sometimes associated with isolated plasma membranes. Nevertheless, its presence in even some

---

Fig. 5. Polyacrylamide gel electrophoresis scans of isolated plasma membrane. India ink was used to mark the location of the tracking dye and accounts for the peaks at a relative mobility of 1.0. (A) Stained with Coomassie blue. (B) Stained with periodic acid–Schiff reagent.

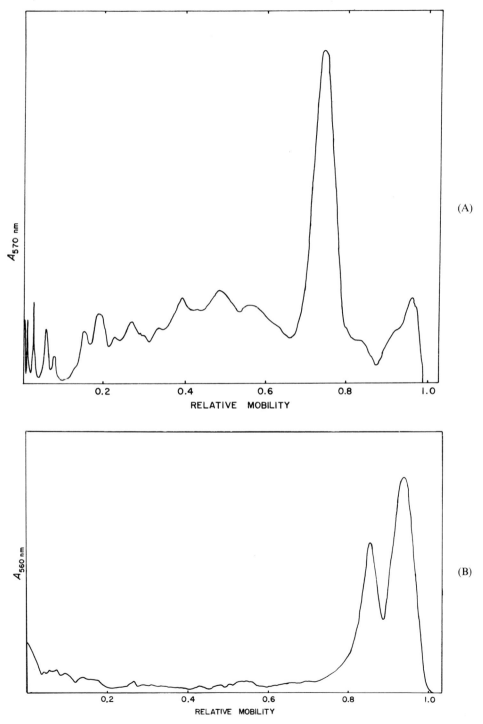

(A)

(B)

preparations is intriguing, and the finding that *Acanthamoeba* also contains myosin (Pollard and Korn, 1972) raises the possibility that these proteins are involved in amoeboid movement and perhaps endocytosis.

Gels stained with periodic acid–Schiff reagent feature two prominent phosphonoglycan bands at R_f values of about 0.85 and 0.95 (Fig. 5B). Phosphonoglycan accounts for 31% of the plasma membrane mass and, insofar as it has been characterized, is known to contain neutral sugars, amino sugars, aminophosphonates, phosphate, and fatty acids (Korn et al., 1974; Dearborn and Korn, 1974).

VII.   Turnover of Plasma Membrane during Phagocytosis

The relatively uncomplicated macromolecular composition of the amoeba plasma membrane, although surprising in the sense that one might expect the surface of a unicellular organism to exhibit complexity, is consistent with the nutritional dependence of *Acanthamoeba* upon endocytosis. A major consequence of nutrient accumulation by this method is the expenditure of large amounts of plasma membrane which must ultimately be replaced. Thus endocytosis can be regarded as a specialized example of the more fundamental phenomenon of cell membrane turnover, and in the face of such extensive turnover it is clearly of benefit to the organism to have a surface membrane of simple composition.

Acanthamoeba can be induced to ingest latex beads by phagocytosis (Weisman and Korn, 1967; Goodall and Thompson, 1971). This technique, in conjunction with the availability of procedures for isolating amoeba plasma membrane, has made possible at least preliminary studies of membrane turnover during phagocytosis (Goodall et al., 1972).

A.   Uptake of Latex Beads

The uptake of latex beads by amoebae is illustrated in Fig. 6. These cells were harvested from exponential-phase cultures, and it is clear that the ingestion of beads is essentially linear as a function of time. This observation was confirmed by examining wet mounts of the cells under a phase-contrast microscope. By contrast, amoebae from stationary-phase cultures are totally incapable of ingesting beads (Chambers and Thompson, 1975). The basis for this loss of phagocytic activity is not known. However, we have recently established that the fluidity of the plasma membrane lipid decreases markedly coincident with the transition to stationary phase (J. A. Chambers

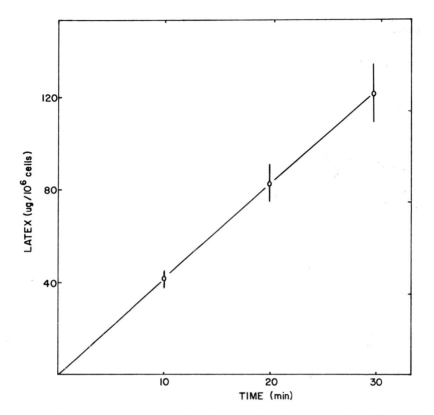

FIG. 6.   Phagocytosis of polystyrene latex beads by *Acanthamoeba castellanii*. Phagocytosis was initiated by adding 33 mg of latex beads (1.01 μm in diameter) to a suspension of amoebae (2×10^7 cells in 9 ml of growth medium) in a 50-ml Erlenmeyer flask. The flask was agitated at 80 cycles per minute in a water bath adjusted to 30°C, and 1-ml samples were taken at 10-minute intervals. Levels of ingested latex were determined as described by Weisman and Korn (1967). Standard errors of the means are indicated; $n = 4$.

and J. E. Thompson, unpublished observations), and it is conceivable that this has some bearing on the capability for phagocytosis.

B.   Visualization of Bead Ingestion by Scanning Electron Microscopy

An examination of bead ingestion by scanning electron microscopy has allowed visualization of the temporal changes in surface membrane topography that accompany phagocytosis, and given as well an indication of the timing of surface membrane replacement relative to actual ingestion. This latter information in particular proved to be helpful in setting up subsequent pulse-chase *in vivo* labeling experiments designed to discover how plasma membrane expended during ingestion is replaced.

A typical asymmetric amoeba, as it appears prior to the addition of latex beads, is shown in Fig. 7a. Several smooth filamentous filopodia are apparent arising from a highly convoluted and deeply folded surface. These extensive convolutions indicate that the surface membrane is not tightly stretched and that there is in fact a surplus of membrane. The relationship between the

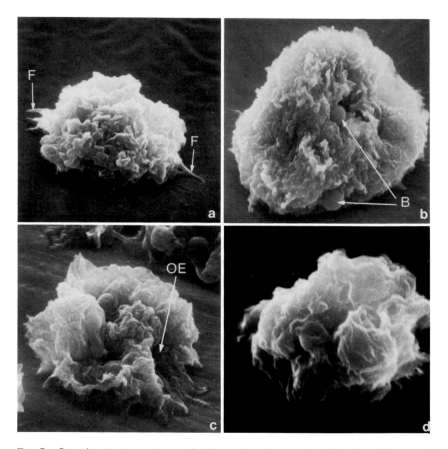

FIG. 7. Scanning electron micrographs illustrating phagocytosis of latex beads by *Acanthamoeba castellanii* (Goodall and Thompson, 1971). Phagocytosis was initiated by adding 100 mg of latex beads (1.95 µm in diameter) to a suspension of amoebae (3 × 10[8] cells in 100 ml of 1.5% proteose peptone in 0.015 M KH$_2$PO$_4$, pH 6.7) in a 250-ml Erlenmeyer flask. The flask was agitated at 80 cycles per minute in a water bath adjusted to 30°C. The samples were processed for scanning electron microscopy as described by Pasternak *et al.* (1970). B, bead; F, filopodia; OE, open end of a membranous tube engulfing a bead. (a) × 3200. (b) × 2900. (c) × 2900. (d) × 2860.

latex bead and the cell during phagocytosis is documented in Figs. 7b and c. The bead becomes attached to the external surface of the cell, and the surface membrane then creeps up around the edge of the bead forming a gulletlike tube of membrane that ultimately engulfs the bead. Large mouthlike openings corresponding to the open external ends of these tubes are clearly evident in Fig. 7c. These open ends apparently fuse once the bead has been engulfed.

By about 30 minutes after the addition of latex beads, marked changes in the surface topology of the amoeba are evident (Fig. 7d). Specifically, the filopodia have virtually disappeared and the deep folds and convolutions in the surface membrane are more shallow, suggesting that the surplus membrane has been expended. If the uningested beads are at this point removed by centrifugation at 800 g for 5 minutes as described by Goodall and Thompson (1971), the amoebae reattain their original surface morphology within about 15 minutes. Thus replacement of plasma membrane seems to occur after rather than during the period of active ingestion.

C. *In Vivo* Labeling of Plasma Membrane during Bead Ingestion

By following the incorporation of labeled precursors into plasma membrane during the ingestion of latex beads, it has been possible to distinguish between *de novo* synthesis and utilization of preexisting membrane as means of replacing surface membrane expended during phagocytosis. [2-^{14}C]glycine or [2-^{3}H]glycerol (specific activity 53 mCi/mmole) were added to a suspension of amoebae (1.6 × 10^9 cells in 200 ml of 1.5% proteose–peptone in 0.015 M KH$_2$PO$_4$, pH 6.7) at concentrations of 0.15 mCi and 0.4 mCi, respectively. The cell suspension was maintained at 30°C in a shaking water bath rotating at 80 rpm. After 30 minutes, phagocytosis was initiated by adding 400 mg of latex beads (2.02 μm in diameter), and after 60 minutes a 100-fold molar excess of unlabeled precursor was added as a chase. One hour later the cells were harvested and plasma membrane was isolated as described in Section III,A. This interval exceeded the time required for the amoebae to return to their prephagocytic morphology as estimated by scanning electron microscopy, thus ensuring that the nature of plasma membrane replenishment was being monitored both during and after the period of most active phagocytosis. Enrichments of radioactivity in the plasma membrane were determined by calculating the ratio of the specific radioactivity in the plasma membrane fraction to that of a corresponding "particulate homogenate" obtained by centrifuging the true homogenate at 133,000 g for 2 hours. Comparisons were made against "particulate homogenates" rather than true homogenates to eliminate contribution to the radioactivity count by unincorporated precursor. In order to monitor the incorporation of label into plasmalemma under conditions in which there was no induced phagocytosis, parallel controls were run in which latex beads were omitted from the incubation mixture.

Enrichment values for both [^{14}C]-glycine and [^{3}H]-glycerol are presented in Table V. Statistical analyses of these data (Student's t test) have confirmed that there is no significant difference between values obtained in the presence of latex beads and those obtained in their absence. This suggests that there was no preferential synthesis of plasma membrane in response to bead ingestion because, otherwise, the enrichment values for phagocytizing cells should have been significantly higher than those for nonphagocytizing cells. It seems reasonable to propose, therefore, that the source of replenishment for surface membrane expended during phagocytosis is preexisting membrane that becomes inserted into the plasmalemma. Indeed, there may be a cytoplasmic pool of such membrane, which in the event of extensive phagocytosis is, as suggested by Chlapowski and Band (1971), derived from a fragmentation of vacuolar membranes. Cycloheximide does not prevent uptake of latex beads by the amoebae, and this also indicates that phagocytosis does not require *de novo* synthesis of membrane.

Clearly, however, labeled precursors of both phospholipid and protein are incorporated into the plasma membrane during this period. In the absence of detectable preferential membrane synthesis this can be interpreted as reflecting routine molecular turnover occurring independently of phagocytosis. It seems likely, therefore, that the ingestion of beads involves turnover of plasma membrane at two levels: cycling of membranes which enables replacement of

TABLE V

INCORPORATION OF RADIOACTIVE PRECURSORS INTO PLASMA MEMBRANE
DURING PHAGOCYTOSIS OF LATEX BEADS[a]

Latex beads	Enrichments[b]	
	[^{14}C]Glycine	[^{3}H]Glycerol
Absent	1.44	0.91
Absent	1.82	1.42
Absent	1.30	1.84
Present	1.72	1.33
Present	1.34	1.68
Present	1.80	1.40

[a] From Goodall *et al.* (1972).

[b] Enrichment values are ratios of the specific radioactivity in the plasma membrane fraction to that of a corresponding "particulate homogenate" obtained by centrifuging the true homogenate at 133,000 g for 2 hours. For enrichments in the presence and in the absence of latex beads $t_4 = 0.48$ for [^{14}C]glycine and 0.28 for [^{3}H]glycerol.

expended surface membrane without *de novo* synthesis; and turnover at the molecular level involving input of newly synthesized material. Warren and Glick (1968) have proposed that molecular turnover of membranes provides a basis for their modification. Such turnover in the case of phagocytosis may be the means whereby membrane fragments newly inserted into the plasmalemma acquire properties characteristic of the cell surface.

REFERENCES

Bosmann, H. B., Hagopian, A., and Eylar, E. H. (1968). *Arch. Biochem. Biophys.* **128**, 51–69.
Bowers, B., and Korn, E. D. (1968). *J. Cell Biol.* **39**, 95–111.
Bowers, B., and Korn, E. D. (1973). *J. Cell Biol.* **59**, 784–791.
Bowers, B., and Olszewski, T. E. (1972). *J. Cell Biol.* **53**, 681–694.
Chambers, J. A., and Thompson, J. E. (1976). *J. Gen. Microbiol.* **92**, 246–250.
Chlapowski, F. J., and Band, N. (1972). *J. Cell Biol.* **50**, 634–651.
Daniels, E. W. (1972). *In* "The Biology of Amoeba" (K. W. Jeon, ed.), pp. 125–169. Academic Press, New York.
Dearborn, E. D., and Korn, E. D. (1974). *J. Biol. Chem.* **249**, 3342–3346.
Fairbanks, G., Steck, T. L., and Wallach, D. F. H. (1971). *Biochemistry* **10**, 2606–2617.
Goodall, R. J., and Thompson, J. E. (1971). *Exp. Cell Res.* **64**, 1–8.
Goodall, R. J., Lai, Y. F., and Thompson, J. E. (1972). *J. Cell Sci.* **11**, 569–579.
Hübscher, G., and West, G. R. (1965). *Nature (London)* **205**, 799–800.
Kellenberger, E., Ryter, A., and Sechaud, J. (1958). *J. Biophys. Biochem. Cytol.* **4**, 671–678.
King, E. J. (1932). *Biochem. J.* **26**, 292–297.
Klein, R. L. (1962). *Exp. Cell Res.* **28**, 549–559.
Korn, E. D., and Weisman, R. A. (1967). *J. Cell Biol.* **34**, 219–227.
Korn, E. D., and Wright, P. L. (1973). *J. Biol. Chem.* **248**, 439–447.
Korn, E. D., Dearborn, D. G., and Wright, P. L. (1974). *J. Biol. Chem.* **249**, 3335–3341.
Lowry, O. H., Rosebrough, N. J., Farr, A. L., and Randall, R. J. (1951). *J. Biol. Chem.* **193**, 265–275.
Lucy, J. A. (1970). *Nature (London)* **227**, 815–817.
Michell, R. H., and Hawthorne, J. N. (1965). *Biochem. Biophys. Res. Commun.* **21**, 333–338.
Neff, R. J. (1957). *J. Protozool.* **4**, 176–182.
Neff, R. J., Ray, S. A., Benton, W. F., and Wilborn, M. (1964). *In* "Methods in Cell Physiology" (D. M. Prescott, ed.), Vol. 1, pp. 55–83. Academic Press, New York.
Pal, R. A. (1972). *J. Exp. Biol.* **57**, 55–76.
Pasternak, J. J., Thompson, J. E., Schultz, T. M. G., and Zachariah, K. (1970). *Exp. Cell Res.* **60**, 290–298.
Pennington, R. J. (1961). *Biochem. J.* **80**, 649–654.
Pollard, T. D., and Korn, E. D. (1972). *Cold Spring Harbor Symp. Quant. Biol.* **37**, 573–583.
Pollard, T. D., and Korn, E. D. (1973). *J. Biol. Chem.* **248**, 448–450.
Schultz, T. M. G., and Thompson, J. E. (1969). *Biochim. Biophys. Acta* **193**, 203–211.
Sottocasa, G. L., Kuylenstierna, B., Ernster, L., and Bergstrand, A. (1967). *J. Cell Biol.* **32**, 415–438.
Ulsamer, A. G., Wright, P. L., Wetzel, M. G., and Korn, E. D. (1971). *J. Cell Biol.* **51**, 193–215.
Vickerman, K. (1962). *Exp. Cell Res.* **26**, 497–519.
Victoria, E. J., and Korn, E. D. (1975). *J. Lipid Res.* **16**, 54–60.
Warren, L., and Glick, M. C. (1968). *J. Cell Biol.* **37**, 729–746.
Weisman, R. A., and Korn, E. D. (1967). *Biochemistry* **6**, 485–497.
Wilkins, J. A., and Thompson, J. E. (1974). *Exp. Cell Res.* **89**, 143–153.

Chapter 18

Induction of Mammalian Somatic Cell Hybridization by Polyethylene Glycol[1]

RICHARD L. DAVIDSON

Division of Human Genetics, Children's Hospital Medical Center, and Department of Microbiology and Molecular Genetics, Harvard Medical School, Boston, Massachusetts

AND

PARK S. GERALD

Division of Human Genetics, Children's Hospital Medical Center, and Department of Pediatrics, Harvard Medical School, Boston, Massachusetts

I. Introduction

The hybridization of somatic cells has become an increasingly important technique for the study of the organization and regulation of the mammalian genome (for reviews, see Ephrussi, 1973; Davidson, 1974; Ruddle and Crea-

[1] This work was supported in part by NICHD Grants HD 04807 and HD 06276.

gan, 1975). In spite of its importance as a technique, however, the mechanisms involved in cell hybridization are not understood. In the absence of this knowledge, progress on improving the efficiency of hybridization has not been rapid and planned studies have been limited sometimes by the ability to isolate hybrid cells. Inactivated Sendai virus has been used widely to induce cell fusion (Harris et al., 1966; Davidson, 1969), and lysolecithin also has been used for this purpose (Croce et al., 1971). However, with both these agents, there are major problems in terms of difficulty of preparation, variations in fusion inducing activity from batch to batch, and absence of reproducibility from one laboratory to another. Clearly, the identification of an agent that is highly active in inducing cell fusion, consistent in fusing activity from batch to batch, easy to use, readily obtainable, and inexpensive would facilitate studies on mammalian somatic cell genetics. Recent results suggest that polyethylene glycol (PEG) may fit all these requirements.

PEG was first shown to be capable of inducing cell fusion in experiments with plant cells (Kao and Michayluk, 1974). Subsequently Pontecorvo (1975) showed that PEG induces fusion and the formation of viable hybrids between mammalian cells. We have confirmed that PEG induces mammalian cell hybridization and have developed procedures for treating cells with PEG that are simple, rapid, and effective (Davidson and Gerald, 1976; Davidson et al., 1976).

II.   Polyethylene Glycol Solutions

The structure of PEG is $HOH_2C(CH_2OCH_2)_n CH_2OH$. PEG grades that have average molecular weights ranging from less than 200 to greater than 6000 have been used in hybridization experiments. The smallest PEG used (PEG 200) has on the average only 3.1 repeating subunits (CH_2OCH_2) per molecule, whereas the largest (PEG 6000) has 152. The molecular weight range and the number of repeating subunits for all the PEG grades used in these experiments are presented in Table I. PEG (all grades) was purchased from the J. T. Baker Co.

PEG solutions for all the grades were prepared as previously described (Pontecorvo, 1975; Davidson and Gerald, 1976). PEG was autoclaved, and warm Eagle's medium without serum was added to the warm liquid PEG. With the higher molecular weight grades of PEG, the PEG solution often begins to solidify after the addition of medium. If this occurs, the PEG can be brought back into solution by heating to 70°C. After the PEG has been redissolved, the solution can be stored at room temperature for long periods of time without solidification.

TABLE I

POLYETHYLENE GLYCOL (PEG) GRADES

PEG grade	Average molecular weight range[a]	Average number of repeating subunits/molecule[b]
200	190–210	3.14
300	285–315	5.41
400	380–420	7.68
600	570–630	12.2
1000	950–1050	21.3
4000	3000–3700	74.7
6000	6000–7500	152

[a] Average molecular weight range as indicated by the supplier.
[b] The average number of repeating subunits/molecule was calculated using the midpoint of the average molecular weight range.

The concentrations of PEG solutions have been expressed on a weight/weight (w/w) basis, assuming that 1 ml of medium has a weight of 1 gm. A 50% solution could be prepared, therefore, by mixing 10 gm of PEG and 10 ml of medium.

III. Hybridization of Cells in Monolayers

A. Cell Lines and Selection of Hybrids

Most of the experiments on the PEG-induced hybridization of cells attached in monolayers involved the fusion of mouse fibroblasts with rat glial cells. The parental cell line in this cross was 3T3-4E, a bromodeoxyuridine-resistant mouse line, and RG6A-TgA, an azaguanine-resistant rat line. In a few experiments, the mouse cell line LM(TK⁻) clone 1D was used. These three lines will be referred to below as 3T3, RG6, and LM, respectively.

The cells were grown in Dulbecco's modified Eagle's medium (E medium) supplemented with 10% fetal calf serum (E-10FCS medium). Hybrids between 3T3 and RG6 cells were selected in E-10FCS medium containing hypoxanthine, aminopterin, and thymidine (E-10HAT medium), in which neither of the parental cell lines can grow because of the enzyme deficiencies associated with their drug resistance (Littlefield, 1964).

In most of the experiments, 60-mm Falcon tissue culture dishes were inoculated with 2.5×10^6 3T3 cells and 2.5×10^5 RG6 cells. This number of cells produces a confluent monolayer. After 24 hours, the cells were treated with PEG as described in Section III,B. Twenty-four hours after the PEG treatment the cells were harvested by trypsinization and plated at lower densities in 100 mm dishes in E-10HAT medium. The selective medium was renewed 4–5 days later. After 7 days in selective medium, the dishes were fixed with methanol and stained with Giemsa, and the hybrid cell colonies were counted in triplicate or quadruplicate dishes.

B.   Treatment of Monolayers with PEG

The protocol for treating cells with PEG was developed using the 3T3-RG6 cross and PEG 6000 (Davidson and Gerald, 1976; Davidson et al., 1976). The parameters that were tested included the method of removal of the medium before the PEG treatment, the length of exposure of the cells to PEG, the temperature of the PEG, and the method of removal of the PEG. Based on a large number of experiments, the following protocol was found to yield the optimal results:

1. Inoculate 60-mm dishes with a mixture of the two parental cell types (the effect of cell density is described in Section III,F) in E-10FCS medium and incubate the dishes for 24 hours at 37°C.

2. Aspirate the medium from the dishes, tilt the dishes on edge, allow the residual medium to drain for 30 seconds, and aspirate the medium that has collected.

3. Cover the cells with 3 ml of PEG solution at room temperature (the effects of PEG concentration and molecular weight are described in Sections III,C and D).

4. Leave the PEG in contact with the cells for 1 minute at room temperature without agitation.

5. Aspirate the PEG and rapidly rinse the dishes three times with 5 ml of E medium (at room temperature).

6. Add 5 ml of E-10FCS medium to the dishes and incubate the dishes for 24 hours at 37°C prior to trypsinizing the cells and plating in selective medium.

Some of the factors tested had no effect on the frequency of hybridization. For example, it made no difference whether or not the cells were rinsed with serum-free medium prior to adding PEG. However, the volume of residual medium in the dishes at the time of PEG addition was found to be very important. The extent of cell fusion is so sensitive to changes in the concentration of PEG (see Section III,C) that even a very small amount of medium remaining in the dishes when the PEG is added can cause a significant reduction in fusion-inducing activity (Davidson et al., 1976). This marked concentration effect is the reason for step 2 above.

The temperature of the PEG during the treatment does not seem to be of significance, since identical results were obtained with PEG solutions either at room temperature (22°C) or at 37°C. However, the length of exposure of the cells to PEG is critical (Davidson and Gerald, 1976). In the 3T3-RG6 cross, exposures to PEG either shorter than 1 minute (e.g., 30 seconds) or longer than 1 minute (e.g., 2 minutes) result in many fewer hybrids than does a 1 minute exposure. The method of removing the PEG after exposure of the cells (step 5 above) is also critical. The highest yield of hybrids is obtained with a series of three quick rinses to remove the PEG. With fewer rinses, or with a slower series of rinses, the yield of hybrids is significantly decreased. It should be noted that a small amount of PEG remains on the cells even after the recommended rinsing, but this residual PEG is apparently not toxic to the cells.

C. Effect of PEG Concentration

The effect of changes in the concentration of PEG on hybridization were investigated initially with the 3T3-RG6 cross and PEG 6000 (Davidson and Gerald, 1976). As shown in Fig. 1A, there is a very marked effect of PEG concentration on the frequency of hybridization. The maximum frequency of hybridization was observed with PEG at a concentration of 50%, and even small changes in PEG concentration resulted in a significant drop in the yield of hybrids. When the concentration of PEG was decreased from 50% to 45%, the yield of hybrids dropped by 80%, and when the concentration of PEG treatment than do 3T3 or RG6.) As shown in Fig. 2, most of the control cultures. Similarly, when the concentration of PEG was increased from 50% to 60%, the yield of hybrids dropped by 90%. (See the legend to Fig. 1 for the actual hybridization frequencies.)

The decreased hybridization with PEG concentrations below 50% seems to be due to a decline in the fusion-inducing activity of PEG at lower concentrations. The effect of changes in PEG concentration on fusion (measured by the formation of multinucleate cells), as distinct from hybridization (measured by the formation of hybrid cell colonies), was investigated with LM cells. (It is easier to study fusion effects with LM cells than with 3T3 or RG6 cells, since LM cells undergo much more extensive fusion following PEG treatment than do 3T3 or RG6.) As shown in Fig. 2, most of the LM cells in a culture treated with 50% PEG fuse into massive multinucleate cells. In contrast, after treatment with 45% PEG, many fewer cells fuse and the size of the multinucleate cells formed is much reduced. The decline in the fusion of LM cells as the concentration of PEG is decreased below 50% parallels the decline in the yield of hybrid cell colonies in the 3T3-RG6 cross with decreased PEG concentrations.

In contrast to the correlation between fusion and hybridization at PEG concentrations below 50%, the results presented in Table II suggest that the decline in the hybridization frequency with PEG concentrations greater than

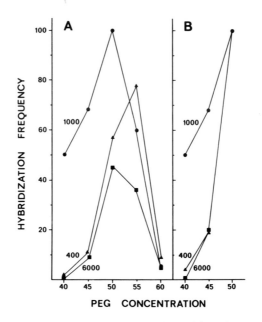

FIG. 1. Effect of polyethylene glycol (PEG) molecular weight and concentration on hybridization. Dishes (60 mm) were inoculated with 2.5×10^6 3T3 and 2.5×10^5 RG6 cells, and the cells were treated with PEG 400, 1000, or 6000 as described in Section III,B. The PEG grades are indicated on the figure. In panel A, the hybridization frequencies for all the PEG grades are expressed relative to that for PEG 1000 at 50%. In panel B, the hybridization frequencies for each PEG grade are expressed relative to the frequency for that grade at 50%. The reference frequencies are assigned values of 100 in both A and B. The actual hybridization frequencies for PEG 400, 1000, and 6000 at 50% are 13.9, 24.4 and 11.0 hybrids per 10^3 RG6 cells, respectively. All values are corrected for the frequency of spontaneous hybridization, which was 0.06 hybrid per 10^3 RG6 cells in this experiment.

50% is not related to a drop in cell fusion. (The experiment presented in Table II was carried out with PEG 1000 rather than 6000. Like PEG 6000, PEG 1000 shows at least a 90% drop in the frequency of hybridization as the concentration is increased from 50% to 60%.) PEG concentrations above 50% become progressively more toxic to the cells, and the decline in the yield of hybrids at PEG concentrations greater than 50% parallels the decrease in the viability of the treated cells. If the reduced viability is taken into account, it is seen that the hybridization frequency remains relatively constant as the PEG concentration is increased above 50%. Furthermore, rather than a drop in fusion, there is actually an increase in cell fusion at the higher concentrations of PEG. It seems, therefore, that the optimal PEG concentration for hybridization is determined by a combination of two factors—cell fusion and cell viability.

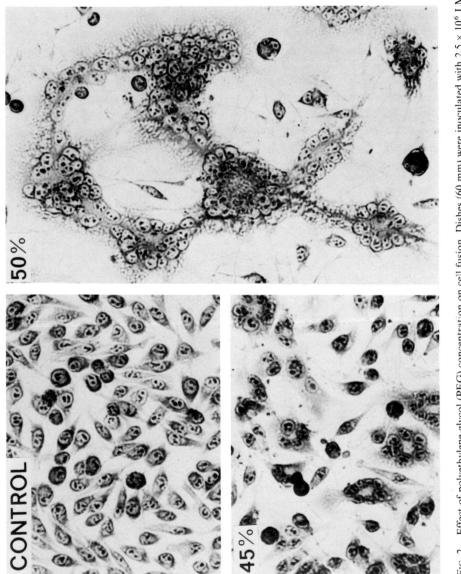

FIG. 2. Effect of polyethylene glycol (PEG) concentration on cell fusion. Dishes (60 mm) were inoculated with 2.5×10^6 LM cells, and the cells were treated with PEG 6000 at the indicated concentration. The cells were fixed and stained 2 hours after the PEG treatment.

TABLE II

EFFECT OF HIGH CONCENTRATIONS OF POLYETHYLENE GLYCOL (PEG) ON HYBRIDIZATION AND VIABILITY

PEG concentration	Viability[a]	Hybridization frequency[b]	
		Per 10^4 cells	Per 10^4 viable cells
50	60.0%	17.5	29.2
55	28.6%	9.12	31.9
60	3.01%	0.800	26.6

[a] Dishes, 60 mm, were inoculated with 10^6 3T3 and 10^5 RG6 cells. The cells were treated with PEG 1000 as described in Section III,B. At 24 hours after the PEG treatment, the plating efficiency of the PEG-treated cells in E-10FCS medium was determined. Viability was defined as the plating efficiency of PEG-treated cells relative to that of control cells not treated with PEG.

[b] Aliquots of the same PEG-treated cell mixtures used to determine viability were plated in E-10HAT medium to determine hybridization frequencies. The hybridization frequencies are expressed as the number of hybrid cell colonies per 10^4 cells inoculated or per 10^4 viable cells (after correcting for relative plating efficiency).

D. Effect of PEG Molecular Weight

The hybridization-inducing activities of PEG 200, 300, 400, 600, 1000, 4000, and 6000 were tested with the 3T3-RG6 cross. All the PEG grades tested were found to be active in inducing hybridization, and, as described in Section III,C for PEG 6000, the hybridization-inducing activity of all the PEG grades was found to be concentration dependent. However, there were major differences among the grades, in terms of peak concentration, maximum hybridization efficiency, and sensitivity to dilution (Davidson *et al.*, 1976).

The effects on hybridization of PEG 400, 1000, and 6000 over the range of concentrations from 40–60% are presented in Fig. 1A. (For simplicity, only the values for these three grades are shown, but these results can be taken as representative of the results obtained with the other grades tested.) The peak concentration for all the grades (except PEG 200) was observed at 50 or 55%. It seems that the grades with molecular weight of 1000 or greater have their peak concentration at 50%, whereas the lower molecular weight grades have their peak concentration at 55%. The highest frequency among all the grades at any concentration was observed with PEG 1000 at 50%. Following treatment with PEG 1000 at 50%, as many as 5% of the RG6 cells have been observed to fuse with 3T3 cells to form viable hybrids.

In addition to variation among the grades in terms of the peak concentration and the maximum hybridization efficiency, differences were observed in the sensitivity of the various grades to changes in concentration. For example, as shown in Fig. 1B, PEG 6000 and 400 lose approximately 80% of their hybridization-inducing activity when diluted from 50% to 45%, whereas PEG 1000 loses only approximately 30% of its activity with the same dilution.

PEG 200, the lowest molecular weight PEG tested, was found to be the least effective in inducing hybridization of all the grades tested. At its peak concentration (60%), PEG 200 caused little more than a 30-fold increase in hybridization (relative to untreated control cultures). All the other PEG grades tested (300–6000) were much more effective at their respective peak concentrations, causing a 180-400-fold increase in hybridization efficiency. Experiments on cell fusion per se revealed that PEG 200 also is much less effective in producing multinucleate cells than are any of the other grades from PEG 300–6000. These results suggest that there is a certain minimum molecular weight (around 300) or minimum number of repeating subunits per molecule (around 5), below which PEG begins to lose its cell-fusing activity. (See Table I for the molecular weight range and number of repeating subunits for the various PEG grades.)

E.   Effect of pH

The effect of the pH of the PEG solution on the efficiency of hybridization was tested with the 3T3-RG6 cross and PEG 1000 at 50%. Over the pH range from 5 to 9, the effect of changes in pH was relatively small (see Table III). It seemed that the frequency of hybridization was the highest at pH 6 and the lowest at pH 9. However, the difference in hybridization frequency between pH 9 and pH 6 was little more than 1.5-fold.

TABLE III

EFFECT OF pH ON THE FREQUENCY OF HYBRIDIZATION[a]

pH of PEG solution	Hybridization frequency
5	38.9
6	49.7
7	34.5
8	37.6
9	30.9

[a] Dishes (60 mm) were inoculated with 10^6 3T3 and 10^5 RG6 cells. The cells were treated with polyethylene glycol (PEG) 1000 at 50% as described in Section III,B. The pH of the PEG solutions was controlled by the addition of HCl or NaOH. The results are expressed as the number of hybrid cell colonies per 10^3 RG6 cells.

F.   Effect of Cell Density

As described above, LM cells show a massive fusion response to PEG. However, PEG-treated mixtures containing LM cells as one of the parental cell types have not yielded hybrids at a high frequency. This discrepancy between cell fusion and the formation of viable hybrids with LM cells raised the possibility that maximal fusion does not necessarily lead to the highest yield of hybrids. Experiments were undertaken, therefore, to determine whether limiting cell fusion (by decreasing cell contacts) in PEG-treated mixtures of 3T3 and RG6 cells would have an effect on hybridization. Dishes were inoculated with varying numbers of 3T3 and RG6 cells (always in a 10 : 1 ratio) so that the cultures ranged from complete confluence to relative sparsity at the time of exposure to PEG. As can be seen in Table IV, as the cell density in the PEG-treated cultures decreased, the frequency of hybridization increased. It thus seems that the highest yield of hybrids is obtained at cell densities that are suboptimal for fusion. This presumably is due to an increased formation of smaller multinucleate cells that can give rise to viable hybrids when fusion is limited by decreasing cell contacts.

G.   Summary

The experiments presented above were undertaken in order to optimize the conditions of PEG-induced hybridization of cells attached in monolayers. Among the parameters considered were: PEG molecular weight, PEG concentration, cell density, pH, length of exposure to PEG, temperature, volume of residual medium at the time of PEG addition, and the method of removing the PEG. The protocol presented in Section III,B seems to represent the optimal conditions, in terms of the last four parameters. It is possible that treatment of cells with PEG solutions at pH 6 gives a slightly higher yield of hybrids than at any other pH between 5 and 9.

For most of the above parameters, the conditions leading to optimal fusion and optimal hybridization are probably the same. However, this is

TABLE IV

EFFECT OF CELL DENSITY ON THE FREQUENCY OF HYBRIDIZATION[a]

Cells/dish		PEG grade and concentration		
3T3	RG6	400–55%	1000–50%	6000–50%
2.5×10^6	2.5×10^5	11	15	6
1.0×10^6	1.0×10^5	25	30	20
5.0×10^5	5.0×10^4	30	41	36

[a] Dishes (60 mm) were inoculated with the indicated number of cells and treated with PEG as described in Section III,B. The results are expressed as the number of hybrid cell colonies per 10^3 RG6 cells.

clearly not the case with the cell density in the PEG-treated cultures, as subconfluent cultures yield significantly more hybrids than do confluent cultures.

At present, the effects on hybridization frequency of changes in either PEG molecular weight or concentration are not understood. However, as a practical point, it seems clear that PEG 1000 at a concentration of 50% represents the best choice. There are two reasons for this choice: (1) PEG 1000 at 50% produces a higher frequency of hybridization than any other grade at any concentration; (2) PEG 1000 in the region of its peak concentration (50%) is the least sensitive of all the grades to dilution effects.

The conditions for PEG-induced hybridization were optimized for the 3T3-RG6 cross. However, we have used the protocol developed with this cross to isolate hybrids from a variety of other crosses, involving human, rat, Syrian hamster, and mouse cells. It seems, therefore, that the basic conditions for the hybridization of cells in monolayers are similar for most cell lines. However, since there clearly are differences among cell lines in sensitivity to PEG (in terms of both frequency of fusion and viability of the treated cells), it may be necessary to alter various conditions somewhat in order to obtain the *maximum* hybridization frequency with each cross.

IV. Hybridization of Cells in Suspension

A. Cell Lines and Selection of Hybrids

The cell lines used for the fusion of cells in suspension were the rodent lines Wg3h (Chinese hamster) and RAG (mouse), both of which are deficient in hypoxanthine–guanine phosphoribosyltransferase activity. The other cell types used were human peripheral leukocytes (WBC) and mouse peritoneal macrophages.

The rodent cells were grown in Dulbecco's modified Eagle's medium (E medium) supplemented with 5% fetal calf serum (E-5FCS medium). A plasma suspension of WBC was obtained by sedimenting for 1–2 hours a mixture of equal volumes of heparinized whole blood and 100,000–200,000 molecular weight dextran solution (3 gm of dextran/100 ml of saline). The WBC were washed 3 times in Hanks' balanced salt solution before use. The mouse peritoneal macrophages were obtained by peritoneal lavage with sterile saline and were washed three times with Hanks' balanced salt solution.

In most of the experiments, mixtures containing 5×10^6 rodent cells (Wg3h or RAG) and 5×10^6 WBC or mouse peritoneal macrophages were treated with PEG as described in Section IV,B. The PEG-treated mixtures were divided among five 60-mm Falcon tissue culture dishes containing E-5FCS medium. Twenty-four hours later, the medium was replaced with E-5FCS medium containing hypoxanthine, aminopterin, and thymidine (E-5HAT medium) (Littlefield, 1964). [The success of this "half-selective" system (Davidson and Ephrussi, 1965) depends upon the inability of WBC or macrophages to divide while the Wg3h and RAG cells are unable to survive in HAT medium.] The E-5HAT medium was renewed every 4–5 days. Fourteen days after fusion, the dishes were fixed with methanol and stained with Giemsa to permit colony counting.

B.   Treatment of Cell Suspensions with PEG

The protocol for fusing cells in suspension with PEG was developed using rodent cell-WBC crosses and PEG 6000. The duration of exposure to PEG was found to be critical. Exposure times shorter or longer than 1 minute reduced the frequency of hybrid colony formation below that obtained with a 1 minute exposure (Davidson and Gerald, 1976). Rapid termination of the exposure to PEG by dilution with E medium was found to be essential to permit precise control of the exposure time. The following simple protocol was developed for Wg3h-WBC fusions and has been utilized for studying the effect of PEG concentration and molecular weight on fusion in suspension (described in Section IV, C):

1. Mix 5×10^6 Wg3h cells and 5×10^6 WBC in a conical-bottom tube.
2. Centrifuge the mixed suspension and aspirate the supernatant (approximately 0.05 ml of fluid still remains in the tube after aspiration).
3. Add 1 ml of PEG solution and mix by pipetting.
4. After 60 seconds, add 9 ml of E medium, centrifuge the suspension, and aspirate the supernatant.
5. Resuspend the cells in 5 ml of E-5FCS medium and divide the suspension among five 60-mm culture dishes in the same medium.

This protocol is less effective with RAG-WBC fusions; this particular cell mixture apparently requires more extensive removal of the PEG in order to obtain the optimum yield of hybrid cell colonies.

C.   Effect of PEG Concentration and Molecular Weight

The experiments with monolayer fusions indicated that the maximum yield of hybrid cell colonies was obtained over a narrow range of PEG concentration. The maximum yield of colonies in suspension fusions similarly was found to be confined to a relatively narrow concentration range when PEG 6000 was used, and, as with monolayer fusions, the peak hybridization frequency was observed with PEG 6000 at 50 % (Table V). However, fusion in suspension of PEG 1000 yielded quite different results (Table V). It is

TABLE V

EFFECT OF POLYETHYLENE GLYCOL (PEG) MOLECULAR WEIGHT
AND CONCENTRATION ON HYBRIDIZATION IN SUSPENSION[a]

PEG concentration (%)	PEG grade	
	1000	6000
35	2.4	0.2
40	1.8	1.0
45	1.0	5.6
50	0.8	5.8
55	0.6	1.6
60	0.0	0.4

[a] Mixtures containing 5×10^6 Wg3h and 5×10^6 WBC were treated in suspension with PEG as described in Section IV,B. The hybridization frequencies are expressed as the number of hybrid cell colonies per 10^6 WBC.

possible that the cells used in the suspension fusions are more sensitive to the toxicity of PEG 1000, and that this increased sensitivity obscures the expected rise in hybridization-inducing activity as the concentration of PEG 1000 is raised.

The protocol presented in Section IV,B for the fusion of Wg3h cells and WBC also has been applied successfully to the hybridization of Wg3h cells with mouse peritoneal macrophages. In the Wg3h-mouse macrophage cross, the yield of hybrids was 2-fold higher than generally seen in Wg3h-WBC crosses.

D. Summary

The experience to date with PEG-induced hybridization of cells in suspension is much less extensive than with cells in monolayer, but a simple procedure giving consistent success with certain cell combinations nonetheless has been developed. Although only a few variables have been examined for hybridization in suspension, it already seems apparent that the principles established with monolayer fusions cannot be extrapolated to suspension fusions without some modification.

REFERENCES

Croce, C., Sawicki, W., Kritchevsky, D., and Koprowski, H. (1971). *Exp. Cell Res.* **67**, 427–435.
Davidson, R. L. (1969). *Exp. Cell Res.* **55**, 424–426.
Davidson, R. L. (1974). *Annu. Rev. Genet.* **8**, 195–218.

Davidson, R. L., and Ephrussi, B. (1965). *Nature (London)* **205**, 1170–1171.
Davidson, R. L., and Gerald, P. S. (1976). *Somat. Cell Gen.* 2, 165–176.
Davidson, R. L., O'Malley, K. A., and Wheeler, T. B. (1976). *Somat. Cell Gen.* **2**, 271–280.
Ephrussi, B. (1973). *"Hybridization of Somatic Cells"* Princeton Univ. Press, Princeton, New Jersey.
Harris, H., Watkins, J. F., Ford, C. E., and Schoefl, G. I. (1966). *J. Cell Sci.* **1**, 1–30.
Kao, K. N., and Michayluk, M. R. (1974). *Planta* **115**, 355–367.
Littlefield, J. (1964). *Science* **145**, 709–710.
Pontecorvo, G. (1975). *Somat. Cell Gen.* **1**, 397–400.
Ruddle, F. H., and Creagan, R. P. (1975). *Annu. Rev. Genet.* **9**, 407–486.

Chapter 19

Preparation of Microcells

T. EGE AND N. R. RINGERTZ

Institute for Medical Cell Research and Genetics,
Medical Nobel Institute, Karolinska Institute,
Stockholm, Sweden

H. HAMBERG

Department of Pathology,
Sabbatsberg Hospital,
Stockholm, Sweden

AND

E. SIDEBOTTOM

Sir William Dunn School of Pathology,
Oxford University,
Oxford, England

I. Introduction

It is a well known and widely exploited characteristic of interspecies somatic cell hybrids that they lose chromosomes (for reviews, see Ruddle and Creagan, 1975; Ringertz and Savage, 1976). The loss is progressive and apparently more or less random as far as specific chromosomes are concerned *except* that the lost chromosomes may come almost entirely from one parental cell type. This preferential elimination, particularly of human chromosomes in human + mouse and human + Chinese hamster hybrids has made it possible to assign genes to almost all the human chromosomes. However, in several hybrid cell systems, in particular with cells from the same or closely related species, chromosome loss may be very slow and unpredictable. To overcome this problem methods have been devised to obtain rapidly cell hybrids containing only a single or a small number of chromosomes from one of the parental cells. With animal cells, the following systems have been investigated:

1. The transfer of isolated chromosomes into cells of a different species. This has been achieved by several groups of workers (Burkholder and Mukherjee, 1970; McBride and Ozer, 1973; Shani *et al.*, 1974; Sekiguchi *et al.*, 1973, 1975; Willecke and Ruddle, 1975; Burch and McBride, 1975; Wullems *et al.*, 1975).

2. Fusion of heavily irradiated cells with intact cells of a different species. The irradiated cells suffer chromosome breaks and, although most chromosomes of the irradiated genome are eliminated (Pontecorvo, 1971, some fragments may be incorporated and replicated in hybrid cells (Goss and Harris, 1975).

3. Fusion of mitotic with interphase cells. The premature chromosome condensation (Johnson and Rao, 1970) induced in the interphase nucleus may lead to chromosome breaks and the incorporation of chromosome fragments into the daughter nuclei, but most of the genetic material from the interphase nucleus is lost (Schwartz *et al.*, 1971; Matsui *et al.*, 1972).

4. Transformation with isolated DNA (Fox *et al.*, 1969) and transduction of a bacterial gene with eukaryotic cells (Merril *et al.*, 1971) have also been claimed, but these observations await further confirmation.

Another method of transferring small amounts of genetic material from one cell into another is to fuse subdiploid cell fragments from the donor cell with the receptor cell (Ege and Ringertz, 1974). The purpose of this article is to outline the techniques recently introduced for producing such subcellular fragments which we have called *microcells*. These fragments consist of a subdiploid micronucleus surrounded by a rim of cytoplasm and a plasma membrane (Ege and Ringertz, 1974). We will describe some of the properties of microcells and discuss the potential value of experiments in which they are fused with normal or mutant cells.

Our method for preparation of microcells depends on two distinct processes (Fig. 1). The first involves the production of populations of micronucleate cells and the second involves the application of enucleation techniques to these micronucleate cells. An alternative method (Johnson *et al.*, 1975) is to subject cells synchronized in mitosis to a cold shock. This induces abnormal cleavage when the cells are returned to normal temper-

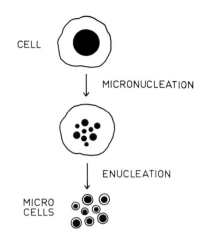

FIG. 1.   Schematic illustration of the generation of microcells. First the cells are induced to undergo micronucleation. The micronucleate cells are then enucleated by cytochalasin treatment and centrifugation.

atures. Microcells also form spontaneously in small numbers in cell cultures treated with colchicine (Cremer *et al.*, 1976; T. Ege and N. R. Ringertz, unpublished observations).

II.   Micronucleate Cells

A.   Formation

Observations of micronucleate cells have been recorded sporadically for many years, especially in relation to studies of abnormal mitosis. Although there have been numerous studies of mitosis and agents that interfere with it, relatively little attention has been focused on the micronuclei that result from such interference and the mechanisms by which these are produced.

In normal mitosis the daughter chromosomes that have separated during anaphase become surrounded by nuclear membranes, either directly as a group or first as single chromosomes with the subsequent fusion of the chromosome-containing vesicles. This is followed by decondensation of the chromosomes to form interphase nuclei. Interference with the separation of chromosomes into an anaphase configuration or with the fusion of the chromosome-containing vesicles can cause the formation of cells containing many micronuclei, each of which contains a different part of the genome. In the most extensively micronucleated cells single chromosomes form separate micronuclei (Levan, 1954; Phillips and Phillips, 1969).

The number of agents that have been found to interfere with mitosis is very large indeed. Such agents, although of widely different natures can be broadly divided into (a) poisons and physical factors that interfere with the assembly of microtubules and the mitotic spindle and (b) chromosome-breaking agents. The best known examples of agents that induce micronucleation are colchicine and X-rays. In our attempts to generate subdiploid cells for fusion experiments, it was clearly neither possible nor necessary to examine the effects of a large number of different agents, and so we concentrated on comparing the effects of colchicine and its deacetyl derivative, Colcemid, with vinblastine in the cell culture systems that were of potential interest for microcell experiments. The effects of these drugs on a series of cell lines of mouse, hamster, rat, and human origin were studied by adding the drug to cell cultures and then observing the cells by phase-contrast microscopy. In some cases time-lapse cinematography was used. In order to study the mechanism of micronucleation, cells were also grown on glass slides, which were fixed and stained by Giemsa or Feulgen at different times after adding the drug.

Addition of colchicine to cells growing *in vitro* arrests dividing cells in metaphase. Stubblefield (1964), in a study of Chinese hamster diploid cells, found that Colcemid blocked cells in metaphase for approximately 12 hours. During this time some cells died, but many were able to progress into interphase as either mononucleate or multinucleate cells. Some cells underwent cytoplasmic division into two cells, but many showed abnormal mitosis and failed to divide. Our own time-lapse studies were performed with a heteroploid SV40 virus-transformed Chinese hamster cell line (C15s) growing in Colcemid (1 μg/ml). These studies showed that cells arrested in metaphase remained rounded and immobile for 8–20 hours. They then formed pseudopodia, assumed an extended form, and underwent vigorous movements. As a result of these movements, pieces of cytoplasm were sometimes pinched off the main cell body. The phase of abnormal movements lasted for 3–10 hours and ended with the cells flattening out onto the substratum. At this stage the cells were usually seen to contain a large number of small micronuclei. Some cells never progressed from the rounded phase into the phase of vigorous movements; others failed to flatten out and died.

The different stages of micronucleation could also be observed in fixed and Giemsa-stained slides. Figure 2 illustrates a mononucleate cell, and Figs. 3 and 5 the micronucleate cells of a polyoma virus-transformed mouse line (Py6). An intermediate stage believed to represent the period of vigorous twitching is shown in Fig. 4. This cell clearly contains several nuclei and has

---

FIG. 2.   Mononucleate Py6 cells. × 675.

FIG. 3.   Micronucleate Py6 cells. × 675.

FIG. 4.   Extended micronucleate cell with pseudopodia of a type observed in time-lapse cinematography to be characteristic of the transition from a rounded, metaphase arrested stage to a flat micronucleate stage. × 675.

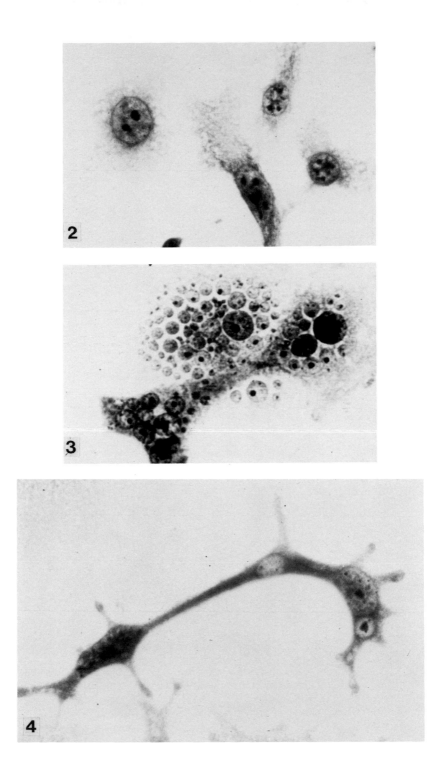

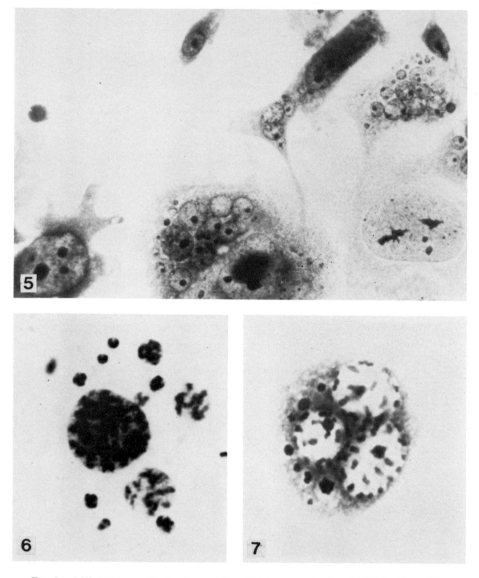

FIG. 5. Different types of cells observed in colchicine-treated cultures of Py6 cells. × 675.
FIG. 6. Micronucleate Cl 5s hamster cell where individual chromosomes can still be observed. × 1350.
FIG. 7. Micronucleate Py6 mouse cell with chromosomelike structures. × 1200.

the extended form with the pseudopodia that are also observed in time-lapse movies. The stained preparations provide some information about the mechanism by which micronuclei form. Figures 6 and 7 show Cl 5s and Py6 cells containing several nuclei in which the chromosomes still seem to be partially condensed. Some micronuclei appear to form from single chromosomes, and others arise from groups of 10–20 chromosomes. These observations confirm those of Levan (1954), who in a study of mitosis in colchicine-treated Landschütz and Ehrlich tumors concluded that "in some cases each single metaphase chromosome forms a telophase nucleus of its own, in other cases two or more chromosomes group together in forming their telophase nucleus." One important observation in these studies is that in most cells the metaphase chromosomes enter telophase directly without separating into two daughter chromatids. Many micronuclei can therefore be expected to contain two identical chromosomes rather than a single chromosome.

B. Properties of Micronucleate Cells

Micronuleate cells obtained by blocking mitosis with microtubular poisons appear to be viable for several weeks. The majority of the micronuclei have a dispersed interphase type of chromatin. In cells that have undergone extensive micronucleation, however, one often finds small pycnotic nuclei or clumps of DNA surrounding a central group of micronuclei with a normal morphology. Like Phillips and Phillips (1969), we found that many micronuclei form nucleoli and incorporate [3H]-uridine into acid-insoluble material suggesting that the genetic information in the micronuclei is actively transcribed into RNA. Many micronuclei also show incorporation of [3H]-thymidine. Further microspectrophotometric measurements on Feulgen-stained micronucleate and untreated normal cells demonstrate clearly that several rounds of DNA replication can occur in the presence of colchicine (Fig. 8). DNA levels 4–6 times higher than the normal G_1 level are found both in micronucleate cells and a few remaining mononucleate giant cells (cf. Fig. 5). The DNA content of individual micronuclei in micronucleate cells could not be measured since the nuclei were aggregated in the center of the cells and sometimes also stacked on top of each other. To overcome this problem, the DNA content of individual micronuclei was measured on micronuclei isolated by a detergent method (Goto and Ringertz, 1974) from parallel cultures. As illustrated in Fig. 8 there is a very wide variation in the DNA content of the isolated nuclei. Some nuclei have a DNA content which is more than 4 times greater than the normal G_1 content while a large group representing approximately 35 % of the total nuclear population contains less than the normal G_1 DNA content. The smallest nuclei in this group have a DNA content approaching that of individual Chinese hamster chromosomes.

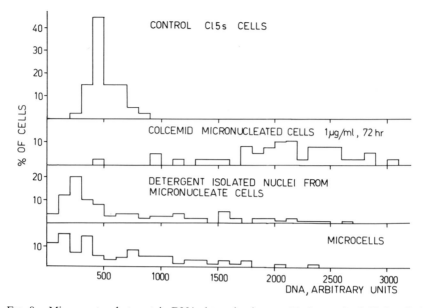

FIG. 8. Microspectrophotometric DNA determinations on Feulgen-stained Cl 5s cells in untreated control cultures and in cultures treated with 1 μg of Colcemid per milliliter for 72 hours. Endoreduplication in the mononucleate cells ensures that these as well as the micronucleate cells in the treated culture have a greater than normal DNA content. The DNA content of individual micronuclei and nuclei was measured on Feulgen-stained preparations of nuclei isolated by a detergent method from a parallel, Colcemid-treated culture. The variation in DNA content in microcells isolated by the cytochalasin and centrifugation method is similar to that found in the detergent isolated nuclei.

An interesting question is whether or not micronucleate cells are able to divide. Some micronuclei show a prophaselike chromatin morphology, and time-lapse cinematography experiments show that micronucleate cells can round up and enter what appears to be mitosis. Similar observations have previously been made by Stubblefield (1964) in Chinese hamster cells treated with Colcemid. There is no evidence, however, to indicate how efficiently micronucleate cells divide to generate viable daughter cells. The observation that the percentage of micronucleate cells and the relative number of micronuclei per cell increases between 48 and 72 hours (see below) suggests that some of the most extensively micronucleated cells arise after more than one "mitosis" in the presence of Colcemid.

C.   Frequency of Micronucleate Cells

All the cell lines we have examined (mouse, rat, hamster, and human) show some degree of micronucleation after drug treatment. Several mouse cell lines, such as Py6 and A9, show extensive micronucleation, whereas others, like 3T3 and 3T6, are much less susceptible. Table I shows an example of the dose dependence of micronucleation in three different cell lines. Colcemid, vinblastine, and colchicine give approximately the same degree of micronucleation in the same cell line, and increasing the drug concentration above a threshold level does not increase the number of nuclei per cell. In Fig. 9 the time course for the induction of micronucleation together with the number of attached cells at the different time points is given for Cl5s Chinese hamster cells. The percentage of micronucleate cells and the mean number of nuclei per cell increases with time but at the same time the number of cells on the slides decreases. The latter effect is due to the fact that many cells which are arrested in metaphase die instead of reentering interphase. The number of micronuclei in micronucleate cells varies and the proportion of cells showing micronucleation varies with the cell line (Fig. 10). Thus for a given dose of drug and time of treatment some cell lines produce a larger proportion of extensively micronucleated cells than do other lines.

III.   Preparation of Microcells from Micronucleate Cells

A.   Spontaneous Formation of Microcells

The literature contains several reports of subdiploid cells with a reduced chromosome number (Levan, 1954; Sinha, 1967) or a DNA content below the normal diploid level, but because they are rare these cells are extremely difficult to study and nothing is known about their properties. The frequency of "spontaneously" formed microcells is increased in cultures exposed to microtubular poisons (Cremer *et al.*, 1976). As described above, these

TABLE I

MEAN NUMBER OF NUCLEI PER CELL IN THREE DIFFERENT CELL LINES TREATED WITH COLCEMID, VINBLASTINE, AND COLCHICINE AT DIFFERENT CONCENTRATIONS FOR 72 HOURS

	Colcemid dose (μg/ml)			Vinblastine dose (μg/ml)			Colchicine dose (μg/ml)	
Cells	0.3	1.0	3.0	0.3	1.0	3.0	1.0	3.0
SV40-transformed Chinese hamster cells Cl 5s	3.0	3.7	3.7	—	2.2	2.9	4.5	3.5
Polyoma-transformed rat cells TCA1	8.0	6.6	7.5	6.5	7.4	8.9	7.3	7.8
Polyoma-transformed mouse cells Py6	8.5	12.4	11.4	12.2	10.9	—	9.2	11.7

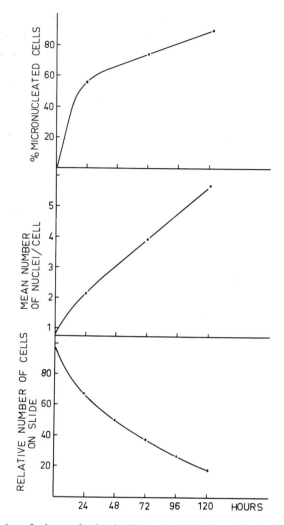

FIG. 9.   Induction of micronucleation in Cl 5s cells (colchicine at 1 μg/ml). The proportion of micronucleate cells increases with time. The number of cells remaining on the slides decreases because many metaphase-arrested cells die or detach.

substances induce the formation of micronuclei at the same time as the cells that are escaping from mitosis engage in the abnormal cell movements which sometimes cause them to pinch off pieces of cytoplasm (Fig. 11). Some of these cell fragments contain micronuclei as well as a small piece of cytoplasm (Fig. 12 c–f) and therefore qualify as microcells. However, even if it were possible to separate out these spontaneously formed microcells from other

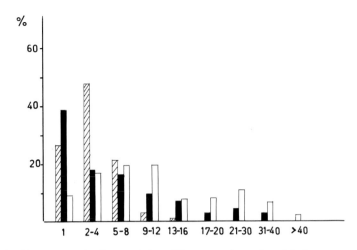

Fig. 10. Proportion of cells containing different numbers of nuclei after treatment for 72 hours with 1 μg of Colcemid per milliliter. Cl 5s (▨) is a transformed Chinese hamster line, TCA1 (■) is a polyoma virus-transformed rat cell line, and Py6 (□) is a polyoma virus-transformed mouse cell line. The Py6 line shows the largest proportion of micronucleate cells and the greatest degree of micronucleation.

cells by exploiting their small size, the quantities obtained would probably be insufficient for cell fusion purposes.

B. Cytochalasin and Centrifugation

In 1972 Prescott *et al.* and Wright and Hayflick reported that large numbers of anucleate cells could be prepared by centrifuging monolayers of cells attached to glass or plastic disks in the presence of cytochalasin B. The same method also yields nuclei that are surrounded by a thin rim of cytoplasm and a cell membrane (Ege *et al.*, 1973, 1974; Prescott and Kirkpatrick, 1973). If the cells have been induced to undergo micronucleation prior to enucleation, a large number of microcells are obtained (Ege and Ringertz, 1974). Later experiments have shown that microcells can be prepared both from monolayers attached to plastic disks and from cells in suspension. The techniques used are as follows:

1. From Plastic Disks

Cells are seeded on 25 mm diameter plastic disks which have been punched out of the bottom of tissue culture dishes. After 12–24 hours of growth, the appropriate mitotic inhibitor is added at a final concentration of 1–3 μg/ml.

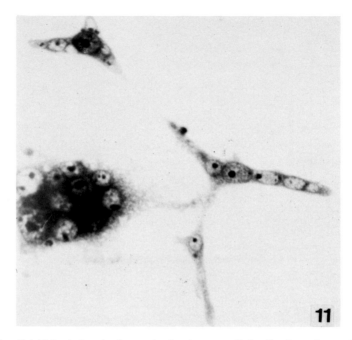

FIG. 11.   Colchicine-induced micronucleation in mouse Py6 cells. One micronucleate cell appears to be splitting up into three microcells. × 765.

After a further 72 hours incubation to allow micronucleation to occur, the disks are removed from the medium and placed with the cells facing downward in centrifuge tubes, each containing about 4 ml of PBS. A plastic plug is placed on top of the plastic disk to stabilize its position in the tube. The disks are then centrifuged at 10,000 rpm (12,000 *g*) for 10 minutes in a Sorvall RC 2B centrifuge prewarmed to 37°C. This centrifugation is a preliminary step to remove loosely attached cells that would otherwise contaminate the microcell fraction After the first centrifugation, the disks are transferred to clean centrifuge tubes containing PBS with 10 % calf serum and 10 μg cytochalasin B per milliliter. The disks are centrifuged at 14,000 rpm (23,500 *g*) for 20 minutes at 37°C.

After the second centrifugation, the pellet material at the bottom of the centrifuge tubes is gently resuspended, pooled, and centrifuged at 500 *g* for 5 minutes. In fusion experiments the supernatant is removed and the pellet is washed and resuspended in Eagle's MEM without serum.

2.   FROM SUSPENSION

Wigler and Weinstein (1975) use a discontinuous Ficoll gradient containing cytochalasin B to enucleate and separate nucleated from anucleate cell

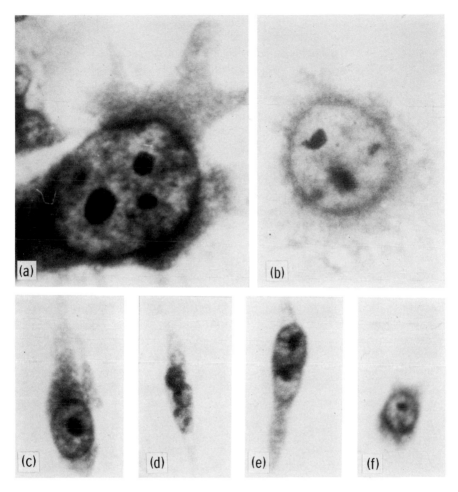

FIG. 12.   (a) Enlarged Py6 mononucleate cell in a colchicine-treated culture. (b) Normal Py6 cell in a control culture. (c–f) Different types of microcells formed by "pinching off" from colchicine-treated Py6 cells. The microcells usually contain only one micronucleus, but some contain 2 (e) or more (d) micronuclei.

fragments. This method has the advantage that more cells can be processed at one time than by the disk method. Up to 6×10^7 cells can be successfully enucleated on one 10-ml gradient. The gradient also has the very important advantage that it is applicable to cells that do not attach firmly to glass or plastic surfaces.

Preliminary experiments have shown that the gradient described by Wigler and Weinstein can be successfully used for enucleation of micronucleated cells, thereby increasing the number of microcells obtained.

C.   Cold Shocks of Mitotic Cells

A different method of producing microcells was recently reported by Johnson *et al.* (1975) and Schor *et al.* (1975). Synchronized cells were arrested in mitosis by exposure to N_2O at 5 atmospheres and then transferred to $+ 5°C$ for 9 hours. When these cells were then transferred to 37°C aberrant chromosome segregation and cytokinesis started within 2 hours. The dividing cells were observed to form structures described as "bunches of grapes." The "grapes" were found to consist of cytoplasmic fragments many of which contained one or several micronuclei. Individual microcells were spontaneously detached from the aggregates or could be freed by shearing. The mechanism underlying this form of abnormal cell division is not known, but since cold treatment is known to interfere with the assembly of microtubules it is possible that the mechanism causing the formation of micronuclei and free microcells after cold shocks is similar to that described above for colchicine-treated cultures.

D.   Fractionation of Microcell Populations

Since it is difficult to induce 100% of the cells in a population to micronucleate extensively, the microcell population will be very heterogeneous, containing both very small and very large nuclei. We are presently attempting to separate the small micronuclei from the large by filtration through Nucleopore filters with pore diameters from 1–5 μm. Figure 13 shows the DNA content of the microcell population before and after filtration through a 5-μm Nucleopore filter. It is evident from the histograms that the larger microcells can be selectively removed in this way.

Another method is to separate the cells on a Ficoll gradient at unit gravity (Scor *et al.*, 1975). The different fractions will then be enriched for microcells of a particular size range and DNA content.

IV.   Properties of Microcells

In a microcell preparation, prepared as above (Fig. 14) usually about 80–90% of the small subcellular fragments exclude trypan blue, suggesting that they have an intact plasma membrane. Direct evidence for the presence of a plasma membrane has been given by Johnson *et al.* (1975), showing by electron microscopy that their microcells consist of a small nucleus surrounded by a layer of cytoplasm and a cell membrane. Feulgen-stained preparations show that some of the fragments contain a DNA-containing micronucleus while others seem to be anucleate cytoplasmic fragments. The fluorescence pattern of acridine orange-stained preparations demonstrates

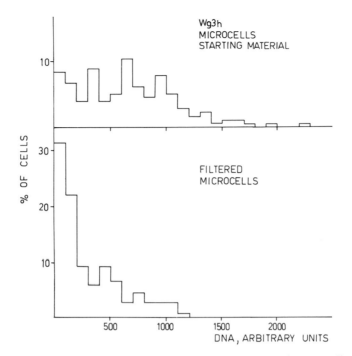

Fig. 13.   DNA content of microcells prepared from Wg3H Chinese hamster cells and of a microcell fraction after passage through a 5-μm Nucleopore filter.

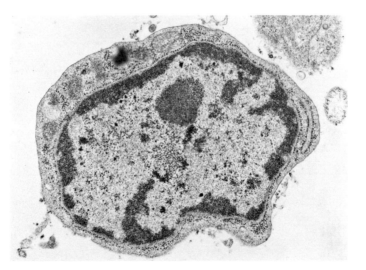

FIG. 14.   Electron micrograph showing the general characteristics of a microcell from Py6. Serial sections revealed a greatest nuclear diameter of approximately 4 μm in this cell. Control cells displayed a fourfold greater nuclear diameter. × 16,500.

that the micronuclei contain a dispersed interphase type of chromatin and that the micronucleus is surrounded by a rim of red fluorescing cytoplasm. Some of the micronuclei also contain small nucleoli. Quantitative DNA and DNA protein determinations on microcells from the L6 myoblast cell line (Fig. 15) and Cl 5s cells (Fig. 8) show that the microcell preparations contain cells with a wide range of DNA values from those similar to the intact, untreated cells down to less than 1/20 of normal G_1 values, which is equivalent to the DNA content of only 1 or 2 chromosomes.

In Fig. 16 the Feulgen values are plotted against the total protein content of intact L6 cells and cells obtained after enucleation of control and micronucleate L6 cells. It is clear from the figure that most of the microcells have a subdiploid DNA content, and also that the ratio of protein to DNA is similar

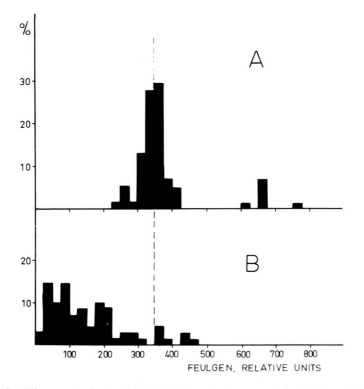

FIG. 15. Microspectrophotometric DNA determinations on minicells (A) and microcells (B) prepared by cytochalasin treatment and centrifugation of normal and micronucleate rat L6 myoblasts. The minicells vary in DNA content from the diploid G_1 value for rat cells (dashed vertical line) to twice this value. The microcells contain subdiploid amounts of DNA.

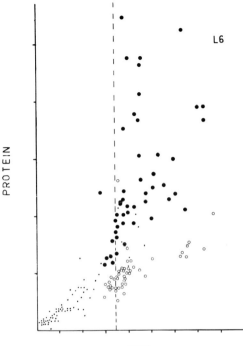

PROTEIN

L6

DNA

FIG. 16. Microspectrophotometric DNA and protein determinations on individual rat L6 cells (●), minicells (○) and microcells (·). The minicell preparation is contaminated with at least one intact cell as judged by its protein content while the microcell preparation contains approximately 10% cells, which may be minicells (from mononucleate cells remaining in the micronucleate cell preparation) or intact cells.

in both diploid and subdiploid cells obtained after enucleation but is far below that of the intact cells.

The properties and survival of microcells have not yet been carefully investigated, but some attach to surfaces and appear to be viable for periods of at least 24 hours. Schor *et al.* (1975) have shown that immediately after preparation some of the microcells are capable of incorporating radioactive precursors into DNA, RNA, and protein. Further indications of viability are given by the fact that microcells can fuse with intact cells when treated with Sendai virus and the incorporated micronuclei are able to synthesize RNA. Microcell heterokaryons can be unequivocally identified if the microcells are prepared from cells which have previously been labeled with [³H]-thymidine (Ege and Ringertz, 1974).

V.   Use of Microcells in Somatic Cell Genetics

Although clones of presumptive hybrid cells have been isolated in the HAT selective system after fusion of HGPRT$^-$, TK$^+$ cells with microcells prepared from HGPRT$^+$, TK$^-$ cells, the hybrid nature of the cells has not yet been rigorously proved. Evidence has recently been presented (Schor *et al.*, 1975) that microcells when fused with mitotic cells will undergo premature chromosome condensation (PCC). The prematurely condensed chromosomes originating from the microcell thereby become visible and can be counted. As expected from the DNA determinations, the small microcells give rise to only a few chromosomes, while the larger ones form a greater number of prematurely condensed chromosomes.

The most obvious potential application for microcell hybrids is in the field of chromosome mapping. The microcell hybrids should provide a direct source of heterogeneous clones of hybrid cells with a reduced chromosome complement from the microcell donor without the inevitable and uncertain delays at present encountered in some systems. Since the mapping of regulatory loci as well as structural loci is now becoming a practical possibility, a wider interest in the field is likely and any method of gene transfer and/or gene segregation will be worthy of evaluation.

Mapping investigations are conducted on mononucleate hybrid cells that have undergone many divisions after the original fusion, but it should not be forgotten that the heterokaryotic stage immediately after cell fusion, but before the first cell division occurs, offers opportunities to study mechanisms of gene expression. Valuable results have already been obtained with heterokaryons constructed from two types of intact cells. Microcell heterokaryons should provide the opportunity of looking at gene expression in cells with incomplete, fragmented genomes. It might be possible to separate structural and regulatory loci with such cells.

It should also be noted that the induction of micronucleation is a process that is not at present understood. An investigation of this process might well shed interesting light on the mechanisms of normal mitosis.

REFERENCES

Burch, J. W., and McBride, O. W. (1975). *Proc. Natl. Acad. Sci. U.S.A.* **72**, 1797.
Burkholder, G. D., and Mukherjee, B. B. (1970). *Exp. Cell Res.* **61**, 13.
Cremer, T., Zorn, C., Cremer, C., and Zimmer, J. (1976). *Exp. Cell Res.* (in press).
Ege, T., and Ringertz, N. R. (1974). *Exp. Cell Res.* **87**, 378.
Ege, T., Zeuthen, J., and Ringertz, N. R., (1973). *In* "Chromosome Identification" (T. Caspersson and L. Zech, eds.), pp. 189–194. Academic Press, New York.
Ege, T., Hamberg, H., Krondahl, U., Ericsson, J., and Ringertz, N. R. (1974). *Exp. Cell Res.* **87**, 365.

Fox, M., Fox, W. B., and Ayad, S. R. (1969). *Nature (London)* **222**, 1086.

Goss, H., and Harris, H. (1975). *Nature (London)* **255**, 680.

Goto, S., and Ringertz, N. R. (1974). *Exp. Cell Res.* **85**, 173.

Johnson, R. T., and Rao, P. N. (1970). *Nature (London)* **226**, 717.

Johnson, R. T., Mullinger, A. M., and Skaer, R. J. (1975). *Proc. R. Soc. London, Ser. B* **189**, 591.

Levan, A. (1954). *Hereditas* **40**, 1.

McBride, O. W., and Ozer, H. L. (1973). *Proc. Natl. Acad. Sci. U.S.A.* **70**, 1258.

Matsui, S., Weinfeld, H., and Sandberg, A. A. (1972). *J. Natl. Cancer Inst.* **49**, 1621.

Merril, C. R., Geier, M. R., and Petricciane, J. C. (1971). *Nature (London)* **233**, 398.

Phillips, S. G., and Phillips, D. M. (1969). *J. Cell Biol.* **40**, 248.

Pontecorvo, G. (1971). *Nature (London)* **230**, 367.

Prescott, D. M., and Kirkpatrick, J. B. (1974). *In* "Methods in Cell Physiology" (D. M. Prescott, ed.), Vol. 7, pp. 189–202. Academic Press, New York.

Prescott, D. M., Myerson, D., and Wallace, J. (1972). *Exp. Cell Res.* **71**, 480.

Ringertz, N. R., and Savage, R. E. (1976). "Cell Hybrids". Academic Press, New York,

Ruddle, F. H., and Creagan, R. P. (1975). *Ann. Rev. Genet.*

Schor, S. L., Johnson, R. T., and Mullinger, A. M. (1975). *J. Cell Sci.* **19**, 281.

Schwartz, A. G., Cook, P. R., and Harris, H. (1971). *Nature (London) New Biol.* **230**, 5.

Sekiguchi, T., Sekiguchi, F., and Yamada, M. (1973). *Exp. Cell Res.* **80**, 223.

Sekiguchi, T., Sekiguchi, F., Tachibana, T., Yamada, T., and Yoshida, M. (1975). *Exp. Cell Res.* **94**, 327.

Shani, M., Huberman, E., Aloni, Y., and Sachs, L. (1974). *Virology* **61**, 303.

Sinha, A. K. (1967). *Exp. Cell Res.* **47**, 443.

Stubblefield, L. (1964). *Symp. Int. Soc. Cell Biol.* **3**, 223.

Wigler, M. H., and Weinstein, I. B. (1975). *Biochem. Biophys. Res. Commun.* **63**, 669.

Willecke, K., and Ruddle, F. H. (1975). *Proc. Natl. Acad. Sci. U.S.A.* **72**, 1792.

Wright, W. E., and Hayflick, L. (1972). *Exp. Cell Res.* **74**, 187.

Wullems, G. J., van der Horst, J., and Bootsma, D. (1975). *Somat. Cell Genet.* **1**, 137.

Chapter 20

Nuclear Transplantation with Mammalian Cells

JOSEPH J. LUCAS AND JOSEPH R. KATES

Department of Microbiology,
State University of New York at Stony Brook,
Stony Brook, New York

I. Introduction

As early as 1967, Carter suggested that the techniques of cytochalasin-induced enucleation would lead to fruitful investigations of the "functional relationship of nucleus and cytoplasm." During the past several years, numerous attempts (Ladda and Estensen, 1970; Poste and Reeve, 1971; Ege *et al.*, 1974; Ringertz, 1974; Sethi and Brandis, 1974; Veomett *et al.*, 1974) were made to fulfill this prediction by constructing cells using Sendai virus-mediated fusion of cytoplasts with either whole cells or karyoplasts.

A particularly significant investigation was reported by Veomett and his colleagues (1974). That paper offered definitive evidence for the reconstruction of mammalian cells by fusion of cytoplasts and karyoplasts formed from mouse L929 cells. Moreover, the observation that some putative hybrids were in mitotic configurations suggested that viable cells were formed.

In developing a generalized approach for nuclear transplantation in cultured cells, we have confirmed and extended these previous results. The techniques required to construct large, viable, homogeneous populations of true cytoplasmic–nuclear hybrids will be described in this review. Other aspects of this work are presented in greater detail elsewhere (Lucas *et al.*, 1976; Lucas and Kates, 1976).

These techniques are now being applied to the construction of novel systems for investigating gene expression and its control in mammalian cells. For this purpose, cytoplasts and karyoplasts prepared from various differentiated and transformed cell lines which exhibit distinct morphological and/or biochemical traits are fused to form true cytoplasmic–nuclear hybrid cells. The hybrids and their progeny are then analyzed for the maintenance, loss, or acquisition of the parental cell traits. Construction of the appropriate hybrid lines will permit direct investigation of a wide variety of problems. Areas of interest now being considered are cytoplasmic control of neuronal differentiation, immunoglobulin synthesis and hormone-dependent growth; the requirement of an integrated viral genome for maintenance of expression of various traits of transformed cells; and the role of the cytoplasm in carcinogen-induced malignancy.

II.   Preparation and Characterization of Karyoplasts

A.   Cell Culture

The enucleation technique described by Prescott and his colleagues (1972) was modified for the production of larger quantities of karyoplasts. Thus, instead of cover slips, plastic sheets cut from the bottom sides of Falcon Tissue Culture Flasks (Cat. No. 3024) were used. The sheets were cut such that two could fit snugly while standing back to back and on edge inside Sorvall 50-ml plastic centrifuge tubes. After each experiment, the sheets were thoroughly washed and dried. They were then soaked in concentrated sulfuric acid for 5 minutes, rinsed extensively with water, and sterilized by washing with 70 % (v/v) ethanol and sterile water, followed by treatment with ultraviolet light.

For making monolayer cultures for enucleation, two sheets were immersed in 15 ml of medium in a 100 mm tissue culture dish. A suspension of cells was placed into the dish, and it was incubated overnight at 37°C in an atmosphere containing 5 % CO_2. For obtaining synchronized cultures, monolayers were incubated for an additional 15–20 hours in medium containing thymidine at a concentration of 5 mM (Xeros, 1962).

Unless stated otherwise, the techniques described here were performed with a clonal derivative, designated D51, of mouse L929 cells. These cells were grown and maintained in Eagle's minimum essential medium (MEM) supplemented with 5% fetal calf serum (FCS). Also used in these studies were the A9 line of strain L929 mouse cells (Littlefield, 1964) and a clonal derivative of mouse neuroblastoma C1300 (Augusti-Tocco and Sato, 1969). Both lines were maintained in monolayer in Dulbecco's modified Eagle's medium (DME) with 10% calf serum (CS).

B. Enucleation of Cells

Sheets containing cell monolayers were placed into centrifuge tubes containing medium with serum and centrifuged (12,000 g) for 10 minutes at 37°C in a Sorvall HB4 rotor. During this procedure, any loosely bound cells were removed from the sheets. The sheets were then transferred to tubes containing fresh medium with 10 μg/ml of cytochalasin B. After a 15-minute incubation at 37°C, the tubes were centrifuged (12,000 g) for 45 minutes at 37°C. Karyoplasts, contained in pellets at the bottoms of the tubes, were suspended in fresh medium with serum and incubated in tissue culture dishes at 37°C in an atmosphere containing 5% CO_2.

C. Purification of Karyoplasts

1. REMOVAL OF WHOLE CELLS

Phase-contrast microscopic observation of the material comprising the pellet obtained after centrifugation of a cell sheet in medium containing cytochalasin B showed that the karyoplasts were contaminated with a large quantity of cytoplasmic fragments and a much smaller amount of what appeared to be whole cells. Techniques were therefore developed for the separation of karyoplasts from these contaminants.

When the pellet material was placed in a tissue culture dish, it was noted that very few of the nucleus-containing bodies could initially adhere to the surface. However, whole cells exposed to the same centrifugation procedure (but without cytochalasin) readily stuck to the plastic vessels. Within 90 minutes, essentially 100% of these "mock-enucleated" cells adhered, whereas only about 4% of the nucleus-containing bodies in the pellet did so. Thus, a short period of incubation will suffice to "absorb out" whole-cell contaminants.

Other control experiments, performed with cells that were first removed from a plastic sheet using trypsin and then pelleted through medium containing cytochalasin, clearly indicate the efficacy of the purification technique. After plating in a tissue culture flask for 2 hours, the supernatant was removed and placed in a second dish. Fewer than 0.01% of the original mock-enucleated cells were found adhered to the second vessel; and, after this second cycle of plating, no detectable viable cells remained in suspension.

2.   REMOVAL OF CYTOPLASMIC FRAGMENTS

After several cycles of plating, the supernatant was centrifuged and the pellet was suspended in 0.5 to 1.0 ml of MEM for suspension culture (MEMS). The sample was then layered over a 26-ml gradient containing 1 to 6% (w/v) of Ficoll in MEMS. After incubation at room temperature for 90 minutes, by which time two bands were observed in the tube, 1-ml fractions were collected. The entire operation was performed aseptically in a vertical flow hood. Equipment was sterilized by washing with 70% (v/v) ethanol and sterile water, followed by treatment with ultraviolet light.

Microscopic observation of the material in fractions obtained from the gradient showed that the upper band was composed almost entirely of enucleated fragments whereas the lower band was much enriched in karyoplasts. It was calculated that 85% of the cytoplasmic fragments were in the top three fractions of the gradient, nearly 100% of the karyoplasts being in the fractions comprising the lower visible band. Thus, by collecting fractions at the peak of the second band, the method permits the elimination of most of the material originally contaminating the karyoplast preparation. Moreover, as indicated by the ability to exclude the dye trypan blue, the viability of the karyoplasts is unaltered by their exposure to the gradient conditions.

3.   REMOVAL OF "NONVIABLE" KARYOPLASTS

Nearly 100% of freshly prepared karyoplasts retain the ability to exclude the dye trypan blue. However, upon further incubation, this ability is rapidly lost. By 5 hours after preparation, about 40% of karyoplasts are stained by the dye. Presumably, many karyoplasts are injured during the process of enucleation. As contamination by such bodies would interfere with many types of experimentation, a method was devised for separation of "living" and "dead" karyoplasts.

In this technique, a preparation of karyoplasts (about 10^7) already freed from whole cells and cytoplasmic fragments, was suspended in 0.9 ml of medium and layered over 4.1 ml of Miles Laboratory Path-O-Cyte 4 albumin preparation. The tube was then centrifuged (740 g) for 35 minutes at 25°C. Nearly all viable karyoplasts (i.e., those that excluded dye) remain at the interface between the sample and the serum. Approximately 70% of nonviable bodies were found in a pellet at the bottom of the tube.

D.   Characterization of Karyoplasts

Figure 1 shows an electron micrograph of a karyoplast and several cytoplasmic fragments. It is readily seen that a karyoplast is composed of a nucleus surrounded by a thin shell of cytoplasm and an outer cell membrane.

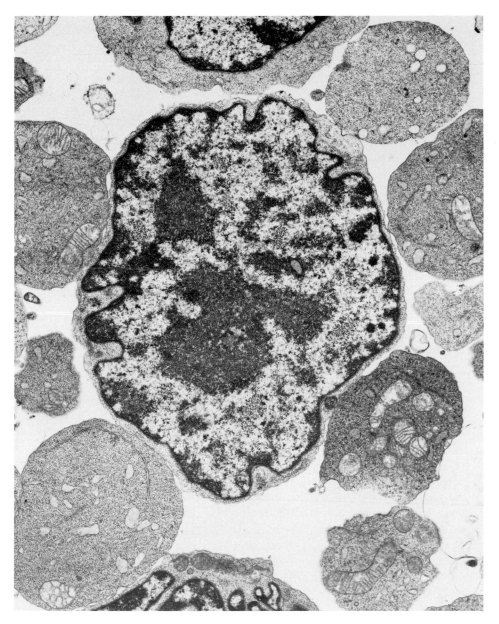

FIG. 1.   An electron micrograph of an L-cell karyoplast surrounded by many cytoplasmic fragments. × 8125. Photograph by G. A. Zorn.

Before performing nuclear transplantation, however, it was essential to determine—with each cell line used—what fraction of the cytoplasm contained within whole cells was retained by karyoplasts. Several methods were devised to determine this quantity.

In one procedure, the amounts of total stable cytoplasmic RNA and, specifically, of ribosomal and transfer RNAs, were taken as measures of cytoplasmic content. First, cells were labeled for 36 hours with [^{3}H]-uridine. Next, the labeled compound was removed and further cell division was prevented by the addition of excess thymidine. After allowing 24 hours for the movement of labeled material from the nucleus to the cytoplasm and for the degradation of unstable RNA species, the thymidine block was removed and, after a further 10-hour incubation, the cells were enucleated.

Small cytoplasmic fragments contaminating the karyoplast pellet were separated by sedimentation through a Ficoll gradient. Cytoplasts, cytoplasmic fragments, and karyoplasts were lysed using Triton X-100 and sonication and the cytoplasmic fraction of each was obtained. The total amount of acid-precipitable labeled material was determined using a portion of each sample. The remaining portions were treated with EDTA and fractionated on a sucrose gradient into ribosomal subunits and free RNA (Girard et al., 1965).

The data showed that of the total cytoplasmic RNA contained within whole cells, approximately 8% was associated with the karyoplasts. Similar figures were obtained when the RNA was fractionated into various components.

In a second method, the numbers of latex spheres ingested into the cytoplasm of cells was taken as a measure of cytoplasmic volume. To apply this technique, monolayers of cells on plastic sheets were incubated for 20 hours in tissue culture dishes with medium containing 10^9 latex spheres (diameter of 1 μm). After enucleation, the numbers of beads in cytoplasts and karyoplasts was determined. Assuming a random distribution of the spheres within the cytoplasm, determination of the number of spheres remaining in cytoplasts and karyoplasts permitted calculation of the fraction of total cytoplasm in each body. Such determinations with L-cell karyoplasts and cytoplasts prepared in the manner described here indicated that the karyoplasts were more than 90% reduced in cytoplasmic volume.

As described above, such bodies initially lack the ability to adhere to the surface of a tissue culture vessel. However, it was noted that upon prolonged incubation under appropriate conditions, a significant fraction of such bodies not only regained the ability to spread upon a surface but also could divide. A description of the conditions required to maximize the extent of karyoplast regeneration and division and to isolate cell lines from such bodies is described elsewhere (Lucas et al., 1976).

III.   Preparation and Characterization of Cytoplasts

The method of preparation of cytoplasts was essentially that described by Prescott and his colleagues (1972). Plastic disks were cut from the bottoms of Falcon tissue culture dishes using a heated No. 7 cork borer. Cell monolayers were formed on sterilized disks by immersing them in tissue culture dishes containing suspensions of cells. Disks containing cell monolayers were placed—cell side down—into centrifuge tubes containing medium with serum and 10 μg/ml of cytochalasin B. After a 15-minute incubation at 37°C, the tubes were centrifuged (11,000 g) for 45 minutes at 37°C in a Sorvall SS34 rotor. The disks, now containing monolayers of enucleated cells, were removed from the tubes, washed vigorously in medium, placed in tissue culture dishes and incubated at 37°C in an atmosphere containing 5 % carbon dioxide. After centrifugation, the cytoplasts appeared more elongated than the parent whole cells. Therefore, before use in nuclear transplantation experiments, the cytoplasts were permitted to recover for about 1 hour, during which time they regained the normal cell morphology. Using the conditions described here, less than 1 % of the culture remained nucleated. This was determined by staining the monolayer with a dye—such as Giemsa stain or crystal violet—which emphasized the appearance of nuclei.

IV.   Nuclear Transplantation

A.   Preparation of Sendai Virus

The Ender's strain of Sendai virus was prepared and assayed using modifications of standard procedures (see, for example, Watkins, 1971). Ten-day-old fertile hen's eggs were inoculated with 0.1 ml of Earle's balanced salt solution (EBS) containing 0.01–0.02 hemagglutinating units of virus. After 3 days' incubation at 37°C, the eggs were chilled and clear allantoic fluid was harvested and pooled. The fluid was heated at 37°C for 15 minutes and then centrifuged (300 g) for 10 minutes at 4°C in the Sorvall SS34 rotor. The supernatant was removed and centrifuged (20,000 g) for 30 minutes at 4°C in the SS34 rotor. The pellet was suspended in EBS and assayed for its ability to agglutinate sheep erythrocytes. Virus stocks, having titers of 2000–10,000 hemagglutinating units per milliliter, were stored frozen at −70°C. Before use in fusion experiments, virus was exposed for 5 minutes, at a distance of 15 cm, to a General Electric G8T5 ultraviolet light bulb.

B. Virus-Mediated Fusion

The desired fusion products contain one cell-equivalent of cytoplasm and one nucleus. The fusion of cytoplasts and karyoplasts in suspension resulted in a wide range of products, including bodies arising from the fusion of many cytoplasts with one or more karyoplasts. It was reasoned that the production of bodies containing more than one cell-equivalent of cytoplasm could be eliminated by fusing karyoplasts to cytoplasts which remained attached to a solid substrate. Therefore, the cell lines used as the cytoplast donors were grown and enucleated on plastic disks, whereas large numbers of karyoplasts were prepared from cultures on plastic sheets.

Each cell line used as cytoplast or karyoplast donor was first characterized as described above. Cell lines were used only after ascertaining that the preparative techniques used yielded cultures containing few or no whole-cell contaminants and that karyoplasts contained no more than 10 % of the cytoplasm found in whole cells.

It should be noted once again that the enucleation procedures described above were used for mouse L929 cells and the A9 derivative of L cells. Procedures employing other concentrations of cytochalasin and durations and speeds of centrifugation were developed to give maximum enucleation for other cell lines.

The disks carrying monolayers of cytoplasts were washed with Earle's balanced salt solution (EBS) at 4°C and then placed into the chambers of a chilled Linbro Multi-Dish Disposo-Tray (Cat. No. FB-16-24-TC). The disks were covered with 0.2 ml of cold (4°C) EBS containing about 100 hemagglutinating units of inactivated Sendai virus and incubated at 4°C for 10 minutes. Control cultures of cytoplasts were incubated in EBS without virus. Next 0.2 ml of a suspension containing about 10^6 karyoplasts was placed into each chamber. After a 20-minute incubation at 4°C, the tray was transferred to a 37°C incubator with an atmosphere containing 5 % CO_2. After 45 minutes, the disks were rinsed vigorously in medium. Since freshly prepared kayoplasts are unable to adhere to a surface, only those karyoplasts fused to the monolayer will remain associated with the disks. Other control experiments showed that Sendai virus itself did not promote the adherence of karyoplasts to a plastic surface. After washing the disks, they were placed in tissue culture dishes for further observation and experimentation.

C. Characterization of Hybrids

The first step in the characterization of the new cultures is the determination of the efficiency of fusion. After permitting 1 hour of incubation as a recovery period, a fused culture and a control (that is, a cytoplast culture exposed to karyoplasts in the absence of virus) were fixed with methanol and treated with Giemsa stain to facilitate the observation of nuclei. Results of an experiment performed with cytoplasts and karyoplasts derived from L cells — that is, the reconstruction of L cells — are shown in Fig. 2. Approximately 23 % of the cytoplasts received nuclei. Upon repeated performance of the experiment, this figure ranged from 10 to 30 %. In control cultures, less than 1 % of bodies enumerated contained nuclei.

Hybridization has been successfully performed using L-cell cytoplasts and A9 karyoplasts and neuroblastoma cytoplasts and A9 karyoplasts. The conditions used were those described above for the reconstruction of L cells. In both cases, the efficiencies of fusion were again between 10 and 30%.

Next, it was essential to determine whether or not the reconstructed or hybrid cells were viable. This was most easily done by following the increase in cell number with time. For example, knowing that the doubling time of L cells is about 24 hours, it was estimated that at least 30 % of the reconstructed cells were capable of division. As it appears that there was some lag before the reconstructed cells assumed the normal doubling time of whole cells, this was a minimum estimate. More accurate determinations can be obtained by performing time lapse cinematography of a developing culture and by cloning the new cells soon after fusion.

D.   Genetic Selection Techniques

A major asset of the transplantation technique described here is that hybrids can be constructed with a previously unobserved frequency. Thus, large viable cultures can be prepared without the use of any genetic selection techniques. Since such cultures contain no more than 1 % contaminating whole cells, they are suitable for immediate biochemical characterization.

However, for certain applications it would be most desirable to eliminate such contaminants entirely. As the major contaminant in fused cultures was whole cells from the cytoplasmic donor cell preparation, a technique was devised specifically to eliminate such cells.

To accomplish this goal, a line of cells resistant to 8-azaguanine was used as the nuclear donor (A9 cells). The cytoplasmic donor remained the normal azaguanine-sensitive line (L cells). Sendai virus-mediated fusion was performed using disc cultures of L-cell cytoplasts and A9 karyoplasts. The efficiency of fusion varied from 10 to 30%. On the second day after fusion, medium containing 8-azaguanine was placed on the cells. The fused cultures continued to grow while control cultures—cytoplasts exposed to karyoplasts in the absence of virus—showed no increase in cell number. It should be noted that treatment with azaguanine did not result in the immediate death and disintegration of sensitive cells. Thus, some whole cells persisted for several

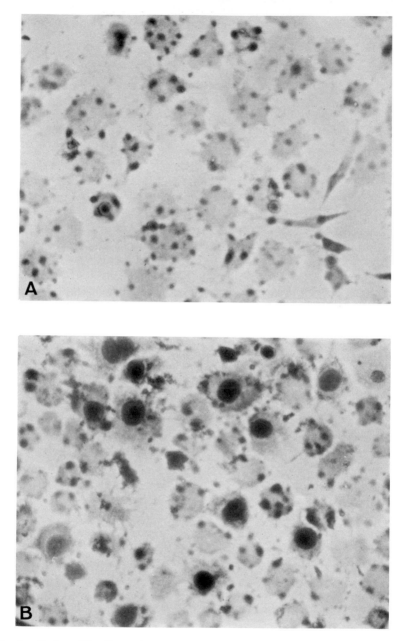

FIG. 2. Fusion of L-cell cytoplasts with L-cell karyoplasts. The photographs show cultures in which karyoplasts were layered over cytoplasts either in the absence (A) or in the presence (B) of Sendai virus. After the appropriate incubations, to permit fusion to occur, excess karyoplasts were removed by vigorous washing. One hour after completion of the experiment, the cultures were fixed with methanol and treated with Giemsa stain to facilitate observation of nuclei. × 340. Photographs are taken from Lucas and Kates (1976).

days in control samples. However, any significant growth of the cultures was prevented. Thus, it was established that the L-A9 cytoplasmic–nuclear hybrid cells were viable and that they retained the genetic marker of the nuclear donor cell type.

The original L-A9 hybrid cells presumably contained the enzyme hypoxanthine–guanine phosphoribosyltransferase (HGPRT) within their cytoplasm. Azaguanine was not added until after a delay of several days in order to permit the loss of this enzymic activity. The success of the experiment described above supported the notion that after 2 days, hybrids no longer contained sufficient enzyme to make them susceptible to killing by azaguanine. This was also shown by the following experiment.

A9 karyoplasts were again fused to a series of L cytoplast cultures. To one culture, azaguanine was added immediately after fusion, while to a second culture, the analog was added after the usual delay of 2 days. After another day of incubation, the number of whole cells in each of the two cultures was determined.

Immediately after fusion was performed, 1250 whole cells were enumerated upon a disk culture. After 3 days of incubation, the culture which received azaguanine immediately after fusion contained 900 cells within an area of the same size, and the culture which received azaguanine after a 2-day delay contained 3140 whole cells. The immediate addition of azaguanine prevented division of the hybrid cells, thus suggesting that they contained active HGPRT contributed by the L-cell cytoplasts. After 2 days of incubation, the hybrids and their progeny no longer contained sufficient amounts of enzyme to make them sensitive to azaguanine.

Such a selection procedure, and others now being developed, should be generally applicable for the selection of cytoplasmic–nuclear hybrids. Thus, before fusing cytoplasts and karyoplasts derived from cell lines of interest, mutants resistant to appropriate drugs and toxins can first be selected.

ACKNOWLEDGMENTS

This research was supported by N.I.H. Grant AI 11839–02. Joseph J. Lucas was supported by a Damon Runyon Memorial Fellowship and a Public Health Service Postdoctoral Fellowship.

We are most grateful to Maureen McLeod, Maja Nowakowski, Elizabeth Szekely, and G. A. Zorn for their valuable contributions to this project.

REFERENCES

Augusti-Tocco, G., and Sato, G. (1969). *Proc. Natl. Acad. Sci. U.S.A.* **64**, 311.
Carter, S. B. (1967). *Nature (London)* **213**, 261.
Ege, T., Krondahl, U., and Ringertz, N. R. (1974). *Exp. Cell Res.* **88**, 428.
Girard, M., Lapham, H., Penman, S., and Darnell, J. E. (1965). *J. Mol. Biol.* **11**, 187.
Ladda, R. L., and Estensen, R. D. (1970). *Proc. Natl. Acad. Sci. U.S.A.* **67**, 1528.
Littlefield, J. W. (1964). *Nature (London)* **203**, 1142.

Lucas, J. J., and Kates, J. R. (1976). *Cell* 7, 397.

Lucas, J. J., Szekely, E., and Kates, J. R. (1976). *Cell* 7, 115.

Poste, G., and Reeve, P. (1971). *Nature (London)* **229**, 123.

Prescott, D. M., Meyerson, D., and Wallace, J. (1972). *Exp. Cell Res.* **71**, 480.

Ringertz, N. R. (1974). *In* "Somatic Cell Hybridization" (R. L. Davidson and F. de la Cruz, eds.), p. 239. Raven, New York.

Sethi, K. K., and Brandis, H. (1974). *Nature (London)* **250**, 225.

Veomett, G., Prescott, D. M., Shay, J., and Porter, K. R. (1974). *Proc. Natl. Acad. Sci. U.S.A.* **71**, 1999.

Watkins, J. F. (1971). *In* "Methods in Virology" (K. Maramorosch and H. Koprowski, eds.), Vol. 5, p. 1. Academic Press, New York.

Xeros, N. (1962). *Nature (London)* **194**, 682.

Chapter 21

Procedure for Preparation and Characterization of Liver Cells Made Permeable by Treatment with Toluene

RICHARD H. HILDERMAN

Department of Biochemistry, Clemson University, Clemson, South Carolina

AND

PETER J. GOLDBLATT

Department of Pathology, University of Connecticut Health Center, Farmington, Connecticut

I. Introduction

In recent years there has been considerable interest in the isolation of individual liver cells to study a variety of cellular processes *in vivo* including protein synthesis (Schreiber and Schreiber, 1972), gluconeogenesis (Veneziale and Lohmar, 1973), lipogenesis (Clark *et al.*, 1974; Capuzzi *et al.*, 1974), and fatty acid synthesis (Craig and Porter, 1973; Goodridge, 1973). A major limitation to many of these studies is the inability to get charged or large exogenous molecules into the cells, making many experiments impossible. Therefore, investigators have developed methods to make cells and organelles permeable to exogenous substrates. Toluene treatment has been used to make bacterial cells permeable to nucleoside triphosohates so that DNA (Moses and Richardson, 1970) and RNA (Peterson *et al.*, 1971) synthesis can be studied in these cells. However, toluene-treated bacteria lose about 85% of their RNA, and protein synthesis does not occur in these bacteria (Jackson and DeMoss, 1965). Recently Atherly (1974) has cold-shocked *Escherichia coli* so that these cells are permeable to an exogenous source of ribonucleoside triphosphates. These permeable cells can synthesize RNA and proteins.

Toluene-treated *Chlamydomonas reinhardi* (Howell and Walker, 1972) incorporated deoxyribonucleoside triphosphates into DNA which is found mainly in chloroplasts. Also, *Ustilago maydis* and *Saccharomyces cerevisiae* made permeable with Brij-58 to deoxyribonucleoside triphosphates synthesize mitochondrial DNA (Banks, 1973). Using a Teflon–glass tissue homogenizer, Burgoyne (1972) made Ehrlich ascites cells permeable to deoxyribonucleoside triphosphates and studied DNA synthesis in these cells. Kumar *et al.* (1974) have treated HeLa cells with amphotericin B so that the cells would take up *E. coli* DNA.

In this report we describe the optimum conditions for toluene treatment of liver cells which render them permeable to exogenous molecules, and we examine some properties of such cells. In addition, we have demonstrated the structural integrity of toluene-treated cells by both electron microscopy and the distribution of various enzyme markers.

II.   Preparation of Liver Cells Made Permeable by Toluene

A.   Isolation of Individual Liver Cells

Hepatocytes were isolated by collagenase and hyaluronidase digestion using the procedure of Berry and Friend (1969), except that the liver was dispersed into a cell suspension by incubating the liver with the enzymes at 19°C for 3 hours. This time and temperature of incubation was chosen because these conditions gave the highest aminoacyl-tRNA synthetase activity per gram wet weight of isolated cells. The cell pellet was resuspended in 10 ml of $CaCl_2$-free Hanks' basic salt solution containing 0.05 M N-2-hydroxyethylpiperazine-N'-2-ethanesulfonic acid, pH 7.5 (buffer A) containing 0.15 M NH_4Cl, and centrifuged at 50 g for 30 sec. Hepatocytes were purified by suspension three times in buffer A and centrifugation at 50 g for 30 sec. These cells, as judged by light and electron microscopy, were 80–90 % hepatocytes. The yield of hepatocytes in different preparations varied from 3 to 7 × 10^7 cells per 5 gm of liver, and 85–90 % of these cells were viable as judged by their ability to exclude trypan blue. Individual hepatocytes retained all their aminoacyl-tRNA synthetase activity when kept for 8 hours in buffer A on ice.

B.   Optimum Conditions of Toluene Treatment

Since we were interested in using the permeable cells in studies of the aminoacyl-tRNA synthetase complex (Hilderman and Deutscher, 1974), the optimal conditions for toluene treatment were determined by assaying the aminoacyl-tRNA synthetase activity for four different amino acids (lysine, arginine, threonine, and alanine). The optimal conditions for toluene treatment were determined to be 7–9 % toluene incubated with the individual liver cells for 2 minutes at 0°C (Hilderman et al., 1975). After toluene treatment the cell are centrifuged at 100 g for 2 minutes, and they are resuspended in buffer A plus 20 % glycerol. Under these conditions about 90 % of the total aminoacyl-tRNA synthetase activity remained associated with the cell (Hilderman and Deutscher, 1974).

C.   Stability of Toluene-Treated Cells

During out initial studies as much as 70 % of the aminoacyl-tRNA synthetase activities leaked out of the cells during our assay of these enzymes for 5 minutes at 37°C. This problem was overcome significantly by the addition of 20 % glycerol to the buffer in which the treated cells were suspended. In the presence of glycerol, 75 % of the activities remained with the cells (Hilderman and Deutscher, 1974). Glycerol was not included during toluene treatment itself, since it decreased the efficiency of this procedure. Despite the presence of glycerol, the toluene-treated cells were relatively unstable to prolonged storage even at 0°C. After about 15 minutes storage on ice, synthetase activities started leaking out of the cells (Hilderman et al., 1975). Thus, in all experiments, the toluene-treated cells were used within 5 minutes of their preparation.

III.   Characterization of the Toluene-Treated Liver Cells

A.   Analysis of Material Extracted from Cells by Toluene Treatment

Although cells after toluene treatment appeared relatively intact (see below), it was important to determine whether any cellular material leaked out during and following toluene exposure. Therefore, we measured the amount of various cell constituents chemically in untreated cells and in the pellet and the supernatant fractions, after toluene treatment. The data in Table I show that no DNA and only about 5% of the cellular RNA were extracted by the toluene treatment. However, about 20% of the protein and 30% of the phospholipids were no longer associated with the cells. The protein released by toluene treatment was examined further by chromatography on Sephadex. Almost all the protein in the toluene supernatant had a molecular weight of 140,000 to 180,000, there being relatively few lower molecular weight components. By isolating *cells injected* with [^{14}C]orotic acid, we were able to show that more than 90% of the labeled RNA released from the cell during toluene treatment cochromatographed with carrier rabbit liver tRNA when analyzed on Sephadex G-150.

These results suggest that toluene-treated cells retain a high degree of structural integrity. However, the appreciable release of phospholipids probably indicates some alterations in the plasma and/or endoplasmic reticulum membranes.

TABLE I

ANALYSIS OF MATERIAL EXTRACTED FROM TOLUENE-TREATED CELLS[a]

	Relative distribution		
Cell constituent	Isolated liver cells	Toluene pellet	Toluene supernatant
Protein	(100)	85	21
Phospholipid	(100)	75	31
RNA	(100)	91	5
DNA	(100)	100	0

[a] Isolated cells were suspended in 5 volumes of buffer A. A portion was treated with toluene as described above; cells were centrifuged and resuspended in 5 volumes of buffer A. Proteins were analyzed by the method of Lowry et al. (1951), phospholipids were analyzed by the method of Chen et al. (1956), DNA and RNA were analyzed by the method of Schneider (1957). Untreated cells contained per gram 189 mg of protein, 0.029 nmole of lipid phosphate, 9 mg of RNA, and 0.9 mg of DNA.

TABLE II

COMPARISON OF AMINOACYL-tRNA SYNTHETASE ACTIVITY (UNITS PER GRAM OF LIVER) IN CELLS MADE PERMEABLE BY TOLUENE TREATMENT OR OPENED BY OTHER METHODS[a]

Amino acid	Homogenization of intact liver		Sonication of individual liver cells		Nonidet P-40 treatment of individual liver cells		Toluene treatment of individual liver cells	
	Supernatant	Pellet	Supernatant	Pellet	Supernatant	Pellet	Supernatant	Pellet
Arginine	6.51	0.74	14.8	0.63	15.0	0.23	1.30	10.2
Lysine	9.08	0.13	11.3	0.19	13.3	1.60	1.21	10.32
Alanine	4.15	0.75	6.21	0.54	6.46	0.48	0.54	4.3
Threonine	7.63	0.86	11.9	0.49	12.6	0.98	1.19	10.1

[a] Isolated cells were suspended in 5 volumes of buffer A containing 20% glycerol. Homogenization of intact liver (0.5 gm) was done in a Dounce homogenizer using 10 strokes of a loose-fitting pestle. Sonication of individual liver cells (0.5 ml) was performed using a Branson Sonifier cell disrupter at a setting of 1 for a total of 60 seconds in 30-second bursts with a 30-second cooling period. Nonidet P-40 treatment involved incubating 0.2-ml aliquots of cells with 50 μl of 10% Nonidet P-40 solution for 10 minutes in ice. Toluene treatment of individual liver cells was performed on a 0.2-ml aliquot as described above. All samples were centrifuged at 100 g for 2 minutes, the pellets were resuspended in 0.2 ml of buffer, and both the supernatant and the pellet fractions were assayed for aminoacyl-tRNA synthetase activities.

B.   Aminoacyl-tRNA Synthetase Activity in Toluene-Treated Cells

In order to determine the efficiency of toluene treatment for making cells permeable, the level of aminoacyl-tRNA synthetases assayable after this procedure was compared to that in cells opened by a variety of other methods. Aminoacyl-tRNA synthetase assay conditions have been described elsewhere (Hilderman and Deutscher, 1974). The results in Table II demonstrate that the total units detectable after toluene treatment were somewhat greater than those found by homogenization of an equivalent weight of intact liver, but slightly lower than that obtained by sonification or Nonidet P-40 treatment of individual liver cells. This lower activity in toluene-treated cells was not due to enzyme inhibition or inactivation since the same concentration of toluene had no effect on aminoacyl-tRNA synthetase activity in a homogenate of intact liver. Also, remixing of toluene supernatant and pellet fractions gave only additive results suggesting that the lower activity was not due to removal of a soluable factor from the cells. In addition, since Nonidet P-40 treatment of toluene-treated cells did not further increase the total units of enzyme detectable, toluene appeared to make all the activity accessible. Also, control mixing experiments of either intact cells or toluene-treated cells with high speed supernatant fractions from liver indicated that the aminoacyl-tRNA synthetases were not bound to the outside of the cells after toluene treatment. These data indicate that this procedure is quite efficient for making cells permeable.

In order to determine whether the aminoacyl-tRNA synthetases actually remained within the cell during the assay, cells were preincubated under the same conditions as the assay (Hilderman and Deutscher, 1974) and centrifuged; both the pellet and supernatant fractions were assayed. As shown in Table III approximately 75% of the total synthetase activity for each amino acid remained within the cell after this treatment. Thus, this suggested that we are measuring the aminoacyl-tRNA synthetase reaction within toluene-treated cells. Also, aminoacylation of tRNA within a toluene treated cell is dependent upon an exogenous source of ATP and tRNA. This indicates that toluene-treated cells are permeable to charged molecules and large exogenous molecules at least the size of tRNA.

C.   Effect of Toluene Treatment on Marker Enzymes

Our studies with aminoacyl-tRNA synthetase indicated that toluene treatment of liver cells could render a presumed cytoplasmic enzyme accessible to exogenous substrates. Thus, it was of interest to determine whether this treatment was also useful for study of enzymes with other subcellular localizations. Accordingly, we measured a variety of marker enzymes to

TABLE III

LOCATION OF AMINOACYL-tRNA SYNTHETASE DURING ASSAY[a]

Amino acid	Supernatant (units/gm cells)	Pellet (units/gm cells)
Arginine	2.48	7.95
Lysine	2.20	5.51
Alanine	1.78	5.14
Threonine	2.62	7.50

[a]After toluene treatment the cells were suspended in 5 volumes of assay mixture without labeled amino acid and preincubated for 5 minutes at 37°C. The cells were then centrifuged at 100 g for 2 minutes, and the cell pellet was resuspended in 5 volumes of the same medium. Ten microliters of the cell and supernatant fractions were assayed as described elsewhere (Hilderman and Deutscher, 1974).

ascertain whether they became accessible to substrates and whether they remained with the cells after toluene treatment. These included tRNA nucleotidyltransferase, presumably located in the cytosol (Deutscher, 1970); 5'-nucleotidase, localized in the plasma membrane (Song and Bodansky, 1967); glucose-6-phosphatase, which is found in the endoplasmic reticulum (Ashmore and Weber, 1959); acid phosphatase, located in the lysosomes (Berthet et al., 1951); and cytochrome c oxidase, a mitochondrial enzyme (Schnaitman and Greenawalt, 1968). The data in Table IV indicate that each

TABLE IV

ENZYMIC ACTIVITIES IN TOLUENE-TREATED CELLS[a]

Enzyme	Relative activity		
	Sonicated cells	Toluene pellet	Toluene supernatant
t-RNA nucleotidyltransferase	(100)	85	16
5'-Nucleotidase	(100)	83	15
Glucose-6-phosphatase	(100)	98	5
Acid phosphatase	(100)	97	7
Cytochrome c oxidase	(100)	100	0

[a]Isolated cells were suspended in 5 volumes of buffer A, and a portion was disrupted by sonication for a total of 60 seconds in two 30-second bursts with a 30-second cooling period. Toluene treatment was performed as described previously. Cells were centrifuged, and the pellet was resuspended in 5 volumes of buffer A. Activities per gram of sonicated cells for the various enzymes were as follows: tRNA nucleotidyltransferase, 1.15 nmoles/minute; 5'-nucleotidase, 8.47 μmoles/15 minutes; glucose-6-phosphatase, 7.94 μmoles/15 minutes; acid phosphatase, 6.00 μmoles/15 minutes; and cytochrome c oxidase, 7.41 units. Activities in toluene-treated cells and supernatant are expressed relative to that in sonicated cells.

marker enzyme could be assayed in toluene-treated cells and that the level of activities made accessible to exogenous substrates were similar to those found by sonification. These data also show that more than 80 % of each enzyme remained associated with the cells. The results indicate that toluene treatment is a useful procedure for enzymes with various subcellular localizations. However, although these marker enzymes remain associated with the treated cells, we do not yet know whether they maintain their original subcellular localization after exposure to toluene.

D.   Electron Microscopic Comparison of Hepatocytes before and after Toluene Treatment

Numerous stages during the isolation and treatment of the cells have been examined by electron microscopy (Hilderman *et al.*, 1975). Even after prolonged exposure to enzymic digestion the majority of the isolated hepatic parenchymal cells retain their basic architecture (Berry and Fried, 1969; Gershanam and Thompson, 1975; Hilderman, *et al.,* 1975). The isolated cells (Fig. 1) become rounded, but retain a few microvilli. Viable-appearing cells have intact plasma membranes, which often have tight junctions with adherent membrane from adjacent cells. They have an external (ectoplasmic) zone in which vacuoles, but few organelles, are seen. The remaining cytoplasm retains its usual density and distribution of organelles. The nucleus and nucleolus are indistinguishable from those seen in intact liver. Some cells, however, show variable damage. Swollen and fragmented endoplasmic reticulum (ER), some swollen mitochondria, and focal extraction of hyaloplasmic density are the most frequent changes. These cells also appear generally swollen with loss of microvilli and frequently show focal breaks in the plasmalemma.

After toluene treatment (Fig. 2) the cells remain intact. Some injury is evident, however, particularly in the endoplasmic reticulum. Rough ER shows loss of parallel arrays, and most appears as small ribosome-studded vesicles. Clumps of free ribosomes are present, but spiral arrays are usually not evident. Some mitochondria in each cell are swollen with loss of matrix density. The cell outline is smooth, and fewer microvilli are present. Some irregular loss of hyaloplasmic density, particularly in the ectoplasmic zone, is also usually present. Nuclear and nucleolar structure is usually not altered by brief toluene treatment.

In summary, although some changes, such as swelling of the cells and their organelles, are seen, the cells remain remarkably intact. Treatment with glycerol (20 %) does not reverse or prevent these changes. Whether some other osmotic agent might protect, or whether these changes can be reversed, has not yet been determined.

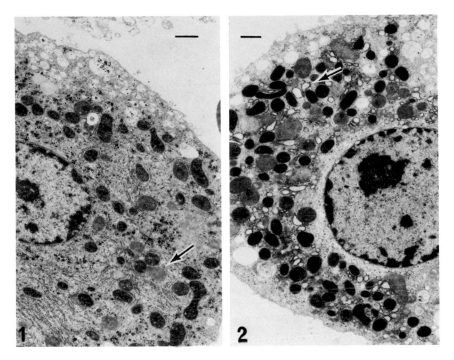

FIG. 1. Isolated hepatocyte prior to toluene treatment. The upper border of the cell shows vacuolization and a zone in which other organelles are not present. Stacks of rough surfaced endoplasmic reticulum are present in the lower left corner of the picture. Some mitochondria have a dense matrix, while others appear slightly swollen (arrow). The bar at upper right is 1 μm. $\times$ 5700.

FIG. 2. Isolated hepatocyte following brief toluene treatment. The cell is relatively intact, but there does appear to be some extraction of hyaloplasmic density, swelling of endoplasmic reticulum (arrow), and accentuation of the mitochondrial swelling. Note that microvilli are still present on the plasma membrane of the upper border of the cell. Bar at upper left is 1 μm. $\times$ 5900.

IV. Conclusions

Isolated individual liver cells were made permeable to charged molecules and macromolecules by treatment with toluene. Toluene treatment was found to be as efficient as various disruptive methods for making internal enzymes accessible to exogenous substrates. Electron microscopic and biochemical examination of the toluene-treated cells indicated that cell structure was not destroyed by the toluene treatment and that most cell constituents remained within the cell.

ACKNOWLEDGMENT

The authors are grateful to Dr. Murray P. Deutscher for guidance and participation in these experiments, most of which were performed in his laboratory.

REFERENCES

Ashmore, J., and Weber, G. (1959). *Vitam. Horm. (N.Y.)* **17**, 91–132.
Atherly, A. G. (1974). *J. Bacteriol.* **118**, 1186–1189.
Banks, G. R. (1973). *Nature (London), New Biol.* **245**, 196–199.
Berry, M. N., and Friend, D. S. (1969). *J. Cell Biol.* **43**, 506–519.
Berthet, J., Berthet, L., Appelman, F., and DeDuve, C. (1951). *Biochem. J.* **50**, 182–189.
Burgoyne, L. A. (1972). *Biochem. J.* **130**, 959–964.
Capuzzi, D. M., Rothman, V., and Margolis, S. (1974). *J. Biol. Chem.* **249**, 1286–1294.
Chen, P. S., Toribara, T., and Huber, W. (1956). *Anal. Chem.* **28**, 1756–1758.
Clark, D. G., Rognstad, R., and Katz, J. (1974). *J. Biol. Chem.* **249**, 2028–2036.
Craig, M. C., and Porter, J. W. (1973). *Arch. Biochem. Biophys.* **159**, 606–614.
Deutscher, M. P. (1970). *J. Biol. Chem.* **245**, 4225–4227.
Gershenson, L. E., and Thompson, E. B., eds. (1975). "Gene Expression and Carcinogenesis in Cultured Liver." Academic Press, New York.
Goodridge, A. C. (1973). *J. Biol. Chem.* **248**, 1924–1931.
Hilderman, R. H., and Deutscher, M. P. (1974). *J. Biol. Chem.* **249**, 5346–5348.
Hilderman, R. H., Goldblatt, P. J., and Deutscher, M. P. (1975). *J. Biol. Chem.* **250**, 4796–4801.
Howell, S. H., and Walker, L. L. (1972). *Proc. Natl. Acad. Sci. U.S.A.* **69**, 490–494.
Jackson, R. W., and DeMoss, J. A. (1965). *J. Bacteriol.* **90**, 1420–1425.
Kumar, B. V., Medoff, G., Kobayashi, G., and Schlessinger, D. (1974). *Nature (London)* **250**, 323–325.
Lowry, O. H., Rosebrough, N. J., Farr, A. L., and Randall, R. J. (1951). *J. Biol. Chem.* **193**, 265–275.
Moses, R. E., and Richardson, C. C. (1970). *Proc. Natl. Acad. Sci. U.S.A.* **67**, 674–681.
Peterson, R. E., Radcliffe, C. W., and Pace, N. R. (1971). *J. Bacteriol.* **107**, 585–588.
Schnaitman, C., and Greenwalt, J. W. (1968). *J. Cell Biol.* **38**, 158–175.
Schneider, W. C. (1957). *In* "Methods in Enzymology" (S. P. Colowick and N. O. Kaplan, eds.), Vol. 3, pp. 680–684. Academic Press, New York.
Schreiber, G., and Schreiber, M. (1972). *J. Biol. Chem.* **247**, 6340–6346.
Song, C. S., and Bodansky, O. (1967). *J. Biol. Chem.* **242**, 694–699.
Veneziale, C. M., and Lohmar, P. H. (1973). *J. Biol. Chem.* **248**, 7786–7791.

Chapter 22

Estimation of Intracellular Fluid Volume[1]

WILLIAM S. RUNYAN

Department of Food and Nutrition, Iowa State University, Ames, Iowa

AND

TERESA H. LIAO

Department of Chemistry, Nutritional Science Program, University of Maryland, College Park, Maryland

I. Introduction

In developing a method for estimating the intracellular fluid volume of L cells (Runyan and Liao, 1974), we were guided by certain extant requirements. Owing to the capacity of our culture facilities, the number of cells available for a given experiment was limited. Therefore, the total volume of intracellular fluid per sample was relatively small. Consequently, the final culture medium containing the cells had to be of a known volume small enough to undergo significant dilution upon release of intracellular fluid. Use of physical means to release the intracellular fluid was necessary if the additional dilutions and manipulations of introducing chemical lysing agents into the samples were to be avoided. All these requirements were met by the following method, which is based on dilution of a known concentration and volume of

[1] This work was supported by the Iowa State University Agriculture and Home Economics Experiment Station, Project 1770, and the Nutrition Foundation, Inc., Grant No. 431.

^{14}C-labeled sucrose solution by extracellular fluid (culture medium) followed by a second dilution due to release of intracellular fluid as a result of ultra-centrifugation.

II. Procedure

The cells were sublines of NCTC clone 929 (strain L) grown as monolayers in 250-ml milk-dilution bottles with Waymouth's 752/1 (Waymouth, 1959) containing 5 % horse serum as the culture medium. Cell harvests were done during logarithmic growth by scraping with rubber policemen. Cell counts were made with a Model B Coulter Counter.

Sucrose solution was prepared by diluting a 100 μCi/ml aqueous solution of [U-^{14}C]-sucrose (S.A. = 1750 μCi/mg) to 5 μCi/ml with culture medium. To compensate for tonicity changes, 4.6 mg/ml of unlabeled sucrose was added to the final solution.

Samples were prepared by scraping cells into the growth medium, which was then pooled, mixed, and distributed to 50-ml conical centrifuge tubes. One milliliter of the pooled cell suspension was appropriately diluted and counted to estimate the number of cells in each tube. The tubes were spun in an International Model HN centrifuge at 800 rpm for 10 minutes at ambient temperature. The growth medium was decanted, saved, and used to resuspend the cells in each tube to a concentration of about 10^7/ml, and the suspensions were again pooled and mixed. Samples containing 5.0 ml of the resultant heavy cell suspension were then distributed to 1 : 2 by 2-inch cellulose-nitrate ultracentrifuge tubes; three 1.0-ml samples were appropriately diluted with the retained growth medium, and each was used for triplicate cell counts. With a forceps, the ultracentrifuge tubes were placed in the bottom of 15-ml centrifuge shields (IEC No. 303) and centrifuged as just described. The supernatant medium was discarded and the tubes inverted briefly on absorbent paper.

To the cells and small volume of growth medium remaining in each ultracentrifuge tube, 0.50 ml of the [^{14}C]-sucrose solution was added from a 1-ml gas-tight syringe. After the cells and solution in each tube were thoroughly mixed with a nichrome wire formed into a 3/8-inch diameter "shepherd's crook," the tubes were again centrifuged in the 15-ml shields. Two-tenths milliliter of the supernatant solution in each tube was removed with a 200-μl pipette and pipette control and placed in a liquid scintillation vial; each micropipette was rinsed into its respective vial with 2 volumes of distilled water, and 20 ml of an ethanol–toluene scintillation cocktail (Buhler, 1962) was added to each vial.

The ultracentrifuge tubes were filled to within 1/8 inch of the top with mineral oil, which was for the primary purpose of preventing collapse of the tubes during centrifugation but also protected the underlying aqueous solution during subsequent steps in the procedure. The samples were centrifuged overnight in a type 50.1 swinging-bucket rotor on a Beckman Model L3-50 ultracentrifuge at a rotor temperature of 4°C. The speed generally used was 50,000 rpm although 40,000 rpm was found to be equally satisfactory in that extensive microscopic examination of the pellet of cell debris resulting from either speed showed no intact cells. After ultracentrifugation, the pellet of cell debris was thoroughly mixed into the aqueous supernatant fluid by inserting the nichrome-wire crook through the mineral oil. This caused a disruption of the oil–aqueous interface, which was removed by centrifugation in the 15-ml shields at 1800 rpm for 10 minutes. The tubes then were allowed to stand overnight at 4°C, after which the stirring and centrifugation at 1800 rpm were repeated. At this point, removal and examination of the supernatant fluid under bright light showed some particulate material that probably was made up of, or arose from, cell organelles and organoids. Microscopic examination did not reveal any large particles, and treatment of the aqueous supernatant and equivalent amounts of culture medium with acid followed by centrifugation in hematocrit capillary tubes produced precipitate volumes which indicated that particulate material made up a small portion (4% or less) of the aqueous supernatant fluid. Moreover, we subsequently found that the aqueous supernatant could be clarified by making the final centrifugation in the ultracentrifuge at 50,000 rpm for 10 minutes, but this did not have a significant effect on the final results. We concluded that the error introduced by the amount of particulate material present was smaller than the inherent error of the method.

After completion of the final stirring and centrifugation, the mineral oil in the ultracentrifuge tubes was carefully drawn off the aqueous supernatant fluid with a low vacuum applied through a fine capillary tube. Care was necessary because during removal of the last few millimeters of oil, the aqueous fluid could easily be aspirated from underneath it, and the sample lost. Finally, 0.20-ml samples of the aqueous fluid were again removed from the tubes with micropipettes and prepared for scintillation counting as before.

Radioactivity was assayed on a Packard Model 3320 liquid scintillation counter; efficiency corrections were made by the channels-ratio procedure. All samples were compared to 0.20-ml volumes of the original sucrose solution that had been added to the tubes. The dilutions of isotope concentration were used as indices of extracellular and intracellular fluid volumes, and calculations were done as follows:

$$(C_1/C_2)(V_1) = V_2; \quad (C_2/C_3)(V_3) = V_4; \quad V_4 - V_3 = \text{intracellular fluid}$$

where V_1 = the volume of $[^{14}C]$sucrose solution added to the intact cells (0.50 ml); C_1 = the concentration of ^{14}C in V_1; V_2 = the volume of extracellular fluid plus V_1; C_2 = the concentration of ^{14}C in V_2 and V_3; $V_3 = V_2$ minus the volume removed to determine C_2 (0.02 ml); C_3 = the concentration of ^{14}C in V_4; and V_4 = intracellular fluid plus V_3.

III.   Results and Discussion

Results are shown in Table I. The relatively large standard errors were probably a result of the fact that differences between small dilutions of small fluid volumes were measured. As described here, the procedure left about 0.1 ml of culture medium with each original cell pellet and released a similar volume of intracellular fluid; consequently, the respective dilutions of $[^{14}C]$-sucrose were between 16 and 25%. Addition of less than 0.5 ml of the original sucrose solution made removal of the 0.20-ml samples difficult and, as stated earlier, use of larger cell samples was beyond the capacity of our facilities. However, we see no reason why this procedure cannot be adapted to the use of angle-head ultracentrifuge rotors and their correspondigly larger ultracentrifuge tubes. That would allow use of cell samples up to seven times larger than those used here, and appropriate adjustment of the volume of $[^{14}C]$-sucrose solution added to the cell samples could result in dilutions of 50% or more, which should reduce the standard error of the method and make the manipulations somewhat less tedious.

Some other factors that might have influenced the results presented here should be mentioned. First, the intracellular fluid volume estimates obtained represent more than just intracellular water. In addition to any particulate

TABLE I

INTRACELLULAR FLUID VOLUME OF L CELLS

Group[a]	Expt. No.	Number of samples	Microliters per million cells	Standard error[b]
1	1	3	1.90	0.29
	2	3	1.93	0.18
	3	4	1.99	0.17
2	1	4	2.39	0.23
	2	4	2.24	0.19
	3	6	2.44	0.13
	4	6	2.37	0.11

[a] Cell sublines in groups 1 and 2 were from different sources.
[b] Based on a t test for small samples.

material, a contribution to fluid volume was undoubtedly made by a number of intracellular solutes, a situation likely common to many partition and dilution methods of estimating biological fluid volumes. A second consideration is the method of cell harvest and any effect it might have had on the results through possible cell damage. Scraping was chosen over trypsinization to avoid removing the cells from serum-containing medium, and chelating agents were avoided because of possible effects on membrane transport or permeability. L cells are relatively tolerant of harvest by scraping; those used in this work were comparable in trypan blue dye exclusion whether scraped or trypsinized, with results similar to those reported for other cell lines harvested with trypsin (Yip and Auersperg, 1972). In this procedure, we consider immoderate trituration a more likely source of cell damage than harvest by scraping. However, chemical harvest should be suitable in application of this method as long as it is kept in mind that the harvesting solution may vary from the culture medium in osmolarity and thereby in osmotic properties (Waymouth, 1970). Application to suspension cultures, of course, would eliminate any concerns about the means of harvest. A third consideration concerns any possible influence on our results by metabolism of the [^{14}C]-sucrose. We are unaware of any ability of L cells to hydrolyze sucrose, but the chance of hydrolysis by microbial contaminants led to the employment of low temperatures during the procedure. Moreover, efforts to trap $^{14}CO_2$ with alkaline-saturated filter-paper fans during overnight storage at 4°C of post-ultracentrifuged samples with the mineral oil removed and to demonstrate ^{14}C that could not be removed from the final cell debris by washing were unsuccessful. We concluded that metabolism of sucrose did not occur during the procedure.

We have compared the intracellular fluid volumes obtained by this method to total cell volumes as estimated by use of a Model J Coulter Automatic Size Distribution Plotter calibrated with ragweed pollen suspended in culture medium, and we found a consistent relationship. Expressed as percentage of total cell volume, the mean value for the intracellular fluid volumes of L cells from six experiments was 72.8 with a standard deviation of 0.9 and a range from 70.4 to 74.2 (Runyan and Liao, 1974). However, we do not know if that relationship would be consistent under nonreplicate culture conditions.

REFERENCES

Buhler, D. R. (1962). *Anal. Biochem.* **4**, 413–417.
Runyan, W. S., and Liao, T. H. (1974). *Exp. Cell Res.* **83**, 406–408.
Waymouth, C. J. (1959). *J. Natl. Cancer Inst.* **22**, 1003–1017.
Waymouth, C. J. (1970). *In Vitro* **6**, 109–127.
Yip, D. K., and Auersperg, N. (1972). *In Vitro* **7**, 323–329.

Chapter 23

Detection of Carbohydrates with Lectin–Peroxidase Conjugates[1]

NICHOLAS K. GONATAS

Department of Pathology (Division of Neuropathology), University of Pennsylvania School of Medicine, Philadelphia, Pennsylvania

AND

STRATIS AVRAMEAS

Immunocytochemistry Unit, Pasteur Institute, Paris, France

I. Introduction

Lectins are agglutinins of invertebrate or plant origin which bind mono- and oligosaccharides specifically and have been used widely in immuno-

[1] This work was supported by United States Public Health Service Grants NS-05572-12 and by Research Grant D.G.R.S.T. No. 72-70298 (France).

logical, biochemical, and biological investigations. Recently, several comprehensive reviews on the isolation, characterization, and application of lectins in biomedical research, have been published (Chowdhury and Weis, 1975; Cohen, 1974; Lis and Sharon, 1973; Nicolson, 1974a; Sharon and Lis, 1972, 1975). The purpose of this chapter is to review information derived from the use of conjugates of the marker enzyme horseradish peroxidase with lectins for the morphological (light and electron microscopic) detection of specific carbohydrates or carbohydrate-containing moieties of the plasma (surface) or intracellular membranes of various types of cells and tissues.

Since the introduction by Graham and Karnovsky of the method for the light and electron microscopic cytochemical demonstration of horseradish peroxidase (HRP) in cells and tissues, conjugates of this enzyme with antibodies have been used for the specific detection of cell and tissue antigens (Gonatas et al., 1972; Graham and Karnovsky, 1966); it has been shown that the specific interactions between antigen and antibody molecules are not impaired by coupling antigens or antibodies with horseradish peroxidase (Avrameas, 1970a; Gonatas et al., 1972); similar specific interactions occur between lectins and carbohydrate moieties, and several lectins labeled with horseradish peroxidase (HRP) have been used for the specific cytological demonstration of carbohydrate moieties on cell surfaces or on intracyto-plasmic membranes (Bernhard and Avrameas, 1971; Bretton and Bariety, 1974; Gonatas and Avrameas, 1973; Wood et al., 1974). The use of lectin-HRP conjugates for the cellular detection of carbohydrates is more advantageous over the existing methods utilizing heavy metals (reviewed by Martinez-Palomo, 1970; Rambourg et al., 1969) because they are more specific and can be applied on fixed and unfixed cells as well (Gonatas and Avrameas, 1973; Gonatas et al., 1975). We shall review the following topics: (1) methods of isolation of lectins and of preparation of lectin-peroxidase (HRP) conjugates; (2) effect of fixation on the application of lectin–HRP conjugate; (3) specificity of binding between lectin-HRP and carbohydrate moieties; (4) topography, mobility, and endocytosis of surface (plasma membrane) lectin–HRP conjugates; (5) use of lectin–HRP conjugates for the demonstration of intracellular carbohydrate moieties in tissues; (6) comparison of lectin–ferritin to lectin–HRP conjugates.

II.   Methods of Isolation of Lectins and Preparation of Lectin–HRP Conjugates

The method of choice for isolation of pure lectins is affinity chromatography, based on the specific and reversible binding of lectins to carbohydrates (Cohen, 1974, p. 232; Avrameas *et al.*, 1973; Lis and Sharon, 1973; Sharon and Lis, 1972, 1975); also, insolubilized lectins have been used for the isolation of soluble and membrane-bound glycoproteins (Avrameas, 1970b; Avrameas *et al.*, 1973; Sharon and Lis, 1975; Zanetta *et al.*, 1975). Elution of the lectin from the appropriate column, or elution of the tissue glycoprotein or polysaccharides from the lectin column, is achieved with the use of specific sugars of buffers at low pH (Avrameas *et al.*, 1973; Gonatas and Avrameas, 1973; Sharon and Lis, 1975; Zanetta *et al.*, 1975). Descriptions of methods of isolation and purification of lectins are beyond the scope of this review, and suitable references are cited by Lis and Sharon (1973), Sharon and Lis (1972, 1975), and in a recent publication of the New York Academy of Sciences on lectins (Cohen, 1974). Lectins are commercially available (Miles-Yeda, Israel, Pharmacia, etc.) Horseradish peroxidase–lectin conjugates have been utilized with the following lectins: (1) concanavalin A (Con A): specificity with α-D-glucopyranosyl, α-D-mannopyranosyl and B-D-fructofurnanosyl residues (Abercrombie *et al.*, 1972; Ackerman and Waksal, 1974; Avrameas, 1970b; Barat and Avrameas, 1973; Bernhard and Avrameas, 1971; Bretton *et al.*, 1972a; Collard and Temminck, 1974; Garrido, 1975; Garrido *et al.*, 1974; Gordon *et al.*, 1974; Huet and Bernadac, 1974; Huet and Hertzberg, 1973; Huet *et al.*, 1974; Martinez-Palomo *et al.*, 1972; Roth and Thoss, 1974; Roth *et al.*, 1972, 1973, 1974; Rowlatt *et al.*, 1973; Torpier and Montagnier, 1973; Wood *et al.*, 1974), (2) wheat germ agglutinin (WGA): specificity with *n*-acetylglucosamine residues (Francois *et al.*, 1972; Garrido, 1975; Garrido *et al.*, 1974; Gonatas and Avrameas, 1973; Huet and Garrido, 1972), (3) ricin: specificity with *d*-galactose residues (Gonatas and Avrameas, 1973; Gonatas *et al.*, 1975); (4) phytohemagglutinin; specificity for *n*-acetylgalactosamine residues (Gonatas and Avrameas, 1973); (5) *Lens culinaris* lectin (LCA): specificity for D-mannose, D-glucose, *N*-acetyl-D-glucosamine residues (Parmley *et al.*, 1973; Roth and Thoss, 1974; Roth *et al.*, 1974, 1975b).

With the exception of Con A, all lectins have been linked covalently with HRP with glutaraldehyde (Avrameas and Ternynck, 1971); also, glutaraldehyde has been used for the cross-linking of ferritin with ricin (Nicolson, 1974a). According to this coupling method, active aldehyde groups are introduced to peroxidase by treating the enzyme for 16–24 hours at room temperature with an excess of glutaraldehyde; the unreacted glutaraldehyde is removed by filtration through a Sephadex G-25 column. The "activated" peroxidase is used immediately for coupling with the lectin (Avrameas and Ternynck, 1971; Gonatas and Avrameas, 1973). The lectin–HRP conjugate is then chromatographed on a Sephadex G-200 column, and fractions are read at 280 and 403 mm to determine their concentration in lectin protein protein and peroxidase, respectively, and to exclude unreacted HRP or glutaraldehyde. Before use, the conjugates are tested for peroxidase and hemagglutinating activities. The two-step coupling method with glutaraldehyde as the cross-linking agent is simple and suitable for coupling lectins to HRP at desirable ratios; furthermore, the absence of bivalent cross-linking agents in the final step of the coupling procedure, prevents the formation of lectin–lectin conjugates (Avrameas and Ternynck, 1971). In the case of Con A, its covalent coupling with HRP is not necessary because the enzyme has a high mannose content and it is bound to Con A after its incubation with unfixed (Bernhard and Avrameas, 1971) or fixed cells or tissues (Wood *et al.*, 1974). According to the original method of Bernhard and Avrameas (1971), Con A in concentrations of 100 μg to 1 mg per milliliter of phosphate-buffered saline (PBS) is incubated with the cell culture, cell suspension, or tissue sections; after several washes in PBS, the preparation is incubated with HRP (1 mg/ml); subsequently, the preparation is washed again, fixed in paraformaldehyde or glutaraldehyde, and stained for HRP with diaminobenzidine tetrahydrochloride (DAB) with the method of Graham and Karnovsky (1966).

A method for covalent coupling with glutaraldehyde of Con A with cytochrome peroxidase +(C-Pox) or ferritin for light and electron microscopic studies is also available (Stobo and Rosenthal, 1972); according to this method, saturation of Con A binding sites with α-methyl-D-glucopyranoside (α-μG) before its exposure to glutaraldehyde, protects them from glutaraldehyde and preserves the mitogenic and binding activities of Con A. Removal of α-μG from Con A after its coupling to ferritin or to C-Pox is achieved by extensive dialysis against PBS at 4°C. Cytochrome peroxidase (C-Pox), unlike HRP, has no carbohydrate binding to Con A, and for that reason it is more suitable for a covalent linkage with Con A; furthermore, precipitation of Con A–C-Pox complexes does not occur during coupling (Stobo and Rosenthal, 1972). The cytochemical demonstration of lectin–HRP conjugates is achieved with the method of Graham and Karnovsky (1966), which is based on oxidation of DAB; 5 mg of DAB is dissolved in 10 ml of 0.05–0.1 M Tris-buffer, pH 7.35, and 1 drop of H_2O_2 (30%) is added to the mixture; paraformaldehyde- or glutaraldehyde-fixed cells or tissue sections are incubated for a few minutes at room temperature. To avoid diffusion of the reaction product, short incubation at 4°–22°C are recommended. The preparations are subsequently washed in an appropriate buffer and fixed in osmium tetroxide. In the light microscope, the peroxidase stain has a light or dark brown color; in the electron microscope, the insoluble oxidized DAB–osmium black precipitates appear as electron dense globules or sheets about 300 Å thick (Fig. 1).

The limitations of the use of HRP conjugates with lectins or other specific ligands are: (1) endogenous peroxidase activity (Norikoff *et al.*, 1971), (2)

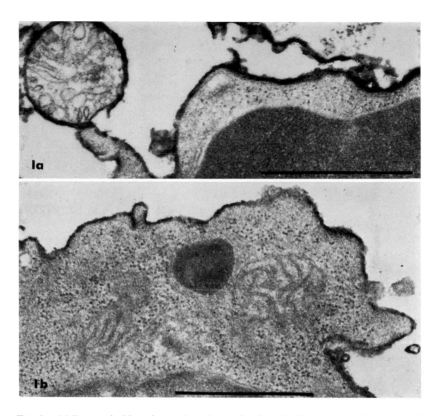

FIG. 1. (a) Rat cervical lymph node lymphocyte incubated with ricin agglutinin–horseradish peroxidase (HRP) conjugate at $^+4°C$ for 1 hour, washed in Earle's balanced salt solution, fixed in 4 % paraformaldehyde in 0.1 M phosphate buffer, and stained with diaminobenzidine (DAB) tetrahydrochloride for 10 minutes at $^+22°C$. Plasma membrane has a heavy deposit of oxidized DAB. Scale bar 1 μm; × 40,000. (b) Same as in 1a but incubation with ricin–HRP was performed after fixation with paraformaldehyde. Note that oxidized DAB deposit is similar to that in 1a. × 36,500. Reproduced from Gonatas and Avrameas (1973, p. 438).

poor penetration of the conjugates in tissues, (3) diffusion of the reaction product. (1) Certain cells and tissues have endogenous peroxidase activity detectable by the DAB method (Novikoff *et al.*, 1971); endogenous peroxidase activity can be easily investigated in DAB-stained control preparations that have not been exposed to lectin–HRP conjugates. (2) Poor penetration of lectin–HRP and diffusion of the peroxidase reaction product are significant limitations and will be discussed in subsequent sections. (3) Diffusion of the reaction product (oxidized DAB); the limitations of this aspect of the method will be discussed in Section VII on the comparison of ferritin–lectin to HRP–lectin preparations.

III.   Fixation

Fixation of cell suspensions or tissues with paraformaldehyde or glutaraldehyde does not affect the binding of Con A, ricin, wheat germ agglutinin or phytohemagglutinin to its sugar-containing receptor molecules (Bretton and Bariety, 1974; Gonatas and Avrameas, 1973; Graham *et al.*, 1974; Wood *et al.*, 1974); however, prefixation of mouse myeloid leukemic cells with glutaraldehyde abolishes their agglutination induced by Con A, WGA, waxbean and soybean agglutinins, although binding per se of lectins to a fixed cell is not significantly affected (Rutishauser and Sachs, 1975); these findings suggest that mobility of lectin receptors, which is abolished by glutaraldehyde fixation, is one of several factors responsible for lectin-induced cell agglutination (Rutishauser and Sachs, 1975).

IV.   Specificity of Binding between Lectin–HRP and Cell Surface or Intracellular Carbohydrate Moieties

Although the interaction between a lectin and its carbohydrate containing receptor is quite specific, special care should be taken to rule out three causes for nonspecific binding: (1) Adsorption of polysaccharides or glycoproteins from the culture medium, serum, etc., on cell surfaces should be avoided. Rowlatt and Wicker (1972) have shown that Con A, visualized by HRP, was bound on the surface of tissue culture bottles containing Eagle's MEM, 10 % calf serum and 10 % tryptose phosphate; thorough washing of cell suspensions or tissue sections in a suitable buffer followed by incubation at 37°C in a buffer (i.e., Eagle's BSS buffered with 10 mM of HEPES N-2-hydroxyethyl-piperazine-N-ethanosulfonic acid; Schwarz-Mann), should be performed before introducing lectins to the system. (2) Various agglutinin preparations may not be pure or monospecific. For example, with the use of affinity chromatography through a galactose-containing polysaccharide, a ricin agglutinin, inhibited by N-acetyl-D-galactosamine, and another agglutinin, inhibited by d-galactose residues, are prepared (quoted in Gonatas and Avrameas, 1973). Therefore, control, inhibition, experiments with several sugars should be performed. Usually, 1 hour of preincubation at 36°C of the lectin–HRP with a 0.1–0.5 M concentration of the appropriate sugar and application of this lectin preparation in cells or tissues prepared in the presence of the inhibiting sugar results in absent or reduced staining (Gonatas and Avrameas, 1973); better inhibition is achieved with fixed rather than unfixed cells (Rutishauser and Sachs, 1975); furthermore, unfixed cells exposed to lectin tend to clump much more easily and diffusion of the lectin is more difficult than in fixed cells, which do not easily agglutinate. It must be noted that removal of the cell-bound lectin with the use of eluents containing the inhibitory sugar is not always feasible (Garrido, 1975; Rutishauser and Sachs, 1975). Therefore in experiments designed to prove the specificity of a preparation, the inhibitory sugar should be used before the lectin is bound on cell surfaces or in intracellular moieties. (3) Doyle *et al.* (1968) have shown that Con A has common binding sites for both neutral polysaccharides containing gluco- or mannopyranosyl residues and various polyelectrolytes (fucoidan, RNA, heparin, bacterial lipopolysaccharides).

V. Topography, Mobility and Endocytosis of Surface (Plasma Membrane), Lectin–HRP Conjugates

Lectin–HRP conjugates have been used extensively for the topographic study of lectin receptors on the plasma membrane of various normal and malignant (transformed) cell lines. These investigations were stimulated by the extensive use of lectins as probes of cell-surface changes associated with malignancy, differentiation, and lectin-induced mitogenicity (Chowdury and Weis, 1975; Cohen, 1974; Collard and Temminck, 1974; Lis and Sharon, 1973; Sharon and Lis, 1972, 1975). The cellular "receptor" that binds the lectin–HRP conjugate is not necessarily identical to the mono- or polysaccharide that inhibits the binding (Lis and Sharon, 1973). Furthermore, morphological differences of the surface or endocytosed lectin–HRP patterns between various cell types cannot be readily related to functional differences in agglutinability, contact inhibition, mitogenicity, and malignancy. For example, it has been strongly suggested that endocytosis (internalization) of a lectin–receptor complex is not a prerequisite for mitogenicity (Lis and Sharon, 1973).

A. Surface Topography of Lectin–HRP "Receptor" Complexes in Normal and Transformed Cells

In 1972, Bretton *et al.* (1972a) and Martinez-Palomo *et al.* (1972) studied the Con A–HRP distribution in normal and SV40-transformed hamster cell cultures; in the majority of the malignant cells the cell coat of Con A–HRP ("glycogram") was uneven and discontinuous when compared to the Con A–HRP stain of normal cells. Similar results were obtained by Garrido *et al.* (1974) in normal and malignant hamster cell lines exposed to Con A–HRP and WGA–HRP; according to these authors, the discontinuous nature of the lectin–HRP stain in malignant cell lines was probably due to an increased mobility of plasma membrane lectin receptors in malignant cells. However, the same authors using identical methods were not able to observe in rat cell lines the differences of surface lectin "receptor" topography found before in hamster cells (Garrido *et al.*, 1974). Torpier and Montagnier (1973) were not able to confirm the topographic differences of Con A–HRP receptors between normal and transformed hamster fibroblasts.

The studies of the topography of various lectin–HRP surface receptors on normal and malignant cells have continued, and the results of these studies are still subject to critical evaluation (Temmink *et al.*, 1975). Garrido in a recent paper (1975) reports that clustering of Con A–HRP and WGA–HRP receptors is greater, and the HRP stain is more discontinuous during mitosis than in the G_1, S, and G_2 phases of cells grown in monolayers; however, if cells grown in suspension were labeled with Con A–HRP or WGA–HRP, there were no differences on surface patterns between cells in mitosis or during interphase. Graham *et al.* (1974) studied the Con A-HRP surface receptors of murine neuroblastoma C-1300 cells grown in suspension or on flat surfaces. Neuroblastoma cells grown in suspension (spinners) showed a continuous pattern of Con A–HRP stain, while extended (differentiated) forms of unfixed neuroblastoma cells exposed to a low concentration of Con A (10–100 μg/liters $\times$ 10^6 cells/ml) showed a patchy surface reaction to HRP; however, at higher concentrations of Con A (250–1000 μg/liter $\times$ 10^6 cells/ml), the HRP stain was continuous and not distinguishable from the stain observed in round cells grown in suspension (Fig. 2). Similar differences of Con A receptors between normal and transformed cells were demonstrated with low concentrations of fluorescent Con A (Shoham and Sachs, 1972). Therefore the pattern of the surface Con A–HRP stain varies depending on the culture conditions (suspension versus monolayer), cell types, and the concentration of the lectin. It is important to note that a continuous surface stain was observed on extended neuroblastoma cells fixed before incubation with low concentrations of Con A (Graham *et al.*, 1974). This observation is in agreement with the conclusion of Rowlatt *et al.* (1973), who noted a continuous Con A–HRP stain on cells fixed before incubation with the lectin; these

FIG. 2.   (a) Extended murine neuroblastoma cells incubated with concanavalin A (Con A), 100 μg/ml per million cells, washed, incubated with HRP, 1 mg/ml per million cells, washed, fixed with glutaraldehyde, and stained with DAB. Note discontinuous reaction product. $\times$ 15,000. (b) Same as 2a. $\times$ 38,000. (c) Same as 2a and 2b, but cells were fixed before incubation with Con A. Note continuous reaction product on plasma membrane. $\times$ 39,000. (d) Neuroblastoma cells grown in spinner (round or undifferentiated form) but treated as in 2a and 2b. Note continuous and thicker reaction product. $\times$ 39,000. PM, plasma membrane. Reproduced from Graham *et al.* (1974), p. 331 and B12.

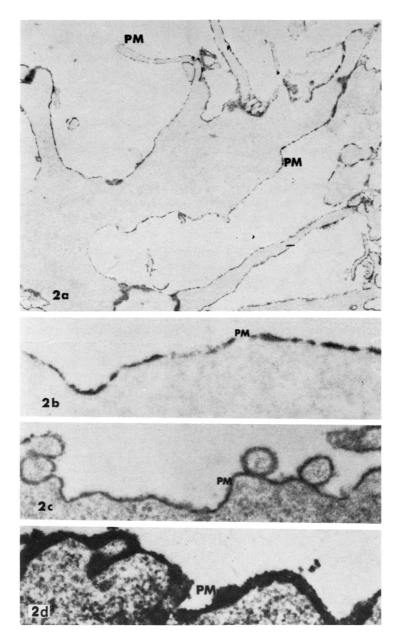

authors attributed the discontinuous Con A–HRP surface stain to a differential loss of material which occurred during the various stages of the Con A–HRP reaction on unfixed cells (Rowlatt et al., 1973). Huet and Bernadac (1974), Huet and Bernhard (1974), and Huet et al. (1974) have combined qualitative morphological studies of the topographic distribution of plasma membrane (surface) lectin–HRP receptor complexes with quantitative investigations of peroxidase activity determined spectrophometrically with O-dianisidine and of Con A binding using ^{63}Ni-labeled Con A; it was found that with the cytochemical demonstration of Con A by the HRP method, only a few lectin molecules are revealed by the enzyme; thus, for each HRP molecule there are 7.7 molecules of Con A on transformed fibroblasts and 3.3 molecules of Con A on normal fibroblasts (Huet and Bernadac, 1974).

Although the surface area of normal and transformed cells cannot be evaluated with great accuracy, the results of Huet and Bernadac (1974), as well as similar results by Collard and Temmink (1974), indicate that with HRP as marker, a small percentage of the total number of Con A molecules binding on the cell surface is revealed and furthermore temperature affects the Con A molecule (Huet et al., 1974). Thus the tetrameric form is predominant at 37°C, but it readily dissociates into dimers at lower temperatures (Huet et al., 1974); therefore the final ultrastructural appearance of the Con A–HRP receptor sites on the plasma membrane ("glycogram") depends on at least four different parameters (Graham et al., 1974; Huet et al., 1974): kinetics of Con A binding, transitions of the Con A molecule, internalization of Con A by the cell, concentration of Con A. Huet and Bernhard (1974) have observed that in malignant (transformed) cell lines, Con A–HRP "receptor" complexes disappeared from cell surfaces faster than in normal cells; interestingly, no differences in the clearance rates of WGA–HRP "receptors" of transformed or normal cells was observed (Huet and Bernhard, 1974).

These studies suggest that (1) Con A–"receptor" complexes segregate selectively in the plasma membrane of transformed cells, and (2) the faster disappearance of Con A–HRP-labeled material from the surface of transformed cells does not necessarily indicate a faster "turnover" of the entire membrane. Collard and Temmink (1974) have confirmed that the cytochemical detection of cell-bound Con A–HRP is not easily quantifiable and that differences in the amount of cell-bound Con A may not always be reflected by differences in the reaction product. In a recent paper, Temmink et al. (1975) have evaluated four different markers to detect Con A binding sites on the surface of normal and transformed cells: horseradish (HRP), hemocyamin (HC), ferritin (Fer), and microperoxidase (MP). In general, despite the significant differences between the four marker molecules used, the qualitative results were quite similar (Temmink et al., 1975).

The effect of enzyme digestion on the cytochemical demonstration of

Con A–HRP has been extensively investigated by Huet and Herzberg (1973), who found that trypsin, pronase, hyaluronidase, neuraminidase, and phospholipase C did not change the surface Con A–HRP stain; this finding indicates that the morphological Con A–HRP method is not sensitive enough to reflect in a quantitative fashion changes on surface Con A "receptors"; perhaps, the HRP activity at room temperature is so great that a relatively small number of HRP molecules may give a diffuse DAB reaction product (Temmink *et al.*, 1975). Brief incubations in the DAB medium at low temperature may result in staining patterns which might reflect more faithfully the amount of bound Con A and HRP.

B. Surface Topography of Lectin–HRP "Receptor" Complexes in Undifferentiated and Differentiated Cells

Ackerman and Waksal (1974) have found thicker Con A–HRP coats of immature human bone marrow cells than in mature neutrophils, basophils, and erythrocytes; in contrast, Parmley *et al.* (1973) found no detectable differences between mature and precursor human marrow cells incubated with Con A–HRP or with *Lens culinaris* hemagglutinins; however, these authors did not perform quantitative studies of the thickness of the oxidized DAB coat. Further studies in immature, mature, and neoplastic narrow cells may clarify the question of differences in lectin-receptor sites on the plasma membranes of these cells.

C. Internalization (Endocytosis) of Lectin–HRP "Receptor" Complexes

When lectin–HRP conjugates are incubated at +4°C with cell suspension or with cells grown on flat surfaces, and are subsequently fixed and stained with DAB, the stain is observed on the plasma membrane (Barat and Avrameas, 1973; Gonatas *et al.*, 1975; Huet and Bernhard, 1974). However, if cells labeled with lectin–HRP at +4°C, are washed and then incubated at +37°C prior to fixation and subsequent stain for HRP, the label is seen in intracytoplasmic vesicles while surface staining is reduced or absent (Barat and Avrameas, 1973; Gonatas *et al.*, 1975; Huet and Bernhard, 1974). This phenomenon has been interpreted to represent endocytosis (internalization) of the plasma membrane complexes of lectin–HRP–"receptor" molecules. In lymphocytes, Con A–HRP vesicles concentrate at one pole of the cell and subsequently appear to fuse with the plasma membrane (Barat and Avrameas, 1973). Endocytosis of plasma membrane lectin–HRP complexes occurred very rapidly in transformed cells in comparison to normal cells (Huet and Bernhard, 1974); fixation in formaldehyde or glutaraldehyde or incubation in dinitrophenol blocked the endocytosis of lectin–HRP complexes, while cytochalasin B (45 gm/ml) only partially suppressed the endocytosis of these complexes (Huet and Bernhard, 1974). Internalization of the presumed neuronal plasma membrane ricin "receptors" was noticed in the cisternae and vesicles of the Golgi apparatus of normal neurons in culture (Gonatas *et al.*, 1975).[2] In neurons, incubated with HRP alone in concentrations similar to or 10-fold higher than those used in the lectin–HRP experiments, surface or internal labeling is not observed (Fig. 3) (Gonatas *et al.*, 1975). A distinction must be made between endocytosis of lectin–HRP-labeled plasma membrane carbohydrate-containing moieties, and the uptake through pinocytic vacuoles of HRP alone, which is not bound to plasma membranes (Barat and Avrameas, 1973; Cornell *et al.*, 1971; Gonatas *et al.*, 1975; Heuser and Reese, 1973; Holtzman and Peterson, 1969; Holtzman *et al.*, 1973; Pelletier, 1973; Teichberg *et al.*, 1975); horseradish peroxidase has been used as a "tracer" molecule for the study of uptake and intracellular transport of proteins (Cornell *et al.*, 1971, Heuser and Reese, 1973; Holteman and Peterson, 1969; Holtzman *et al.*, 1973; Teschberg *et al.*, 1975); in these studies, HRP has been introduced into the extracellular space in high concentrations (3% or 10 mg/ml) whereas in the lectin–HRP experiments, the lectin-bound enzyme was introduced in low concentrations ranging from 50 gm to 400 gm per milliliter (Barat and Avrameas, 1973; Gonatas and Avrameas, 1973; Gonatas *et al.*, 1975); furthermore, the initial binding of lectin to receptor is done under temperature conditions (4°C) virtually inhibiting pinocytic activity; cells with labeled plasma membrane lectin "receptor" are subsequently transferred to and allowed to stay at 36°–37°C for various periods of time in a medium not containing HRP or lectin. The continuous association of lectin with the marker molecule (HRP) in the intracytoplasmic sites has not been formally demonstrated, and the shedding off of receptor–lectin–HRP molecules, and the subsequent pinocytic reuptake of these receptor–lectin–HRP complexes has not been ruled out. Therefore, more work is required for the clarification of the nature and significance of the intracytoplasmic translocation or "recycling" of the presumed lectin–HRP–receptor complexes (Barat and Avrameas, 1973; Gonatas *et al.*, 1975).

There are indications supporting the view that the intracytoplasmic transfer of lectin–HRP may not be analogous to the uptake of the "tracer" molecule HRP. The "tracer" HRP was found often in "coated" vesicles or in

---

[2] Since this review was written, it has been found that sites of internalized ricin–HRP, on PHA–HRP in cultured neurons correspond to GERL (Golgi-endoplasmic reticulum-lysosome) system [Gonatas *et al.*, *J. Cell Biol.* (in press)].

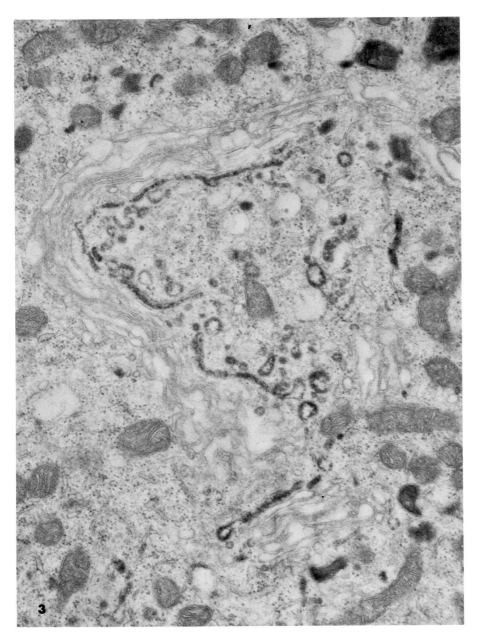

Fig. 3.   Rat dorsal root ganglion cell in culture exposed to ricin–HRP for 1 hour at −4°C, washed and incubated in full culture medium for 3 hours at −37°C prior to fixation and stain with DAB. Note peroxidase-positive cisterns and vesicles of the Golgi apparatus. × 36,000. (N. K. Gonatas, S. U. Kim, and A. Stieber., unpublished observations.)

multivesicular or dense bodies, while the endocytozed lectin–HRP was found in smooth vesicles and cisterns of the Golgi apparatus (Gonatas *et al.*, 1975; Heuser and Reese, 1973; Holtzman and Peterson, 1969; Holtzman *et al.*, 1973; Teichberg *et al.*, 1975). Occasionally, absorptive intestinal cell, neurons in organotypic cultures, somatotrophs and mammotrophs of rat pituitary, Golgi vesicles and saccules contained "tracer" HRP (Cornell *et al.*, 1971; Holtzman *et al.*, 1973; Pelletier, 1973). However, in one study (Pelletier, 1973) the existence of endogenous peroxidase was not "excluded completely" (Pelletier, 1973), and uptake of HRP was noted after the *in vitro* incubation of cells in N^6-monobutyryl cyclic AMP (Pelletier, 1973) or after stimulation with strychnine (Holtzman *et al.*, 1973) which are not necessary for the uptake of lectin–HRP complexes. Studies on the mobility of plasma membrane lectin receptors have been performed on unassociated cells. In tissues, similar studies are probably not feasible because (1) the ligand does not penetrate easily (see below) and (2) plasma membrane mobility may be limited because of cell-to-cell interactions and generally in cells that have assumed permanent fixed positions. Matus *et al.* (1973) have used subcellular fractions from brain (myelin, synaptosomes) for the study of mobility of Con A surface receptors. In these studies Con A–ferritin and a second ligand, rabbit antiferritin, were used (Matus *et al.*, 1973); in unfixed synaptosomes or myelin fragments exposed to these two ligands, redistribution (patching) and limited endocytosis of Con A–ferritin within some synaptic vesicles was seen (Matus *et al.*, 1973). Thus in synaptosomes and myelin, Con A "receptors", are mobile and can be redistributed after cross-linking with bi- or multivalent ligands. Whether similar segregations of Con A "receptors" occur in intact tissues remains to be proved.

VI.   The Use of Lectin–HRP Conjugates for Demonstration of Intracellular Carbohydrate Moieties in Tissues

Despite the known poor penetration of lectin–HRP conjugates in highly organized mammalian tissues, a successful stain of intracytoplasmic moieties has been accomplished in kidney and central nervous system (Fig. 4) (Bernhard and Avrameas, 1971; Bretton and Bariety, 1974; Roth, 1973; Wood *et al.*, 1974). Use of frozen or chopper sections and other manipulations to enhance the penetration of lectin–HRP conjugates are necessary for the effective use of lectin–HRP for morphological ultrastructural studies in intact tissues. Although at present the problems of lectin–HRP penetration in intact tissues have not yet been resolved, the successful use of these conjugates in two different complex tissues should stimulate further investigations in this important area of lectin–HRP applications (Bretton and Bariety, 1974; Wood *et al.*, 1974).

VII.   Comparison of Lectin–Ferritin to Lectin–HRP Conjugates

Because of its smaller size, HRP is considered a better label for demonstration of intracellular antigens than ferritin; on the other hand, ferritin because of its discrete, pointlike, ultrastructural image, has been used extensively for mapping of surface antigens (Nicolson, 1971, 1974a,b; Nicolson and Singer, 1971; Nicolson *et al.*, 1975). Nicolson (1971) and Nicolson and Singer (1971) have developed a technique that allows the topographic study of Con A–ferritin "receptor", on the outer or inner surface of red blood cell or fibroblast plasma membranes. With this technique, the outer surface of red blood cell plasma membrane binds Con A–ferritin while the inner surface has no Con A binding sites (Nicolson and Singer, 1971). Using the same technique, Nicolson (1971) has found that the outer cell surface of normal 3T3 mouse fibroblasts has randomly distributed Con A "receptors" while in the outer surface of the plasma membrane of SV40 transformed mouse fibroblasts the Con A "receptors" are in patches. The number of Con A receptor sites was equal between normal and transformed cells at saturation doses of Con A. Although for surface lectin binding studies, has been apparently preferred over HRP, a thorough comparative study between these two markers has not been performed. Stobo *et al.* (1972) using Con A–HRP or Con A–ferritin for the visualization of Con A binding on lymphoid cell populations, found no significant differences between these two labels.

In 1972, Bretton *et al.* (1972b) compared ferritin with HRP for labeling of cell surface immunoglobulins with appropriate antibody–ferritin or antibody–HRP conjugates. According to this study, more antigenic sites were labeled with HRP than with ferritin, and the continuous labeling obtained with HRP conjugates was due to heavier labeling, and not to diffusion of the reaction product. A highly diluted peroxidase conjugate resulted in a discontinuous, random labeling, similar to the labeling obtained with undiluted ferritin conjugates (Bretton *et al.*, 1972b).

Lectin–ferritin conjugates have been used for the localization of Con A receptors in the synaptic cleft (Cottman and Taylor, 1974); ferritin–Con A was found on the postsynaptic membrane but not in the intact cleft or on the postsynaptic densities (Cottman and Taylor, 1974). These topographic

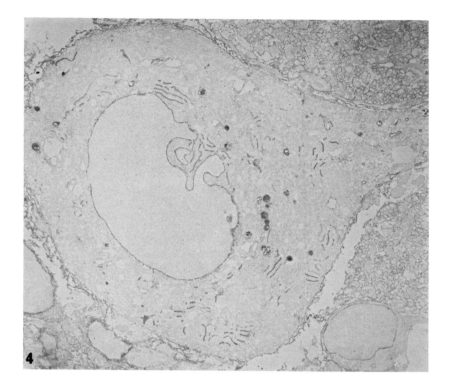

FIG. 4.   Rat Purkinje cell stained with Con A–HRP according to Wood *et al.* (1974). Note stain on certain endoplasmic reticulum membranes and on nuclear membranes; the stain of Con A–HRP is less ubiquitous than in preparations by Wood *et al.*, who noticed Con A–HRP stain also in the Golgi cisterns. Paucity of intracytoplasmic stain may be due to poor penetration of Con A in fixed solid tissues. × 4300. (Courtesy of Dr. A. Tessler, unpublished observations).

differences may be due to fewer or absent Con A binding sites in these areas or to the large size and poor penetration of the Con A–ferritin conjugate. Because of their apparent precise localization, lectin–ferritin conjugates have also been used for the study of relationships between surface lectin receptors and intramembranous particles (Pinto de Silva and Nicolson, 1974; Roth *et al.*, 1974; Tillack *et al.*, 1972; Triche *et al.*, 1975). In addition to the study of surface topography, conjugates of the toxin ricin with ferritin have been used for the ultrastructural visualization of its entry into mammalian cells (Nicolson, 1974b); according to the report, after 30–60 minutes of incubation of ricin–ferritin, endocytosis of the complexes had begun, and after incubations of 60–90 minutes ferritin (and presumably ricin) was seen in endocytotic vesicles and free in the cell cytoplasm (Nicolson, 1974b). It has been shown that in contrast to the hemagglutinin of *Ricinus communis,* which is nontoxic, the toxin ricin is a potent protein synthesis inhibitor and its effect on mammalian cells becomes apparent after some time, which presumably is required for the entry of the toxin in the cytoplasm (Olsnes and Pihl, 1973; Olsnes *et al.,* 1974; Refsnes *et al.,* 1974); this lag time at 37°C is 30 minutes, and at 20°C, 6 hours (Refsnes *et al.,* 1974); in cell-free systems, protein synthesis inhibition occurred immediately after the addition of ricin or abrin (Refsnes *et al.,* 1974). Conjugates of the hemagglutinin ricin with HRP internalize in normal neurons much faster than toxin ricin/ferritin (Gonatas *et al.,* 1975, and unpublished observations). In normal nerve cells in culture, the internalization time of ricin hemagglutinin (MW 120,000)–HRP conjugates, judged by electron microscopy, is 15–30 minutes at 37°C (personal observation); this time is similar to the "lag" time required for the toxin ricin (MW 65,000) to inhibit protein synthesis in intact cells (Refsnes *et al.,* 1974). The faster internalization time of ricin hemagglutinin–HRP over ricin–ferritin may be due to a higher sensitivity of HRP detection, to the smaller size of ricin–HRP than ricin–ferritin, or to differences in the cell systems used (Gonatas *et al.,* 1975; Nicolson, 1974b).

VIII. Summary

The specific binding of lectins with various sugars has been exploited for the cellular demonstration of carbohydrate-containing moieties. Wheat germ agglutinin, phytohemagglutinin, ricin agglutinin, or *Lens culinaris* lectin, covalently linked with horseradish peroxidase (HRP), have been used for the light and electron microscopic cytochemical (HRP) detection of surface and intracellular carbohydrates.

Horseradish peroxidase has been used for the detection of concanavalin A (Con A) receptor sites of unfixed or fixed cells and tissues. Because of the high mannose content of HRP, its covalent coupling to Con A (specificity to α-D-mannopyranosyl residues) is not required and incubation of HRP with cells or tissues previously exposed to Con A is sufficient for a Con A–HRP linkage.

Lectin–HRP conjugates offer the following advantages over existing metallic impregnation methods for the cytological detection of carbohydrates: (1) lectin–HRP conjugates can be used in unfixed cells under culture conditions; (2) with the use of radioactive lectins and assays of HRP activity, the interactions between lectin–HRP and cells can be studied quantitatively; (3) several lectin–HRP-cell interactions can be inhibited with appropriate carbohydrates (specificity of reaction).

Methods of lectin isolation and of preparation of lectin–HRP conjugates have been reviewed briefly.

The following topics have been discussed: (1) effect of fixation on the use of lectin–HRP (2) specificity of binding; (3) topography, mobility, and endocytosis of plasma membrane lectin–HRP conjugates in normal and transformed cells; (4) use of lectin–HRP conjugates for detection of intracellular carbohydrates in tissues; and (5) comparison of lectin–ferritin to lectin–HRP conjugates.

The application of lectin–HRP conjugates has been useful for the morphological mapping of surface lectin receptors in normal or transformed cells. In transformed cells, the surface stain ("glycogram") is less regular and continuous than in normal cells. These observations are consistent with the hypothesis of increase mobility of lectin receptors in transformed cells.

Internalized plasma membrane lectin–receptor conjugates have been traced in the vesicles and cisterns of the Golgi apparatus, and the significance of these observations is under current investigation (Gonatas *et al.*, 1975).

Con A receptor sites in kidney (Bretton and Bariety, 1974) and in cerebellum (Wood *et al.*, 1974) have also been studied with the use of HRP; due to poor penetration of lectins, these studies have been limited.

The information derived from the use of the lectin–HRP conjugates on cells or tissues, although still in a formative stage, is promising; the application of these conjugates for the study of problems of cell biology and cytology, deserves further exploration.

REFERENCES

Abercrombie, M., Heaysman, J. E. M., and Pegrum, S. M. (1972). *Exp. Cell Res.* **73**, 536.
Ackerman, C. A., and Waksal, S. D. (1974). *Cell Tissue Res.* **150**, 331.
Avrameas, S. (1970a). *Int. Rev. Cytol.* **27**, 349.
Avrameas, S. (1970b). *C. R. Hebd. Seances Acad. Sci.* **270**, 2205.
Avrameas, S., and Ternynck, T. (1971). *Immunochemistry* **8**, 1175.
Avrameas, S., Gonatas, N. K., and Guilbert, B. (1973). *Colloq. C. N. R. S.* **221**, No. 11, 733.
Barat, N., and Avrameas, S. (1973). *Exp. Cell Res.* **76**, 451.
Bernhard, W., and Avrameas, S. (1971). *Exp. Cell Res.* **64**, 232.
Bittiger, H., and Schnebli, H. P. (1974). *Nature (London)* **249**, 370.
Bretton, R., and Bariety, J. (1974). *J. Ultrastruct. Res.* **48**, 396.
Bretton, R., Wicker, R., and Bernhard, W. (1972a). *Int. J. Cancer* **10**, 397.
Bretton, R., Ternynck, T., and Avrameas, S. (1972b). *Exp. Cell Res.* **71**, 145.
Chowdhury, T. K., and Weis, A. K., eds. (1975), "Concanavalin A,"*Adv. Exp. Med. Biol.*, Vol. 55. Plenum, New York.
Cohen, E., (1974). "Biomedical Perspectives of Agglutinins of Invertebrate and Plant Origins," Annals, Vol. 234. N.Y. Acad. Sci., New York.
Collard, J. C., and Temminck, H. M. (1974). *Exp. Cell Res.* **86**, 81.

Cornell, R., Walker, W. A., and Isselbacher, K. J. (1971). *Lab. Invest.* **25**, 42.
Cottman, C. W., and Taylor D. (1974). *J. Cell Biol.* **62**, 236.
Doyle, R. T., Woodside, E. E., and Fishel, C. W. (1968). *Biochem. J.* **106**, 35.
François, D., Vu-Van, T., Febvre, H., and Haguenau, F. (1972). *C. R. Hebd. Seances Acad. Sci.* **274**, 1981.
Garrido, J. (1975). *Exp. Cell Res.* **94**, 159.
Garrido, J., Burglen, M. J., Samoly, K. D., Wicker, R., and Bernhard, W. (1974). *Cancer Res.* **34**, 230.
Gonatas, N. K., and Avrameas, S. (1973). *J. Cell Biol.* **59**, 436.
Gonatas, N. K., Antoine, J. C., Stieber, A., and Avrameas, S. (1972). *Lab. Invest.* **26**, 253.
Gonatas, N. K., Stieber, A., Kim, S. U., Graham, D. I., and Avrameas, S. (1975). *Exp. Cell Res.* **94**, 426.
Gordon, N., Dandekar, P. V., and Bartoszewicz, W. (1974). *J. Reprod. Fertil.* **36**, 211.
Graham, D. I., Gonatas, N. K., and Charalampous, F. C. (1974). *Am. J. Pathol.* **76**, 285.
Graham, R. C., and Karnovsky, M. J. (1966). *J. Histochem. Cytochem.* **14**, 201.
Heuser, J. E., and Reese, T. S. (1973). *J. Cell Biol.* **57**, 315.
Holtzman, E., and Peterson, E. R. (1969). *J. Cell Biol.* **40**, 863.
Holtzman, E., Teichberg, S., Abrahams, S. J., Citkowitz, E., Crain, S. M., Kawai, N., and Peterson, E. R. (1973). *J. Histochem. Cytochem.* **21**, 349.
Huet, C., and Bernadac, A. (1974). *Exp. Cell Res.* **89**, 429.
Huet, C., and Bernhard, W. (1974). *Int. J. Cancer* **13**, 227.
Huet, C., and Garrido, J. (1972). *Exp. Cell Res.* **75**, 523.
Huet, C., and Hertzberg, M. (1973). *J. Ultrastruct. Res.* **42**, 186.
Huet, C., Longchamp, M., Huet, M., and Bernadac, A. (1974). *Biochim. Biophys. Acta* **365**, 28.
Lis, H., and Sharon, N. (1973) *Ann. Rev. Biochem.* **42**, 541.
Martinez-Palomo, A. (1970). *Int. Rev. Cytol.* **29**, 29.
Martinez-Palomo, A., Wicker, R., and Bernhard, W. (1972). *Int. J. Cancer* **9**, 676.
Matus, A., DePetris, S., Raff, M. C. (1973). *Nature (London) New Biol.* **244**, 278.
Nicolson, G. L. (1971). *Nature (London) New Biol.* **233**, 244.
Nicolson, G. L. (1974a). *Int. Rev. Cytol.* **39**, 89.
Nicolson, G. L. (1974b). *Nature (London)* **251**, 628.
Nicolson, G. L., and Singer, S. J. (1971). *Proc. Natl. Acad. Sci. U.S.A.* **68**, 942.
Nicolson, G. L., Lacorbiere, M., and Eckhart, W. (1975). *Biochemistry* **14**, 172.
Novikoff, A. B., Beard, M. E., Albala, A., Sheid, B., Quintana, N., and Biempica, L. (1971). *J. Microsc. (Paris)* **12**, 381.
Olsnes, S., and Pihl, A. (1973). Biochemistry **12**, 3121.
Olsnes, S., Saltvedt, E., and Pihl, A. (1974). *J. Biol. Chem.* **249**, 803.
Parmley, R. T., Martin, B. J., and Spicer, S. S. (1973). *J. Histochem. Cytochem.* **21**, 912.
Pelletier, G. (1973). *J. Ultrastruct. Res.* **43**, 445.
Pinto de Silva, D., and Nicolson, G. L. (1974. *Biochim. Biophys. Acta* **363**, 311.
Rambourg, A., Hernandez, W., and Leblond, C. P. (1969). *J. Cell Biol.* **40**, 395.
Refsnes, K., Olsens S., and Pihl, A. (1974). *J. Biol. Chem.* **249**, 3557.
Roth, J. (1973). *Exp. Pathol.* **85**, 157.
Roth, J., and Thoss, K. (1974). *Experientia* **30**, 414.
Roth, J. Meyer, H. W., Bolck, F., and Stiller, D. (1972). *Exp. Pathol.* **6**, 189.
Roth, J., Meyer, H. W., Neupert, G., and Block, F. (1973). *Exp. Pathol.* **8**, 19.
Roth, J., Wagner, M., and Meyer, H. W. (1974). *Pathol. Eur.* **9**, 31.
Roth, J., Thoss, K., Wagner, M., and Meyer, H. W. (1975a). *Histochemistry* **43**, 275.
Roth, J., Wagner, M., and Meyer, H. W. (1975b). *Blut* **30**, 31.
Rowlatt, C., and Wicker, R. (1972). *J. Ultrastruct. Res.* **40**, 145.

Rowlatt, C., Wicker, R., and Bernhard, W. (1973). *Int. J. Cancer* **11**, 314.
Rutishauser, U., and Sachs, L. (1975). *J. Cell Biol.* **65**, 247.
Sharon, N., and Lis, H. (1972). *Science* **177**, 949.
Sharon, N., and Lis, H. (1975). *In* "Methods in Membrane Biology" (E. D. Korn, ed.), Vol. 3, p. 147. Plenum, New York.
Shoham, J., and Sachs, L. (1972). *Proc. Natl. Acad. Sci. U.S.A.* **69**, 2479.
Stobo, J. D., and Rosenthal, A. S. (1972). *Exp. Cell Res.* **70**, 443.
Stobo, J. D., Rosenthal, A. S., and Paul, W. E. (1972). *J. Immunol.* **108**, 1.
Teichberg, S., Holtzman, E., Crain, S. M., and Peterson, E. R. (1975). *J. Cell Biol.* **67**, 215.
Temmink, J. H. M., Collard, J. G., Spits, H., and Roos, E. (1975). *Exp. Cell Res.* **92**, 307.
Tillack, T. W., Scott, R. E., and Marchesi, V. T. (1972). *J. Exp. Med.* **135**, 1209.
Torpier, G., and Montagnier, L. (1973). *Int. J. Cancer* **11**, 604.
Triche, T. J., Tillack, T. W., and Kornfeld, S. (1975). *Biochim. Biophys. Acta* **394**, 540.
Wood, J. G., McLaughlin, B. J., and Barber, R. P. (1974). *J. Cell Biol.* **63**, 541.
Zanetta, J. P., Morgan, I. G., and Gombos, G. (1975). *Brain Res.* **83**, 337.

Chapter 24

Measurement of the Growth
of Cell Monolayers in Situ

GERARDO MARTINEZ-LOPEZ

Virus Laboratory, Instituto Colombiano Agropecuario, Bogota, Colombia

AND

L. M. BLACK

Department of Genetics and Development, University of Illinois, Urbana, Illinois

I. Introduction

Measurement of the increase in number of cells has been the parameter most used in determinations of the effect of different conditions on the growth of cell monolayers, but the lack of a simple and accurate means for such measurements has been one of the main problems in quantitative work with cell monolayers (Martínez-López and Black, 1973).

Several procedures have been developed for estimating cell growth in monolayers. Most of them involve the removal of the cells from the culture followed by dilutions and direct enumeration of cell number with a hemacytometer (Sanford *et al.*, 1951; Harris, 1959; Parker, 1961; Ørstavik and Mykrovold, 1967; Black, 1969; Paul, 1970) or an electronic cell counter (Harris, 1959; Parker, 1961; Ørstavik and Mykrvold, 1967; Eagle and Levine, 1968; Amborski and Moskowitz, 1968; Helleman, 1969; Helleman and Benjamin, 1969a,b; Paul, 1970; Martínez-López, 1971; Martínez-López and Black, 1971). Some other methods measure the packed cell volume (Parker, 1961; Paul, 1970), dry or wet weight (Parker, 1961; Paul, 1970), or some cellular constituent, such as protein (Parker, 1961; Eagle and Levine, 1967; Paul, 1970; Eagle, 1971, 1973), or deoxyribonucleic acid (Parker, 1961; Paul, 1970; Gautschi *et al.*, 1971). Enumeration of the total number of cells in each of a number of microscopic fields (Jamieson *et al.*, 1969) or the enumeration of the cells in ten squares on a diagonal line of a grid ocular micrometer in different microscopic fields (Mitsuhashi *et al.*, 1970) have been used. Castor (1971), Froeze (1971), and Gerschenson *et al.* (1972) have also utilized microscopic methods.

We (Martínez-López and Black, 1973) developed a procedure for accurate, quick, and simple measurement of the growth of cell monolayers with the use of an inverted phase-contrast microscope to enumerate *in situ* those cell nuclei whose circumferences include any of the 25 points of a reticule in the eyepiece. The counts were used to determine relative numbers of cells in flasks, or by taking into consideration the average diameter of the nuclei being counted, to determine estimates of the absolute number of cells in a flask.

II.   Cells Used in Development of the Technique

The counting procedure was developed in work with the cell line AC20, established by Chiu and Black (1967), from the leafhopper *Agallia constricta*, a vector of certain plant viruses. However, it should be applicable to other cell types growing as monolayers in containers having optical properties satisfactory for observation of cells or nuclei by phase-contrast microscopy.

III.   Handling of Cell Monolayers

The monolayers were grown in 25-cm^2 disposable plastic flasks (Falcon Plastics, Los Angeles). Special care must be taken to obtain an even distribution of the seeded cells on the bottom surface of the flask. A homogeneous suspension of AC20 cells was obtained by using a Vortex-Genie mixer to stir the cells in a centrifuge tube at least 5 times at maximum speed for 2–3 seconds each time. The degree of homogeneity could be judged by the subsequent distribution of cells on the bottom of the surface of the flask to which the cells were transferred. Four milliliters of cell suspension were added to each new flask, and an even distribution of the cells on the bottom of the flask was obtained by moving the flask constantly to prevent cell attachment until it was put on a leveled rack in an incubator, where it was left undisturbed until cell attachment was completed. The constant movement was designed to keep the longitudinal axis horizontal while rotating the flask back and forth about the axis through a short arc. Each freshly seeded flask contained between 1.5×10^6 and 2.5×10^6 cells, a cell density at time of subculture that gave the best growth.

IV. Enumeration of Cell Nuclei

Cell nuclei were enumerated in an inverted phase-contrast Zeiss microscope, with a body magnification of 1.25, equipped with a Ph 2 Plan 40/0.60 objective lens, and a focusable Kpl. 10 × eyepiece having a reticule with 25 points distributed systematically over a circle and connected by straight lines to facilitate orientation during counting (Fig. 1). The total magnification used during counting was × 500. The objective lens has a diaphragm, and it is important to have its opening set optimally for counting the cells readily.

Experimental flasks to be studied were held in the mechanical stage of the microscope. Cell nuclei that coincided with any one of the 25 points in the reticule were counted in the number of fields that the investigator considered sufficiently accurate for the purpose. The positions of the fields (Fig. 2) were

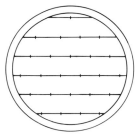

FIG. 1.   Reticule with 25 points distributed systematically within a circle and connected by straight lines to facilitate orientation during counting. (See Addendum on p. 419.)

FIG. 2.   Outline of flask with numbered points indicating the location of microscope fields in which nuclei were counted.

selected by displacement of the mechanical stage in the two coordinate directions to fixed values in the longitudinal and transverse scales.

V.   Number of Fields in Which Counts Should Be Made

On the basis of counts in microscopic fields 1 to 5, 1 to 10, 1 to 20, and 1 to 40 (Fig. 2), Martínez-López and Black (1973) determined experimentally the number of microscopic fields in which counts should be made when working with high and low cell densities. No significant differences were observed between the different numbers of counts when the variance was analyzed. Obviously one of the main factors making this possible was the uniform distribution of the cells in the monolayers. The 95 % confidence limit (Lp) as a percentage of the mean [Lp = $(1.96 \; \sigma/\mu)/\sqrt{n}$, where σ/μ = coefficient of variation], showed that it was similar for both low and high cell densities (Table I). The enumeration of nuclei in 10 microscopic fields per flask usually provided sufficiently accurate information for many of our purposes.

VI.   Use of the Technique for Testing the Suitability of Different Batches of a Component for a Medium

The technique was used by Martínez-López and Black (1973) for comparing the capacity of eight different batches of lactalbumin hydrolyzate to support the growth of AC20 cells in media that were otherwise the same. Comparisons were made between each two flasks of a pair containing the same lot of lactalbumin hydrolyzate, and between pairs containing different batches of this component. Microscope fields 1 to 10 were counted in each flask 4 days after seeding cells in the medium. When compared statistically by the F test, no significant differences were observed between any members of

TABLE I

THE 95% CONFIDENCE LIMITS (Lp) AS PERCENTAGES OF
MEAN NUMBERS OF NUCLEI ON RETICULE POINTS IN
VARIOUS NUMBERS OF MICROSCOPIC FIELDS EXAMINED
PER FLASK

Number of microscopic fields counted	Lp values (%) when the mean number of nuclei on reticule points was:	
	3.2	6.0
5	36	31
10	25	22
20	18	15
40	13	11

a pair having the same batch of lactalbumin hydrolyzate, but significant differences were demonstrated between a number of different batches (Table II). Details of the tests (Martínez-López and Black, 1973) showed that batch B appeared to be best although no significant difference was demonstrated when it was compared with batch F.

VII.   Estimation of Absolute Numbers of Cells

A.   Determination and Use of Constant K

Other things being equal it seems reasonable to assume that the chances of any nucleus coinciding with a reticule point are proportional to the area of the nucleus and to the number of nuclei growing on the bottom of a flask. As before, the area of a nucleus was taken as the area within its circumference when its equator was in focus. With our cell lines a 1:1 relationship between cells and nuclei was assumed. If the number of nuclei coinciding with the reticule points and the average area of the nuclei were determined for a flask with an unknown cell population it should be possible to calculate its cell population by reference to a standard flask *with a known cell population* for which these two values had also been determined.

In our first paper (Martínez-López and Black, 1973, p. 6) we devised such a method for estimating the absolute number of cells in a monolayer in a flask. The standard flask contained 3.04×10^6 AC20 cells with an average nuclear diameter of 12.88 μm and with 7.68 nuclei coinciding with the

TABLE II

F CAL VALUES FOR COMPARISONS BETWEEN COUNTS OF NUCLEI IN CELL MONOLAYERS
GROWN IN MEDIA CONTAINING DIFFERENT COMMERCIAL BATCHES OF LACTALBUMIN
HYDROLYZATE

[a] →	A	B	C	D	E	F	G	H
A	0.00							
B	120.10*	0.64						
C	15.58*	48.70*	0.00					
D	9.09*	62.33*	0.64	0.64				
E	9.09*	62.33*	0.64	0.00	0.64			
F	86.36*	2.59	28.57	38.96*	38.96*	0.64		
G	22.07*	38.96*	5.84*	2.59*	0.00	20.77*	0.00	
H	13.63*	48.70*	0.81	1.94	1.94	23.77*	0.00	0.64

[a] Batch of lactalbumin hydrolyzate. Each batch was tested in 2 flasks.

* Any value for F cal greater than 3.95 means that the difference between two samples is significant at the 95% confidence level.

reticule points. In this first study the square of the diameter was taken as a useful index of nuclear area. The nuclei in an unknown could, therefore, be calculated from the following equation:

$$\text{nuclei/flask} = 3.04 \times 10^6 \times [(12.88)^2/7.68] \times A/B \qquad (1)$$

A and B were determined for the unknown and, in the formula, A was the average number of nuclei counted on the points of the reticule and B was the square of the average nuclear diameter in micrometers. The above equation was then reduced to: nuclei/flask $= 65.8 \times (A/B) \times 10^6$, and 65.8 became a standard reference number. Because of the importance of the reference number 65.8 in the calculation, we indicated the desirability of further work on this value.

This we have since attempted to do in two experiments—one with AC20 cells and another with AS2 cells derived from the leafhopper *Aceratagallia sanguinolenta*. The average nuclear area of AS2 cells is about 10% greater than that of AC20 cells, and it was important to know whether derivation of the standard reference number from cells with this difference in nuclear area would give essentially the same standard reference number. In both experiments data were recorded in such a way that the standard error for each term could be calculated.

In the work now reported one flask of AC20 cells was selected for use as one standard for determining the absolute number of cells in other flasks. The flask, with a bottom surface of 25 cm^2 was carefully prepared to have a uniform monolayer which was studied at a stage in its development most suitable for the purpose. A circular glass plate was mounted in the eyepiece with the ocular lens, and this was marked to delimit a square area in the microscopic field. With the lens system used in our study and with a total magnification of

× 500, the square measured 192.4 μm on a side and delimited 37.0×10^{-5} cm² in the field of the microscope. The average number of cells, 45.1 ± 1.50, in the square of 37.0×10^{-5} cm² was determined from counts in 25 different squares, each in a different field. Cells in contact with the upper and left-hand edges of the square were not counted, whereas those in contact with the remaining two sides were counted. The average number of nuclei, 7.68 ± 0.340, appearing on the reticule points was also determined for 25 different fields. However, in this study the area of the nucleus was measured more accurately than in our first paper. A micrometer scale in a × 12.5 eyepiece was used giving a total magnification of × 600. The longest and the shortest diameter of each nucleus was measured to the nearest scale unit, and the area of each nucleus was calculated according to the formula πab, for the area of an ellipse (in this case $a = \frac{1}{2}$ of one diameter and $b = \frac{1}{2}$ of the other). Such data were collected for each of five nuclei selected at random in each of ten different microscopic fields. The average area of the nuclei was 131.13 ± 4.91 μm². Following the same reasoning as in our original paper, this standard flask was calculated to contain:

$$\frac{(45.1 \pm 1.50) \times 25}{37 \times 10^{-5}} \text{ cells}$$

and the reference value was calculated from

$$\frac{(131.13 \pm 4.91)}{(7.68 \pm 0.340)} \times \frac{(45.1 \pm 1.50) \times 25}{37 \times 10^{-5}}$$
$$\text{or } 52.03 \pm 3.48$$

All the data needed for the calculation were collected during one interval of 4 hours.

A similar experiment was carried out, using as a second standard flask one with AS2 cells. The following results were obtained for the same terms:

$$\frac{(145.62 \pm 5.44)}{(7.40 \pm 0.300)} \times \frac{(41.72 \pm 1.17) \times 25}{37 \times 10^{-5}}$$
$$\text{or } 55.47 \pm 3.43$$

The original data of these two experiments are given in the Appendix.

The standard errors indicate that there is no significant difference between 52.03 and 55.47. We conclude that they are measurements of the same parameter, which we now designate as a constant "K" and that they can be averaged to give a better estimate of K as 53.75 ± 2.44. The standard error of K in this determination is 4.5%.

We propose that our original formula (Martínez-López and Black, 1973, p. 6) be revised as follows:

$$\text{cells per 25 cm}^2 \text{ of monolayer} = (53.75 \pm 2.44) \times (NP/NA) \times 10^6 \quad (2)$$

For mnemonic reasons, in the new form of the equation we have substituted NP (nuclei on points) for A and NA (nuclear area) for B. For greatest accuracy the average nuclear area (NA) should be calculated by the formula πab and expressed in square micrometers.

B.   Discussion of the Constant K

It should be noted that there is an error, presumably a small error, in our K caused by the fact that there is always a higher density of cells in a narrow band around the perimeter of the bottom of the flask. We attribute this to cells settling into this band from the meniscus against the sides of the flask. There are experimental approaches to the measurement of this error, but we have not as yet dealt with the problem. This error means that our K is somewhat smaller than it actually should be. The meniscus error would affect measurements of absolute numbers of cells per flask, but it would seem not to affect measurements of relative concentrations of cells in two or more comparable flasks. If the nuclei are essentially spherical the formula πr^2 gives essentially the same result, and indeed in the case of the work on the AC20 and AS2 standard flasks above, calculation of the average nuclear areas by πr^2 directly from half of the average of the 100 diameters in each case, gave 130.25 μm^2 and 145.91 μm^2 for the AC20 and AS2 nuclei, respectively. These results are virtually identical with those calculated by the more laborious πab formula above. In such cases $K_{\pi r^2}$ is essentially the same as K (which is really $K_{\pi ab}$). However, with the use of πr^2 the more ellipsoidal the nuclei the more inaccurate both the average area and the standard error become, and in the extreme very large errors in both would result.

In our original paper, nuclear areas in both standard and unknown were approximated by simply squaring the average diameter $(2r)^2$. This gave us an original standard reference number of 65.8, an estimate of a value we now symbolize as K_{d^2}. The ratio between d^2 and πr^2 [or $(2r)^2/\pi r^2$] $= 1.27$ and $K \times 1.27 = K_{d^2}$. A better estimate of K_{d^2} than that in our original paper is therefore $(53.75 \pm 2.44) \times 1.27$ or 68.26 ± 3.10, a value that differs from our original one by 4 %.

Calculations may often be simplified by not converting micrometer scale units to micrometers. In *our* case the micrometer scale unit was 1.88 μm. To convert areas in μm^2 to areas in *our* scale units (su) squared, we divided by 1.88^2. To calculate cell numbers for unknowns in this manner, with our scale unit, K must be modified by dividing by 1.88^2 to become 15.21 ± 0.69. Such K values could be designated by K_{su} to indicate that all nuclear areas are based on scale units. We have used this method of calculation extensively.

Our K is based on the assumption of a $1:1$ relationship between cells and nuclei. We believe that this is essentially a true relationship with our two cell lines, but this assumption may have introduced a small error in our K and, with some materials, could introduce a serious error in cell numbers, though not necessarily in growth.

Our K is specific for our reticule and Fig. 1 represents the positions of its points. It seems desirable to record here that at a magnification of $\times 500$ with our equipment the diameter of the inner reticule circle containing the points (Fig. 1) is 230 μm and that of the microscope field is 347 μm. Also our determinations of nuclei on points and of nuclei in the square of 37×10^{-5} cm^2 were done with a $10 \times$ eyepiece, whereas those of nuclear diameters were done with a $12.5 \times$ eyepiece. However, because nuclear diameters and the size of the square in which cells were counted were both determined in micrometers, no error was thereby introduced.

Although our K of 53.75 is also specific for a monolayer area of 25 cm^2 adjustments for other areas of monolayer can readily be made by multiplication of K by $1/25 \times$ the area of the other monolayer in square centimeters. However, the meniscus error in a different monolayer might be different because of a different ratio of perimeter to area, a different depth of cell suspension over the area seeded and different meniscus properties of the medium suspending the cells.

Additional studies of K are desirable to adduce more evidence of its being a constant and to ascertain more exactly what it is. Such studies would be better than those we were able to carry out if they utilized perfectly circular nuclei (or other selected bodies of suitable optical properties), in populations of less variation in nuclear size and in populations with considerably greater and smaller diameters than those of our two species of nuclei. It would be an advantage if other chosen bodies were readily available in known concentrations and had the property of taking fixed positions in the microscope field after being randomly distributed.

Possibly mathematics alone could be used to determine K; however, our mathematical abilities are inadequate to this task.

The evidence here provided that K is a constant for two species of cells in monolayers supports the proposal in our first paper that it would seem to be applicable independently of species and conditions of culture. It also justifies corrections for differences in nuclear areas in certain purely relative comparisons of cell populations.

VIII.   Estimates of Specific Cell Populations from Nuclei on Reticule Points Alone

This section represents a useful application of corrections for nuclear areas in work with a specific cell line under standardized conditions.

Investigation of an AC20 cell monolayer during growth after subculture showed that average nuclear area changed with time and that it could be related to the average number of nuclei appearing on the reticule points. Figure 3, in which nuclear areas are based on calculations of πr^2 (not πab), shows this relationship according to data collected from many experiments. We have assumed that data based on corrections for nuclear area would give more accurate estimates of cell populations than would the uncorrected data. Accordingly, Table III was prepared converting various average areas for various numbers of nuclei appearing on reticule points to cells per flask. By the use of K, the cell populations were calculated from Fig. 3 on the basis of average nuclear areas read from the curve and the corresponding number of nuclei on reticule points. This table, specific for AC20 cells grown under our conditions, eliminated repetition of much calculation.

IX.   General Discussion

The methods described provide simple quick procedures for indexing changes in the number of cells in a monolayer culture. The time required per treatment is short, and because of this many experiments can be run simul-

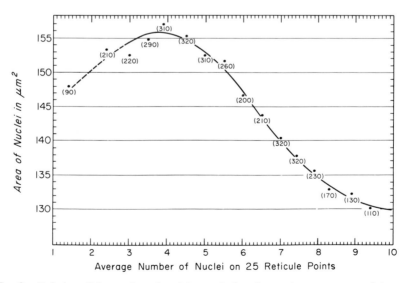

FIG. 3.   Relation of the number of nuclei on reticule points to the average area of the nuclei during our standard conditions of growth of an AC20 monolayer subsequent to cell transfer. Numbers in parentheses give the number of measurements from which the average area was obtained.

TABLE III

ESTIMATES OF MILLIONS OF AC20 CELLS PER FLASK OF 25 CM^{2a,b}

	0.0	0.1	0.2	0.3	0.4	0.5	0.6	0.7	0.8	0.9
1.	0.37	0.41	0.44	0.48	0.51	0.55	0.58	0.61	0.65	0.68
2.	0.72	0.75	0.78	0.81	0.85	0.88	0.91	0.95	0.98	1.01
3.	1.04	1.08	1.11	1.14	1.17	1.21	1.24	1.28	1.31	1.35
4.	1.38	1.42	1.45	1.49	1.53	1.56	1.60	1.64	1.68	1.72
5.	1.76	1.80	1.84	1.88	1.92	1.97	2.01	2.05	2.10	2.14
6.	2.19	2.24	2.29	2.34	2.39	2.43	2.49	2.54	2.58	2.63
7.	2.69	2.74	2.79	2.84	2.89	2.94	2.99	3.04	3.09	3.15
8.	3.19	3.24	3.29	3.34	3.39	3.44	3.49	3.54	3.59	3.64
9.	3.68	3.73	3.78	3.82	3.87	3.92	3.96	4.00	4.05	4.10

[a] Based on average numbers of nuclei observed on reticule points, average areas derived from Fig. 3, and calculations from K.

[b] Average numbers of nuclei on 25 reticule points are registered by the digit (e.g., 7.) in the first vertical column and by decimal (e.g., 0.7) horizontally across the top. The box indicated by the average of 7.7 nuclei gives an estimate of 3.04×10^6 cells per flask.

taneously. In our hands the accuracy of the procedure was better than that obtained in our earlier studies using a hemacytometer (Black, 1969) or a Coulter counter (Martínez-López, 1971; Martínez-López and Black, 1971). The use of fixed positions on the mechanical stage made it possible to make the counts more rapidly than would have been the case if the positions had been picked at random for each flask examined. However, if the distribution of the cells in the flask were to be uneven, it is possible that it would be necessary to use positions selected at random and many more of them.

It seems possible that the enumeration of nuclei on reticule points will be sufficiently accurate for many purposes. Where more precise information is desirable, it is necessary to take into consideration the average area of the nuclei being enumerated. Such information makes it possible to correct estimates of relative numbers of nuclei or to estimate the absolute number of cells per flask (Martínez-López and Black, 1973).

Because there is no damage or disturbance of the cells in the monolayer under study, it is often possible to obtain more information from each flask than in procedures necessitating the removal of the cells from the flask. With this technique, counts may be made as often as desired on a flask without introducing the variations that take place when a large number of replicate flasks must be used, as in techniques in which harvesting of the cells is a requirement. Such variations may be introduced by difficulties in dissociating cells, by the production of debris in the suspension or in other ways during the preparation of the many dilutions required in some counting techniques.

The direct enumeration of cell nuclei by phase-contrast microscopy coincidentally results in observations on cell appearance, proportion of mitotic cells, and cytological changes.

The resolution and high magnification now available in the phase-contrast microscope, with a working distance of 4 cm between phase-contrast disk and the object, has been one of the more important factors in the ulilization of this procedure. When lower magnifications are used, it becomes more difficult and less accurate to enumerate the cell nuclei that coincide with the reticule points. This high magnification is also a requirement when the diameters of nuclei must be measured.

APPENDIX

Original Data of Experiments on AC20 and AS2 Cells for the Determination of K

AC20 cells				AS2 cells			
NP[a]	C/S[b]	$a \times b$[c]		NP[a]	C/S[b]	$a \times b$[c]	
8	54	8 × 7	7 × 6	12	47	7 × 7	8 × 8
11	42	6 × 8	6 × 6	8	48	7 × 6	6 × 10
8	51	8 × 8	6 × 7	6	32	6 × 8	8 × 7
9	50	10 × 7	5 × 7	7	44	7 × 10	5 × 7
5	54	7 × 7	6 × 8	8	45	6 × 8	11 × 8
7	41	7 × 8	5 × 6	9	35	8 × 8	8 × 5
7	42	6 × 6	6 × 8	8	49	6 × 7	6 × 7
5	38	6 × 6	7 × 6	6	32	8 × 9	9 × 6
6	36	5 × 7	8 × 7	6	39	8 × 8	8 × 5
6	26	7 × 6	7 × 7	7	40	5 × 6	11 × 5
9	49	5 × 6	6 × 8	10	46	8 × 7	8 × 8
6	49	10 × 6	6 × 8	6	28	5 × 5	5 × 9
12	51	6 × 5	5 × 8	7	48	6 × 6	6 × 7
10	57	6 × 6	9 × 8	8	44	4 × 9	8 × 7
8	57	7 × 7	6 × 6	6	44	6 × 6	6 × 7
9	49	6 × 7	5 × 7.	8	47	8 × 7	8 × 9
7	45	6 × 6	5 × 8	8	46	8 × 7	8 × 9
7	38	6 × 6	8 × 10	8	39	6 × 8	10 × 9
8	45	7 × 8	6 × 6	7	42	6 × 8	6 × 7
8	35	8 × 7	6 × 6	6	38	7 × 6	5 × 6
8	49	6 × 7	8 × 6	6	43	8 × 6	8 × 7
7	43	7 × 9	7 × 6	7	43	7 × 7	8 × 8
6	45	7 × 7	8 × 9	9	48	8 × 7	7 × 7
7	37	6 × 6	8 × 6	6	43	7 × 9	8 × 8
8	45	9 × 8	8 × 8	6	33	7 × 8	9 × 7

[a] Numbers of nuclei on reticule points in each of 25 microscopic fields (NP = nuclei on points).
[b] Numbers of cells in a square with an area of 37×10^{-5} cm^2 in each of 25 microscopic fields (C/S = cells in square).
[c] Longest and shortest diameters (a and b) in scale units, of 1.88 μm, for each of 50 nuclei selected at random.

ACKNOWLEDGMENT

We wish to thank Dr. S. G. Carmer for his counsel on the statistics and for his calculation of the standard error of K.

REFERENCES

Amborski, R. L., and Moskowitz, M. (1968). *Exp. Cell Res.* **53**, 117–128.

Black, L. M. (1969). *Annu. Rev. Phytopathol.* **7**, 73–100.

Castor, L. N. (1971). *Exp. Cell Res.* **68**, 17–24.

Chiu, R. J., and Black, L. M. (1967). *Nature (London)* **215**, 1076–1078.

Eagle, H. (1971). *Science* **174**, 500–503.

Eagle, H. (1973). *J. Cell. Physiol.* **82**, 1–8.

Eagle, H., and Levine, E. M. (1967). *Nature (London)* **213**, 1102–1106.

Eagle, H., and Levine, E. M. (1968). *Nature (London)* **220**, 266–269.

Froese, G. (1971). *Exp. Cell Res.* **65**, 297–306.

Gautschi, J. R., Schindler, R., and Hürni, C. (1971). *J. Cell Biol.* **51**, 653–663.

Gerschenson, L. E., Okigaki, T., Andersson, M., Molson, J., and Davidson, M. B. (1972). *Exp. Cell Res.* **71**, 49–58.

Harris, M. (1959). *Cancer Res.* **19**, 1020–1024.

Helleman, P. W. (1969). *Scand. J. Haematol.* **6**, 160–165.

Helleman, P. W., and Benjamin, C. J. (1969a). *Scand. J. Haematol.* **6**, 60–76.

Helleman, P. W., and Benjamin, C. J. (1969b). *Scand. J. Haematol* **6**, 128–132.

Jamieson, C. W., Russin, D., Bencs, E., De Witt, C. W., and Wallace, J. H. (1969). *Nature (London)* **222**, 284–285.

Martinez-López, G. (1971). M.S. Thesis, University of Illinois, Urbana.

Martinez-López, G., and Black, L. M. (1971). *Phytopathology* **61**, 902. (abstr.)

Martinez-López, G., and Black, L. M. (1973). *In Vitro* **9**, 1–7.

Mitsuhashi, J.; Grace, T. D. C., and Waterhouse, D. F. (1970). *Entomol. Exp. Appl.* **13**, 327–341.

Ørstavik, J., and Mykrvold, V. (1967). *Acta Pathol. Microbiol. Scand.* **70**, 341–348.

Parker, R. C. (1961). "Methods of Tissue Culture," 3rd ed. Harper, New York.

Paul, J. (1970). "Cells and Tissue Culture," 4th ed., pp. 280–284. Williams & Wilkins, Baltimore, Maryland.

Sanford, K. K., Earle, W. R., Evans, V. J., Waltz, H. K., and Shannon, J. E. (1951). *J. Natl. Cancer Inst.* **11**, 773–786.

Addendum

The reticule used in our work is a Carl Zeiss Integrating Eyepiece, catalog No. 47 40 36, which was discontinued in about 1972. Because our constant K is specific for this reticule we have arranged to lend it to Merz Optical Instruments, Inc., 2519 West Peterson Avenue, Chicago, Illinois 60659, which company is able to provide exact duplicates of the reticule. It also sells the focusable eyepiece we used. The authors have no financial interest in any of these arrangements.

Chapter 25

Peptones as Serum Substitutes for Mammalian Cells in Culture

WILLIAM G. TAYLOR

*Cell Physiology and Oncogenesis Section, Laboratory of Biochemistry,
National Cancer Institute, Bethesda, Maryland*

AND

RAM PARSHAD

Department of Pathology, Howard University College of Medicine, Washington, D.C.

A chemically characterized protein-free and reproducible medium is not only desirable but also ultimately necessary in order to study the biosynthetical capacities of cells *in vitro* . . . growing mammalian cells in chemically defined media may be a matter of gradually learning to handle a specific cell strain under the more demanding culture conditions of protein-free media rather than a matter of always basically altering the cell strain.—Evans *et al.* (1964).

I. Introduction

Since, *in vivo*, mammalian cells have evolved in a milieu of blood and lymph, it is appealing to postulate *in vitro* cellular requirements for plasma or serum. The apparent serum dependency of most primary cultures and low passage lines reinforces this hypothesis. In fact, the broad spectrum usefulness of serum as a medium supplement obscures certain problems associated with its use (Taylor, 1974). Biochemical variability between lots of serum (Esber *et al.*, 1973; Honn *et al.*, 1975), serum enzymes—such as arginase (Kihara and de la Flor, 1968), asparaginase and glutaminase (Wein and Goetz, 1973), and peptidase (Jones *et al.*, 1975)—which deplete cell-free medium of essential nutrients and may, thereby, contribute to chromosome instability (Parshad and Sanford, 1968, 1972; Freed and Schatz, 1969), and the potential introduction of adventitious microbial contamination (Fogh, 1973) are ways in which the serum supplement contributes to variability in the culture system.

Approaches to replacing serum as a medium supplement include the use of: (a) biologically active macromolecules such as purified serum proteins (Temin *et al.*, 1972); (b) plasma extenders such as dextrans (Mizrahi and Moore, 1971), starch and cellulose derivatives (Kuhler *et al.*, 1960; Bryant, 1970), and polyvinylpyrrolidone; and (c) digests and hydrolyzates such as peptones. When the yield and growth-promoting activity of fractionated serum proteins are compared with the growth responses in untreated serum, the time and expense involved elimate this approach for other than research purposes. Dextrans and methylcellulose protect cells in suspension culture from sheer forces exerted by turbulence, but are of marginal benefit for surface substrate-dependent cells. Peptones are inexpensive, have an indefinite shelf life, and promote rapid, continuous, proliferation of several lines from a variety of species (Section III,A).

Cells adapted to growth in serum-free peptone containing medium can be used for several purposes. They support production of infectious virus in the absence of immunogenic serum proteins (Pumper *et al.*, 1960; Medzon and Merchant, 1971; Taylor *et al.*, 1974b; Keay, 1975). An endogenous "growth factor" (GF) from conditioned medium of peptone-dependent cells has been isolated (Alfred and Pumper, 1960) and characterized (Alfred and Pumper, 1962); the similarity between GF and other growth-promoting agents isolated from conditioned medium (Shodell, 1972; Smith and Temin, 1974) has not been established. Monkey kidney (LLC-MK$_2$) cells (Hull *et al.*, 1962) adapted to serum-free peptone-supplemented medium (Taylor *et al.*, 1974b) form "domes" (unpublished data) which are morphologically similar to those observed in canine and mouse cells (Abaza *et al.*, 1974; Visser and Prop, 1974). This dome formation shows that maintenance of differentiated structure can be achieved in serum-free medium. Earlier studies on the effect of trypsin on peptone-dependent cells (Pumper *et al.*, 1971) demonstrated the exquisite protease sensitivity of cells grown in serum-free as compared to serum-supplemented medium (Sefton and Rubin, 1970; Burger, 1970). Peptone-containing medium, therefore, represents a useful intermediate between serum-containing and serum-free medium, an "interim compromise" to be used, while improved medium formulations and adequate serum substitutes are developed.

II.   Peptones as Serum Substitutes

A.   Selection

Peptones are enzymic or acid hydrolyzates of plant or animal tissue prepared under rigidly specified production conditions (Bridson and Brecker, 1970). Proteolytic digests such as Witte's peptone (Friedr. Witte Chemische. Fabrik, Frankfort, Germany) (Carrel and Baker, 1926; Wilmer and Kendal, 1932) and Bacto-peptone (Difco Laboratories, Detroit, Michigan) (Waymouth, 1956; Pumper, 1958; Merchant and Hellman, 1962) have been used most frequently, but several other enzymic digests will promote rapid and continuous proliferation in serum-free medium (Taylor *et al.*, 1972a).

In this laboratory a 0.5–2.0 % w/v peptone supplement can be used with virtually any tissue culture medium. The cell lines discussed in this chapter grow in serum-free, peptone-supplemented medium but will *not* survive and proliferate in: (a) isotonic tissue culture medium devoid of peptone or peptone dialyzate (PD, Section II,C) as shown by growth in PD-containing as compared with PD-free medium (Tables I and III) and growth of NCTC 8096 as compared with 8207 and 8081 (Fig. 1); or (b) isotonic saline containing otherwise adequate concentrations of peptone or PD. Short-term survival and/or growth may be observed in peptone-free media such as NCTC 135 or Dulbecco–Vogt's modified minimum essential medium (Morton, 1970; Rutzky and Pumper, 1974), but most frequently the cultures degenerate.

Peptones prepared by acid digestion support generally lower cell yields (Taylor *et al.*, 1972a; Keay, 1975). During production the pH of the hydrolyzate must be neutralized, and as a result the salt content of the hydrolyzate increases to 30–40 % w/w (Bridson and Brecker, 1970). Media containing such peptones are frequently hyperosmotic.

B.   Preparation of Whole Peptone

A 10 % w/v peptone stock should be prepared in the highest quality distilled water available (Pumper, 1973). Warm or boiling water may be necessary, although Bacto-peptone and Peptic Peptone (NBCo, Cleaveland, Ohio) are quite soluble at room temperature and promote good growth; similar characteristics have been observed for Bacteriological Peptone, Code L37 (Oxoid Ltd., available from Flow Laboratories, Rockville, Maryland) (H. J. Morton, personal communication). Stock peptone solutions must be sterilized by autoclaving; filtered preparations promote significantly lower yields of peptone-dependent cells under otherwise identical conditions. An autoclave should be used which: (a) is not employed for decontamination purposes, and (b) has a minimum of volatilized filming amines and heavy

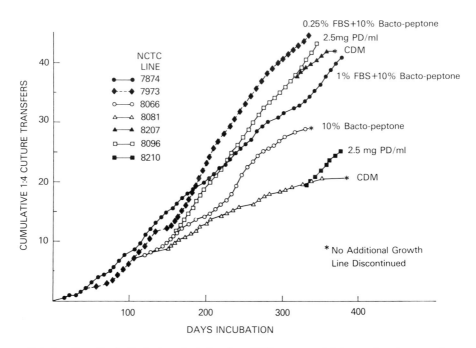

FIG. 1.   Growth of cells during adaptation from 10% serum-containing medium to serum-free and chemically defined medium (CDM). Because all cultures were transferred when they became confluent, the slopes represent relative growth rates. Parent line NCTC 5405, grown in serum containing NCTC 135 was subcultured weekly at a 1:3 split ratio, and the slope for 5405 was comparable to line NCTC 8066 in peptone-containing medium. Note that the growth rate in CDM (NCTC 8081) gradually diminished; by microscopic examination the general cell appearance became progressively less satisfactory. These effects were reversed if a 2.5 mg peptone dialyzate (PD) per milliliter supplement was introduced (line NCTC 8210). Similarly line NCTC 8096, which was grown continuously in NCTC 135 containing 2.5 mg PD/ml, immediately exhibited a diminished growth rate when subcultured in PD-free CDM (line NCTC 8207). In contrast to the rapid adaptation obtained with a combined peptone-serum supplement, over a year was required to adapt NCTC 5405 to 1/64% FBS by serial, 2-fold dilution of the serum supplement; these cells remained serum-dependent and could not be grown continuously in CDM (F. Price, personal communication).

metals in the steam (Kahn *et al.*, 1974), since these can "chemically contaminate" the serum substitute.

Cell yields may be improved by addition of 1 gm of glucose per 100 ml of peptone solution prior to autoclaving (Pumper *et al.*, 1965; Lasfargues *et al.*, 1973). Alternatively, a separate, aqueous 10% w/v glucose stock can be sterilized and 1 ml added per 100 ml of peptone solution. In either case a 10% v/v peptone supplement will increase the glucose concentration of the medium by 1 mg/ml. Although convenient for bulk production of peptone-dependent cells (Taylor *et al.*, 1975), additional glucose appears unnecessary if periodic medium renewals are used. In quantitative nutritional studies with glucose,

sodium pyruvate, and certain of the tricarboxylic acid cycle (TCA) inter-
mediates, no differences in cell yield were observed between experimental and
control cultures (Table I). In PD-free medium 199 marginal growth enhancement with 1 mM sodium pyruvate was anticipated (Neuman and McCoy,
1958), and slightly higher yields were also obtained with 1 mM succinic acid
in medium 199 containing. 0.5 mg PD/ml. The data show that these metabo-
lites, at 1 mM, neither supplant the peptone requirement of rabbit heart (RH)
cells (Pumper et al., 1965) in PD-free medium nor amplify RH cell growth in
medium containing 0.5 mg PD/ml.

C. Preparation of Peptone Dialyzate

Unlike serum, the growth-promoting activity of Bacto-peptone (Pumper
et al., 1965), Peptic peptone, and Witte's peptone (R. A. Dworkin, V.
J. Evans, and W. G. Taylor, unpublished data) is dialyzable. The procedure

TABLE I

GROWTH OF RABBIT HEART CELLS IN MEDIUM 199 CONTAINING ADDITIONAL GLUCOSE,
SODIUM PYRUVATE, OR A TRICARBOXYLIC ACID CYCLE INTERMEDIATE

Peptone dialyzate (mg/ml)	Carbohydrate supplement[a]	Experiment, average cell number $\times 10^5$/T-15[bc]		Medium osmolality (mOsm/kg H_2O)
		NCTC 8256	NCTC 8265	
2.5	None	45.5 ± 1.5	35.4 ± 1.0	292
	+ Glucose, 1 mg/ml	43.4 ± 1.4	35.0 ± 0.7	298
0.5	None	20.4 ± 0.8	17.5 ± 0.7	ND[d]
	Sodium pyruvate	19.8 ± 0.7	17.0 ± 0.4	281
	Oxaloacetic acid	21.7 ± 0.6	12.8 ± 0.8	281
	α-Ketoglutaric acid	21.7 ± 0.6	15.6 ± 0.9	281
	Disodium fumarate	18.5 ± 0.3	14.9 ± 0.5	282
0.0	None	8.15 ± 0.6	3.91 ± 0.1	ND
	Sodium pyruvate	10.2 ± 0.7	6.77 ± 0.4	278
	Oxaloacetic acid	9.90 ± 0.5	5.44 ± 0.1	278
	α-Ketoglutaric acid	8.79 ± 0.3	6.75 ± 0.3	276
	Disodium fumarate	$8.70 + 0.2$	$4.90 + 0.5$	280

[a]Sodium pyruvate, oxaloacetic acid, α-ketoglutaric acid, and disodium fumarate (Calbiochem, La Jolla, California) were filtered as 20 mM stock solutions and tested in medium 199 at a
final concentration of 1 mM. The addition of 1 mg of glucose per milliliter of medium yielded a
glucose concentration of $\sim$ 200 mg/100 ml.

[b] Five replicate T-15 cultures per variable per experiment were prepared (Taylor and Evans,
1974) and incubated 5 days with a complete medium change on day 2. On day 5, the cell populations were dispersed with 0.25 % trypsin and counted with a Coulter Model B counter. Standard
error was estimated by the method of Mantel (1951). Osmolality determinations were made on
an Advanced Instrument Osmometer (Advanced Instruments; Needham, Massachusetts).

[c]Inoculum size (cells $\times 10^5$/T-15 flask): NCTC 8256, 1.34; NCTC 8265, 1.24.

[d]ND, not done.

used in this laboratory is a scaled-up modification of the method described by Pumper *et al.* (1965). One hundred fifty grams of peptone are dissolved with stirring (2-inch magnetic stirring bar) in 1000 ml of warm triple glass-distilled water in a 2-liter graduated cylinder. The cylinder is capped loosely with absorbent cotton and oil-free aluminum foil. Simultaneously, four 21-inch strips of 1.21-inch flat width dialysis casing (Fisher Scientific, Cat. No. 8-667C) are wetted in triple glass-distilled water, then transferred to a 1-liter beaker containing 600–700 ml of fresh triple glass-distilled water and boiled for 5 minutes. The casings are again rinsed with several volumes of fresh triple-glass-distilled water and knotted at one end; excess casing below the knot is cut off, and each casing is filled with 75 ml of triple glass-distilled water. A thin nylon cord and the open end of the dialysis casing are tied, and the 4 casings are placed in a 2-liter graduated cylinder containing ~ 1600 ml of triple glass-distilled water, and the cylinder is capped as above. The nylon cords should extend outside the neck of the graduate. The peptone, dialysis casings, and an additional 700–900 ml fresh glass-distilled water are sterilized by autoclaving.

After equilibration of all reagents at 5°–8°C, the volume of peptone is brought to 1.5 liter with sterile distilled water and stirring. The dialysis casings are aseptically transferred to the 10% w/v peptone by means of the nylon cords, the cords are cut just above the knotted casing, and dialysis with stirring is carried out for 24 hours at 5°–8°C (Fig. 2). Subsequently, casings are removed and washed with cold, sterile distilled water, and the fluid *inside* the dialysis casings is harvested by (a) aseptically withdrawing the peptone dialyzate (PD) with a needle and syringe; or (b) pooling and evaporating the PD (Taylor *et al.*, 1972b). PD harvested as a fluid is used without additional processing at 5% v/v; if lyophilized, it is convenient to reconstitute 2 gm of PD per 40 ml of triple glass-distilled water and use this solution at 5% v/v or 2.5 mg PD per milliliter of medium. However, the optimal PD concentration varies with the cell line (Fig. 3), the culture medium, and whether or not fluid renewal(s) is performed during the growth period. The nondialyzable fraction (Pumper *et al.*, 1965) and the excluded high-molecular-weight fraction assayed by Amborski and Moskowitz (1968) support lower cell yields than do the whole peptone or PD; this cytostatic effect may explain the variability in growth-promoting activity sometimes observed between lots of whole peptone (H. J. Morton, personal communication).

FIG. 2.   Schematic representation of peptone dialyzate preparation, as described in Section II,C.

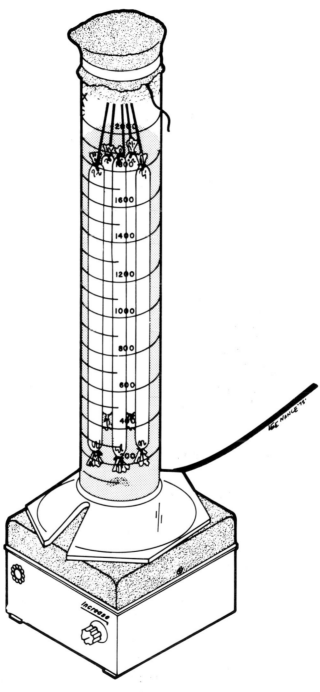

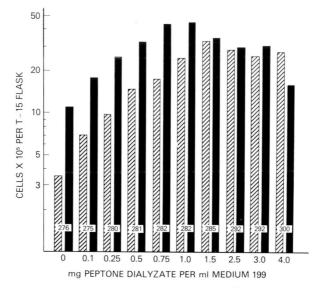

FIG. 3.   Growth of rabbit heart and L-M cells in medium 199 containing different peptone dialyzate (PD) concentrations. Replicate cultures were prepared and handled as described for Table I. Yields of rabbit heart cells (Pumper *et al.*, 1965), shown by the crosshatched bars, were consistently lower at < 1.0 mg PD/ml than yields of L-M cells, shown by the solid bars. This difference in yield demonstrates a greater peptone dependency or requirements in rabbit heart cell populations. The numbers set in the bar graphs are the medium osmolality values, expressed in mOsm/kg H_2O. Values increased with increasing PD concentration since PD contains ∼ 6.4% w/w inorganic salts, but the variation among the media used would not significantly influence growth of these cells. Inoculum size ($\times 10^5$ cells/T-15 flask): rabbit heart, 1.52; L-M, 1.73.

III.   Cell Growth in Peptone-Containing Medium

A.   Adaptation Procedures

Although many established cell lines exhibit a serum dependence, they frequently can be adapted to continuous growth in serum-free medium. In addition to strain L cells, NIH Swiss Mouse Lung (Pumper, 1958), rabbit (Pumper *et al.*, 1965; Taylor *et al.*, 1974b), hamster (Keay, 1975), rhesus and African green monkey (Taylor *et al.*, 1974a; R. W. Pumper, unpublished data), and human KB (R. W. Pumper, personal communication) and mammary tumor (Lasfargues *et al.*, 1973) cell lines can be grown continuously in medium supplemented with 0.5–1.0% w/v whole Bacto-peptone or 1.0–2.5 mg PD per milliliter.

There are two general ways of adapting cells to growth in serum-free medium. First, the serum can be abruptly replaced with peptone or PD. Even at high inoculum sizes, an obvious growth lag or "crisis" may be observed after 2–3 culture transfers (15–30 days) in serum-free medium; during this growth lag periodic changes of medium are useful though not necessary. Subsequently, populations "adapted" to growth in serum-free medium will proliferate and form confluent monolayers at rates comparable to those of parent cultures in serum-containing medium. The second method involves progressive, stepwise reduction of the serum concentration and concomitant addition of peptone or PD to the growth medium, as described in Table II and shown in Fig. 1. Serum-dependent mouse line NCTC 5405 exhibited no lag or crisis phase during adaptation by the latter procedure, and only 159 days were required to adapt this line from medium NCTC 135 (Evans *et al.*, 1964) containing 10 % v/v serum to growth in serum-free, peptone-free chemically defined medium (CDM) (NCTC 8081). Further reduction in the peptone-serum concentration(s) was attempted only after experimental lines could be regularly subcultured at a minimum 1 : 3 ratio. Best attachment and outgrowth frequently were obtained by mechanically dispersing a newly formed cell sheet in ~ 1 ml of spent medium, gently aspirating the suspension to dissociate large cell clusters, and diluting the appropriate volume of cell suspension in fresh experimental growth medium. It is important to note that the absence of serum-antitryptic proteins precludes the use of proteolytic enzymes to disperse confluent monolayers. A cellophane mop (Evans *et al.*, 1964) or a sterile bacteriological loop is used most frequently; chelating agents such as EDTA or EGTA have not assessed adequately, although Pumper reports that 5 % EDTA is useful for this purpose (personal communication).

TABLE II

RELATIONSHIP OF CELL STRAINS DURING ADAPTATION OF LINE NCTC 5405 TO GROWTH IN SERUM-FREE MEDIUM

Parent culture[a]	Derivative strain	Adaptation period[c] (days)	Fetal bovine serum (% v/v)	Bacto-peptone (% w/v)	Peptone dialyzate (mg/ml)
NCTC 5405–333	7873[b]	—	5	0.5	0.0
	7874	49	1	1.0	0.0
7874–4	7938[b]	—	0.5	1.0	0.0
7874–6	7973	68	0.25	1.0	0.0
7973–8	8066	42	0.0	1.0	0.0
8066–3	8081	—	0.0	0.0	0.0
8066–5	8096	—	0.0	0.0	2.5

[a] The number following the hyphen indicates the number of serial culture transfers before further reduction in serum–peptone supplement was attempted.

[b] Cultures were not used for further adaptation studies.

[c] The total time required to adapt NCTC 5405 to growth in serum-free, peptone-free chemically defined medium was 159 days.

B. Chromosome Analyses

Because it is frequently suggested that extensive karyotypic alterations occur during the adaptation period, chromosome analyses were performed on each cell strain developed. Cells from T-15 cultures grown in each of the culture media were planted into T-30 flasks. At the peak of mitotic activity, when the growth surface was 75–80% covered by cells, dividing cells were arrested at metaphase by addition of a final concentration of 0.1 μg of Colcemid (Ciba Pharmaceutical Co., Inc., Summit, New Jersey) per milliliter of culture medium for 3 hours. The cells were then scraped from the flask and centrifuged at 500 g for 10 minutes. The cell pellet was suspended in 0.9% aqueous solution of sodium citrate for 12–15 minutes at 37°C. The cell suspension was again centrifuged at 500 g for 8 minutes and, without disturbing the cells, the pellet was fixed in absolute methanol–glacial acetic acid (3:1) at 4°C for 30 minutes. After fixation, the methanol–glacial acetic acid was replaced twice, after intervals of 30 minutes followed by centrifugation at 500 g for 10 minutes. Subsequently, the volume of cell suspension was brought to 0.5 ml with fixative, and to spread the cells a few drops of suspension were allowed to fall onto a wet, cold slide from a distance of about 12–18 inches with blowing. Slides were dried overnight and stained with 2% aqueous solution of Giemsa (Harleco, Philadelphia, Pennsylvania) for 3 minutes. The stained slides were rinsed in water, air dried, cleared in xylene, and mounted in Permount.

Despite the varied serum and peptone concentrations in which derivatives of NCTC 5405 were grown, no significant difference in chromosome number of frequency of abnormal chromosomes was observed between cells adapted to growth in test media and the parent culture (Fig. 4). The stable aneuploid chromosome mode presumably acquired by line NCTC 5405 with time in may have contributed to these karyologic similarities; obviously, continued growth in the different media required continued modifications in cellular responses to the changing environment. Additional studies, particularly with primary and low-passage cells, are needed.

IV. Limitations and Prospects

As stated in Section I, peptone-containing, serum-free medium may represent and "interim compromise"; as such it is useful for some purposes and unsatisfactory for others. For example, primary cultures have been initiated and established in peptone-supplemented NCTC 135, but better and more reproducible results were obtained with serum-supplemented

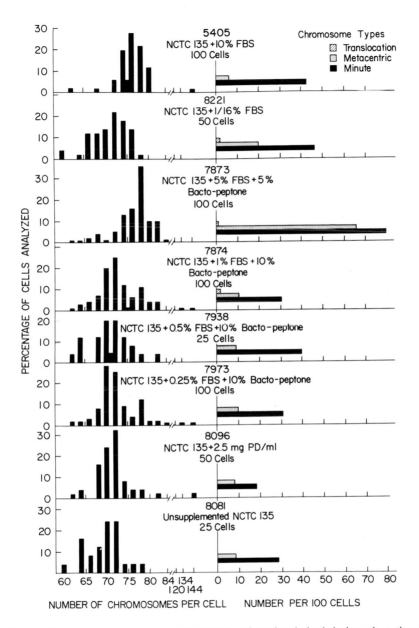

FIG. 4.   Chromosome analyses of NCTC 5405 and strains derived during adaptation to serum-free and chemically defined medium. Despite variations in the medium supplement(s), no significant change in modal chromosome number or frequency of abnormal chromosomes was observed. For comparison, line NCTC 8221 was adapted by gradual, serial dilution of the serum supplement rather than with a combined peptone–serum supplement.

TABLE III

GROWTH RESPONSE OF PEPTONE-DEPENDENT RABBIT HEART CELLS TO 6,8-DIHYDROXYPURINE

Medium 199 containing		Average cell number × 10^5/T-15				Medium osmolality (mOsm/kg H_2O)
Peptone dialyzate (mg/ml)	6,8-Dihydroxypurine[a] (μg/ml)	NCTC 8483		NCTC 8492		
		Day 2	Day 5	Day 2	Day 5	
2.5	—	3.71 ± 0.14	19.1 ± 1.7	5.53 ± 0.13	31.5 ± 0.6	291
0.0	—	2.31 ± 0.14	2.16 ± 0.67	3.34 ± 0.10	4.47 ± 0.89	277
	0.01	2.70 ± 0.18	3.27 ± 0.40	3.87 ± 0.08	6.27 ± 0.28	270
	0.03	1.81 ± 0.16	2.95 ± 0.12	3.53 ± 0.17	6.01 ± 0.65	274
	0.10	2.27 ± 0.37	4.91 ± 0.40	3.76 ± 0.15	7.34 ± 0.27	277
	0.30	2.63 ± 0.35	3.74 ± 0.85	3.86 ± 0.12	5.85 ± 0.90	271
	0.60	2.35 ± 0.46	4.24 ± 1.1	3.66 ± 0.07	5.64 ± 0.21	276
0.5	—	3.25 ± 0.26	12.26 ± 1.2	4.87 ± 0.47	17.2 ± 0.7	282
	0.01	2.93 ± 0.20	13.23 ± 0.47	3.80 ± 0.35	14.0 ± 2.7	278
	0.03	3.22 ± 0.10	13.40 ± 1.3	4.58 ± 0.46	16.0 ± 1.2	278
	0.10	2.94 ± 0.16	13.1 ± 1.3	4.61 ± 0.60	16.7 ± 0.7	278
	0.30	3.03 ± 0.10	11.5 ± 0.9	4.15 ± 0.42	14.4 ± 0.6	276
	0.60	3.13 ± 0.35	12.8 ± 0.6	5.06 ± 0.33	15.13 ± 0.6	277

[a] An aqueous 2 mM solution of 6,8-dihydroxypurine (Aldrich Chemical Co., Milwaukee, Wisconsin) was prepared and sterilized by autoclaving. Growth media were supplemented at 10–600 μg/liter after appropriate dilution of the 2 mM stock solution with sterile triple glass-distilled water.

[b] Experimental procedures were identical to those described for Table I. Values represent the mean population observed in 3–4 T-15 flasks per variable per incubation period.

[c] Inoculum size (cells × 10^5/T-15): NCTC 8483, 1.58; NCTC 8492, 1.66.

NCTC 135. Established lines can be adapted to growth in peptone-containing medium (Section III,A), but more physiological means of dispersing monolayers must be developed. Although many infer that the growth-promoting activity is associated with low-molecular-weight peptide(s) (C. Waymouth, personal communication; Hsueh and Moskowitz, 1973), the chemical nature of the biologically active component(s) has not been firmly established (Taylor et al., 1974a). Recent reports attribute the growth-promoting activity of Bacto-peptone to 6,8-dihydroxypurine (Yamane and Murakami, 1971, 1973). However, studies in this laboratory show that this purine neither supplants the peptone requirement of RH cells nor amplifies growth in medium containing a growth-limiting PD concentration, 0.5 mg/ml (Table III). Little is known about how peptones support continued cell survival and proliferation, though it is plausible that peptones and PD promote production of an endogenous growth factor such as described by Alfred and Pumper (1960, 1962).

The multifunctional role of serum and the diverse activities—hence requirements—of cells in culture will continue to frustrate oversimplified efforts to supplant the in vitro serum requirements. Continued efforts must be directed at formulating a milieu that will maintain the explanted cell in a genotypic and phenotypic state comparable to that of its in vivo progenitor.

REFERENCES

Abaza, N. A., Leighton, J., and Schultz, S. G. (1974). In Vitro 10, 172.
Alfred, L. J., and Pumper, R. W. (1960). Proc. Soc. Exp. Biol. Med. 103, 688.
Alfred, L. J., and Pumper, R. W. (1962). Biochem. Biophys. Res. Commun. 7, 284.
Amborski, R. L., and Moskowitz, M. (1968). Exp. Cell Res. 53, 117.
Bridson, E. Y., and Brecker, A. (1970). Methods Microbiol. 3A, 229–295.
Bryant, J. C. (1970). Biotechnol. Bioeng. 12, 429.
Burger, M. M. (1970). Nature (London) 227, 170.
Carrel, A., and Baker, L. E. (1926). J. Exp. Med. 44, 503.
Esber, H. J., Payne, I. J., and Bogden, A. E. (1973). J. Natl. Cancer Inst. 50, 559.
Evans, V. J., Bryant, J. C., Kerr, H. A., and Schilling, E. L. (1964). Exp. Cell Res. 56, 439.
Fogh, J. (1973). "Contamination in Tissue Culture." Academic Press, New York.
Freed, J. J., and Schatz, S. A. (1969). Exp. Cell Res. 55, 393.
Honn, K. V., Singley, J. A., and Chavin, W. (1975). Proc. Soc. Exp. Biol. Med. 149, 344.
Hsueh, H. W., and Moskowitz, M. (1973). Exp. Cell Res. 77, 376.
Hull, R. N., Cherry, W. R., and Tritch, O. J. (1962). J. Exp. Med. 115, 903.
Jones, J. E., Naider, F., and Becker, J. M. (1975). In Vitro 11, 41.
Kahn, R. H., Burkel, W. E., and Perry, V. P. (1974). J. Natl. Cancer Inst. 53, 1471.
Keay, L. (1975). Biotechnol. Bioeng. 17, 745.
Kihara, H., and de la Flor, S. D. (1968). Proc. Soc. Exp. Biol. Med. 129, 303.
Kuhler, R. J., Marlowe, M. L., and Merchant, D. J. (1960). Exp. Cell Res. 20, 428.
Lasfargues, E. Y., Coutinho, W. G., Lasfargues, J. C., and Moore, D. H. (1973). In Vitro 8, 494.
Mantel, N. (1951). Am. Stat. 5, 26–27.
Medzon, E. L., and Merchant, D. J. (1971). In Vitro 7, 46.

Merchant, D. J., and Hellman, K. B. (1962). *Proc. Soc. Exp. Biol. Med.* **110**, 194.
Mizrahi, A., and Moore, G. E. (1971). *Appl. Microbiol.* **21**, 754
Morton, H. J. (1970). *In Vitro* **6**, 89.
Neuman, R. E., and McCoy, T. A. (1958). *Proc. Soc. Exp. Biol. Med.* **98**, 303.
Parshad, R., and Sanford, K. K. (1968). *J. Natl. Cancer Inst.* **41**, 767.
Parshad, R., and Sanford, K. K. (1972). *J. Natl. Cancer Inst.* **49**, 1155.
Pumper, R. W. (1958). *Science* **128**, 363.
Pumper, R. W. (1973). *In* "Tissue Culture: Methods and Applications" (P. F. Kruse and M. K. Patterson, Jr., eds.), pp. 674–677. Academic Press, New York.
Pumper, R. W., Alfred, L. J., and Sackett, D. L. (1960). *Nature (London)* **185**, 123.
Pumper, R. W., Yamashiroya, H. M., and Molander, L. T. (1965). *Nature (London* **207**, 662.
Pumper, R. W., Fagan, P., and Taylor, W. G. (1971). *In Vitro* **6**, 266.
Rutzky, L. P., and Pumper R. W. (1974). *In Vitro* **9**, 468.
Sefton, B. M., and Rubin, H. (1970). *Nature (London)* **227**, 843.
Shodell, M. (1972). *Proc. Natl. Acad. Sci. U,S.A.* **69**, 1455.
Smith, G. L., and Temin, H. M. (1974). *J. Cell Physiol.* **84**, 181.
Taylor, W. G., and Evans, V. J. (1974). *In* "Methods in Cell Biology" (D. M. Prescott, ed.), Vol. 8, pp. 47–73. Academic Press, New York.
Taylor, W. G., Dworkin, R. A., Pumper, R. W., and Evans, V. J. (1972a). *Exp. Cell Res.* **74**, 275.
Taylor, W. G., Taylor, M. J., Lewis, N. J., and Pumper, R. W. (1972b). *Proc. Soc. Exp. Biol. Med.* **139**, 96.
Taylor, W. G. (1974). *J. Natl. Cancer Inst.* **53**, 1449.
Taylor, W. G., Evans, V. J., and Pumper, R. W. (1974a). *In Vitro* **9**, 278.
Taylor, W. G., Pumper, R. W., and Rutzky, L. P. (1974b). *In Vitro* **10**, 370.
Taylor, W. G., Evans, V. J., Fox, C. H., Camalier, R. F., and Sanford, K. K. (1975). *Biotechnol. Bioeng.* **17**, 1847.
Temin, H. M., Pierson, R. W., Jr., and Dulak, N. C. (1972), *In* "Growth, Nutrition and Metabolism of Cells in Culture" (G. H. Rothblat and V. J. Cristofalo, eds.), Vol. 1, pp. 50–81. Academic Press, New York.
Visser, A. S., and Prop, F. J. A. (1974). *J. Natl. Cancer Inst.* **52**, 293.
Waymouth, C. (1956). *J. Natl. Cancer Inst.* **17**, 315.
Wein, J., and Goetz, I. E. (1973). *In Vitro* **9**, 186.
Willmer, E. N., and Kendal, L. P. (1932). *J. Exp. Biol.* **9**, 149.
Yamane, I., and Murakami, O. (1971). *Biochem. Biophys. Res. Commun.* **45**, 1431.
Yamane, I., and Murakami, O. (1973). *J. Cell Physiol.* **81**, 181.

Chapter 26

In situ Fixation and Embedding for Electron Microscopy

MARLENE SABBATH

Department of Pathology, Lenox Hill Hospital, New York, New York

AND

BARBRO ANDERSSON

Department of Pathology, School of Medicine, Leahi Hospital, Honolulu, Hawaii

I. Introduction

Most methods generally available to the electron microscopist for the fixation and embedding of tissue culture cells involve repeated manipulation of cell suspensions by centrifugation. Such procedures do not lend themselves to studies aimed at elucidating the orientation and interaction of cells *in vitro*. Any investigation requiring the maintenance of the relationship between the cell and its environment would be greatly facilitated by a methodology permitting *in situ* fixation and embedding. Several such techniques allowing the ultrastructural examination of cultured cells without the disruption of their natural interrelationships have been developed. These techniques can be divided into two major categories determined by whether the cells are grown on unsectionable or sectionable substrates.

II.   Cells on Unsectionable Substrates (Flat-Faced Embedding Technique)

A.   Glass Substrates

Initial approaches to *in situ* fixation for electron microscopy involve the fixation and embedding of cells on the glass surface upon which they were grown. Borysko and Sapranauskas (1954) were the first to report the embedding of whole colonies of normal rat fibroblasts for sectioning in a plane perpendicular to the substrate. Similar techniques were applied by Gay (1955) to the study of *Drosophila* chromosome smears. A modification of the two previous procedures was introduced by Howatson and Almeida (1958), who utilized inverted gelatin capsules filled with partially polymerized methacrylate to embed selected areas of a slide. The removal of the glass slide after polymerization was accomplished by rapid cooling with solid CO_2. After separation, the embedded cells were sectioned in a plane parallel to the glass surface. This procedure was shown to be advantageous in that it allowed the embedding of selected cells for detailed study. Various mechanical refinements of this procedure were introduced by Nebel and Minick (1956), Latta (1959), Nirshiura and Rangan (1960), Persijn and Daems (1964), Sparvoli *et al.* (1965), and Gorycki (1966).

The removal of embedded cells from the glass substrate by rapid cooling was often complicated by the inability to separate cells from their substrate, particularly in cultures grown for several days. With the advent of epoxy resins the problem of separating embedding media from glass substrates became more difficult. In an attempt to overcome these problems, various methods involving the pretreatment of the glass surface have been employed.

B.   Coated-Glass Substrates

In studying the effects of microbeam irradiation on single cells in culture, Bloom (1960) grew them on glass or quartz cover slips which were later fixed, dehydrated, and embedded *in situ* with methacrylate. To facilitate the separation of the monolayer from glass, Bloom suggested the cultivation of cells on carbon-coated cover slips, which would separate more easily. Subsequently, several investigators reported the treatment of glass supports with carbon (Davies and Wallace, 1958; Robbins and Gonatos, 1964; Robbins and Jentzsch, 1967), collagen (Heyner, 1963), silicone (Rosen, 1962; Micou *et al.*, 1962), silica (Kushida and Suzuki, 1968), 1% nitrocellulose (Flaxman *et al.*, 1968), or polytetrafluorethane (Buckley, 1971). These methods have been subject to much controversy, since the effects of such substrates on tissue culture cells are unknown.

Another approach to the coating of glass surfaces was developed by Chang (1971) and proved to be nontoxic to all tissue culture lines tested. Cells were grown on cleaned cover glasses that were lightly sprayed with Teflon. All selected cover slips, with the monolayer attached, were fixed and dehydrated in small staining dishes and later embedded with Epon, or other plastics, utilizing a silicone rubber embedding mold (Fig. 1). The Teflon-coated cover glasses were then easily separated from the resin block by sudden immersion in liquid nitrogen. After separation, the Epon blocks could be viewed and photographed by light microscopy.

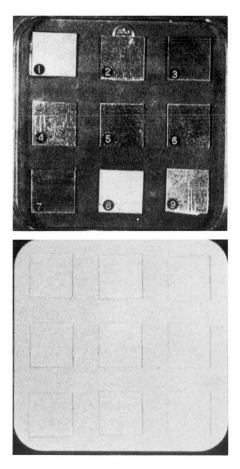

FIG. 1.   Master template for preparing silicone rubber embedding molds (top). Completed silicone rubber embedding mold (bottom). From Brinkley and Chang (1973), p. 441. By permission of Academic Press, New York.

Two important features of this technique are: (1) numerous pretreated cover slips from a single culture flask can be studied; (2) selected cells can be cut vertically and horizontally for ultrastructural examination. In addition, this method not only is applicable to cell monolayers but can also be used for fresh imprints, cell suspensions, ascites fluids, blood and cell smears, as well as cryostat sections.

C.   Synthetic Substrates

As an alternative to glass supports, cells were cultured directly on substrates easily detached from polymerized embedding resins. Such supports as plastic sheets (Kjellen et al., 1955), mica (Persijn and Scherft, 1965; Yardley and Brown, 1965), cover slips of polyester sheeting (Melinex) (Firket, 1966; Sykes and Basrur, 1971), vinyl plastic cups (Anderson and Doane, 1967), and silicone rubber membranes (Shahar et al., 1973) have been used with satisfactory results. The most widely applied technique has been that introduced by Brinkley et al. (1967; Brinkley and Chang, 1973), who used Falcon plastic petri dishes and flasks as substrates for cell growth. After fixation in glutaraldehyde and dehydration in the water-miscible resin hydroxypropyl methacrylate, the cells were embedded in situ with a thin layer of Epon (Luft, 1961). When polymerization of the resin was completed, individual cells could be observed and photographed by light microscopy. After such examination, selected groups of cells were mounted on BEEM capsules and subsequently serially sectioned for electron microscopic analysis. The chief advantage of this technique is that large quantities of cells may be prepared at any one time and easily screened by light microscopy. The capability of selecting cells at specific states of the mitotic cycle makes this method particularly useful in the ultrastructural study of cell division (Fig. 2).

All the previously discussed techniques require the removal of cells from their growth supports. Such procedures inherently risk the loss or damage of cells during separation of resin and substrate. This difficulty led to a search for procedures in which the growth substrate remained as an integral part of the sectioning process.

III.   Cells on Sectionable Substrates

A.   Polystyrene

In order to avoid the necessity of separating substrate from cells, culture supports that could either be sectioned with the cells or else dissolved during

FIG. 2.   A light micrograph (× 1530) and an electron micrograph (× 8900) of the same female rat kangaroo cell in late prophase. From Brinkley *et al.* (1967) p. 281; by permission of Rockefeller University Press, New York.

preparation of the specimen were sought. Several investigators used polystyrene petri dishes (Nelson and Flaxman, 1972) or cover slips (Richters and Valentin, 1973). In these methods both the embedded cells and the substrate are sectioned together, avoiding any disruption of cell-to-cell relationships. The use of sectionable substrates allows cells to be cut both parallel to and at right angles to the plane of attachment. A limitation noted by Nelson and Flaxman (1972) was the wrinkling of the substrate during sectioning due to differences in the cutting properties of polystyrene and resin. This difficulty was overcome by spreading the thin sections with xylene vapors.

B.   Millipore Filters

A method that allows embedding of both cell monolayers and substrate was developed by McCombs *et al.* (1968) and Dalen and Nevalainen (1968). This technique is based on the growth and adsorption of tissue culture cells onto Millipore filters and the use of these filters in manipulating the cells during the embedding process. McCombs *et al.* (1968) demonstrated the usefulness of such filters as a substrate by the growth of lymphoblastoid and fibroblastic cell cultures. The filter-grown fibroblasts retained their characteristic elongated shape, and excellent preservation of plasma membranes and microvilli was obtained in lymphoblastoid cells. In addition to cell cultures, this technique can be used to study the development of viruses that bud from cell membranes (Fig. 3), as well as of viral particles that have been concentrated directly on Millipore filters (McCombs *et al.*, 1968; Lemcke, 1971).

C.   Thin-Agar Layer Technique

A technique valuable for the study of specimens which are very fragile or available in such small quantities that standard embedding methods cannot be readily applied is the thin-layer agar method. Specimens to be examined are placed in a pure agar jelly (which protects and supports them), and subsequently processed according to the methods of Van Noord *et al.* (1973). Both human and mouse hemopoietic cells have been successfully examined by the thin-agar layer technique, and it was suggested that one particularly useful application would be in experimental hematology, where agar cultures of human hemopoietic tissue are a routine procedure.

D.   Sandwich-Embedding Technique

A number of other substrates, such as nucleopore polycarbonate filters (Cornell, 1969) and BEEM capsules (Tan *et al.*, 1975), were found suitable for *in vitro* growth with subsequent processing for electron microscopy.

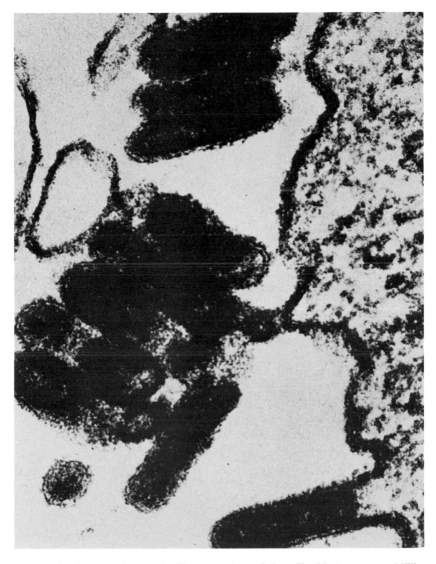

FIG. 3.   An electron micrograph of human embryonic lung fibroblasts grown on Millipore filters and infected with vesicular stomatis virus (× 115,500). From McCombs *et al.* (1968), p. 238. By permission of Rockefeller University Press, New York.

Egeberg (1965) described a sandwich method for embedding monolayers of cells grown on thin plates of polymerized resin. The procedure was used in examining leukocytes fixed in different phases of phagocytosis and proved to be more practical and less time-consuming than the flat-faced embedding methods. Both Epon (Egeberg, 1965; Zagury *et al.*, 1966) and Araldite (Smith *et al.*, 1968) have been used in this technique with the limitation that some cell lines did not proliferate on Epon substrates (Zagury *et al.*, 1968), and on Araldite higher cell concentrations were required for adequate seeding as compared to glass supports (Smith *et al.*, 1968).

In our recently published report (Sabbath *et al.*, 1973), we have adapted the newly discovered Spurr medium (Spurr, 1969), with qualities of low viscosity and high penetration, to the sandwich-embedding technique. The procedure involves the *in situ* fixation and embedding of cell monolayers grown on Spurr disks (Fig. 4). These disks were found to be particularly suitable as a substrate for cell growth since their ingredients have not proved to be cytotoxic nor do they seem to alter cellular morphology. In addition,

FIG. 4.   Polypropylene caps (upper left), Spurr disk removed from cap after polymerization (upper right). Cultures grown on Spurr disk placed in petri dish (bottom). From Sabbath *et al.* (1973), p. 487. By permission of Academic Press, New York.

this technique permits the viewing and selection of cells in both the living and fixed state, as well as the photographing of specific cells by phase contrast and light microscopy. It also eliminates the coating of the cellular substrate with an antiadhesive substance, which is used in some *in situ* techniques and might have an undesirable effect on ameboid movement and phagocytic processes (Egeberg, 1965).

REFERENCES

Anderson, N., and Doane, F. W. (1967). *Stain Technol.* **42**, 169–173.
Bloom, W. (1960). *J. Biophys. Biochem. Cytol.* **7**, 191–193.
Borysko, E., and Sapranauskas, P. (1954). *Bull. Johns Hopkins Hosp.* **95**, 68–73.
Brinkley, B. R., and Chang, J. P. (1973). *In* "Tissue Culture: Methods and Applications" P. F. Kruse, Jr. and M. K. Patterson, Jr., eds.), pp. 438–443. Academic Press, New York.
Brinkley, B. R., Murphy, P., and Richardson, L. C. (1967). *J. Cell Biol.* **35**, 279–283.
Buckley, I. K. (1971). *Lab. Invest.* **25**, 295–301.
Chang, J. P. (1971). *J. Ultrastruct. Res.* **37**, 370–377.
Cornell, R. (1969). *Exp. Cell Res.* **58**, 289–295.
Dalen, H., and Nevalainen, T. J. (1968). *Stain Technol.* **43**, 217–220.
Davies, C., and Wallace, R. (1958). *J. Biophys. Biochem. Cytol.* **4**, 231–232.
Egeberg, J. (1965). *Stain Technol.* **40**, 343–346.
Firket, H. (1966). *Stain Technol.* **41**, 189–191.
Flaxman, B. A., Lutzner, M. A., and Van Scott, E. J. (1968). *J. Cell Biol.* **36**, 406–410.
Gay, H. (1955). *Stain Technol.* **30**, 239–242.
Gorycki, M. A. (1966). *Stain Technol.* **41**, 37–42.
Heyner, S. (1963). *Stain Technol.* **38**, 335–336.
Howatson, A. F., and Almeida, J. D. (1958). *J. Biophys. Biochem. Cytol.* **4**, 115–118.
Kjellen, L., Lagermalm, G., Svedmyr, A., and Thorsson, K. G. (1955). *Nature (London)* **175**, 505–506.
Kushida, H., and Suzuki, K. (1968). *J. Electron. Micros.* **17**, 350–352.
Latta, H. (1959). *J. Biophys. Biochem. Cytol.* **5**, 405–410.
Lemcke, R. M. (1971). *Nature (London)* **229**, 492–493.
Luft, J. H. (1961). *J. Biophys. Biochem. Cytol.* **9**, 409–414.
McCombs, R. M., Benyesch-Melnick, M., and Brunschwig, J. P. (1968). *J. Cell Biol.* **36**, 231–243.
Micou, J., Collins, C. C., and Crocker, T. T. (1962). *J. Cell Biol.* **12**, 195–198.
Nebel, B. R., and Minick, O. T. (1956). *J. Biophys. Biochem. Cytol.* **2**, No. 4, Suppl. 61.
Nelson, B. K., and Flaxman, B. A. (1972). *Stain Technol.* **47**, 261–265.
Nishiura, M., and Rangan, S. R. S. (1960). *J. Biophys. Biochem. Cytol.* **7**, 411–412.
Persijn, J. P., and Daems, W. T. (1964). *Stain Technol.* **39**, 125–128.
Persijn, J. P., and Scherft, J. P. (1965). *Stain Technol.* **40**, 89.
Richters, A., and Valentin, P. L. (1973). *Stain Technol.* **48**, 185–188.
Robbins, E., and Gonatas, N. K. (1964). *J. Cell Biol.* **20**, 256–259.
Robbins, E., and Jentzch, G. (1967). *J. Histochem. Cytochem.* **15**, 181–182.
Rosen, S. (1962). *Stain Technol.* **37**, 195–197.
Sabbath, M., Andersson, B., and Ioachim, H. L. (1973). *Exp. Cell Res.* **80**, 486–490.
Shahar, A., Monzain, R., and Staussman, Y. (1973). *Tissue & Cell* **5**, 691–696.
Smith, W., Gray, E. W., and Mackay, J. M. K. (1968). *J. Microsc. (Oxford)* **89**, 359–363.
Sparvoli, E., Gray, H., and Kaufmann, B. P. (1965). *Stain Technol.* **40**, 83–88.
Spurr, A. R. (1969). *J. Ultrastruct. Res.* **26**, 31–43.

Sykes, A. K., and Basrur, P. K. (1971). *In Vitro* **7**, 68–73.
Tan, C. K., Chan, H. L., and Tan, C. H. (1975). *Experientia* **31**, 253–255.
Van Noord, M. J., Blansjar, N., and Nakeff, A. (1973). *Stain Technol.* **48**, 239–246.
Yardley, J. H., and Brown, G. D. (1965). *Lab. Invest.* **14**, 501–512.
Zagury, D., Zeitoun, P., and Viette, M. (1966). *C. R. Hebd. Seances Acad. Sci.* **262**, 1458–1459.
Zagury, D., Pappas, G. D. and Marcus, P. I. (1968) *J. Microsc.* **7**. 287–292.

Chapter 2 7

Genetic and Cell Cycle Analysis of a Smut Fungus (Ustilago violacea)

JOSEPH E. CUMMINS AND ALAN W. DAY

The University of Western Ontario, Department of Plant Sciences, London, Ontario, Canada

I. Introduction

The Ustilaginales have proved very useful in genetic studies (Holiday, 1961; Day and Jones, 1968, 1969). The anther smut, *Ustilago violacea*, provides another advantage in its relatively simple and synchronous sexual morphogenesis. This morphogenesis consists of the cooperative construction of a cylindrical fusion tube between a part of uninucleate, yeastlike cells of opposite mating type. In this basidiomycete, copulation and nuclear fusion are temporally separated and the mechanisms governing each can be analyzed separately (Day and Jones, 1968). Studies employing this organism have begun to provide new insights into the genetic regulation of the interaction between the cell cycle and morphogenesis. These findings are summarized as follows:

1. Two alleles of the mating-type locus that regulate assembly of the copulatory organelle have different cell cycle controls; allele a_1 allows mating only in G_1 (a stringent control) and allele a_2 allows mating during the entire cell cycle (a relaxed control) (Cummins and Day, 1973).

2. These cell cycle controls are allele specific and cis is dominant. It is suggested that when both alleles are active simultaneously they "cancel out" and the cell does not mate (Day and Cummins, 1973). Thus in the heterozygous diploid mating is inducible only during G_2 and S and a_2 is dominant.

3. Spontaneous changes in the heterozygote affecting the cell cycle control system of at least the a_2 allele appear to arise frequently. An example was reported in which allele a_2 changed from relaxed control to stringent control during subculture of the diploid. Such a change had the effect of rendering the diploid incapable of mating even though both alleles were functional when segregated into haploid strains (Day and Cummins, 1975). These observations, particularly the finding that the cell cycle properties of an allele can be altered without affecting its function, provide the first clear-cut evidence for cell cycle control genes, which serve to limit the temporal expression of other genes controlling morphogenetic steps. Furthermore, the molecular biology and genetics of this synchronous relatively uncomplicated morphogenetic process should provide valuable source material for students of molecular biology and genetics.

The purpose of this chapter is to describe genetic, cytological, and biochemical techniques that have been successfully applied in our laboratory to the study of the cell cycle and development in the anther smut. An earlier review concerned with the cell cycle and development in the anther smut (Cummins and Day, 1974a) should also be consulted.

II. Biology and Culture

A. Life Cycle

The life cycle of *U. violacea* involves the haplophase, dikaryophase, and diplophase (Fischer and Holton, 1957). The haplophase is saprophytic, and the dikaryophase is obligately parasitic. In nature, the dikaryophase is initiated by conjugation of compatible sporidia and parasitizes host plants in the family Caryophyllaceae. The formation of diploid nuclei occurs following fusion of the nuclei in dikaryotic mycelium infecting anthers of the host. Thick-walled diploid brandspores then replace pollen in the anthers of the host plant. The haplophase is reestablished after meiosis in the promycelia which develop from mature brandspores when they are provided with water and air. The three-celled promycelia and the brandspore form a tetrad, each cell of which produces a uninucleate haploid sporidium. These sporidia are yeast-like and reproduce by budding, and can be maintained indefinitely on nutrient medium.

The mating system in *U. violacea* is bipolar, that is, there are two mating-type alleles a_1 and a_2 of a single locus. Sporidia-bearing diploid or even higher ploidy levels can be produced artificially (Day and Jones, 1968; Day, 1972). Polyploid strains are capable of mating, and the mating-type determinant a_2 is clearly dominant. Polyploids heterozygous for the mating-type locus are solopathogenic, that is they are capable of infecting the host plant in the absence of sporidia bearing the complementary mating-type, and they can complete meiosis in the anthers of an infected plant (Day, 1968). It has not yet been clearly established whether or not diploids homozygous for mating type are solopathogenic and whether or not they are able to complete meiosis.

B. Culture

Since the dikaryotic and diplophases of this smut's normal life cycle are obligately parasitic, the discussion below refers to the growth of sporidial cultures. The techniques for handling smut cultures are essentially similar to those for maintaining and cultivating yeast (see Volumes 10, 11 of this series). It is only in the obligately parasitic phases of this organism's life cycle that the methodology differs significantly from that involved in cultivating yeast.

1. MEDIA FOR *Ustilago violacea* (DAY, 1968)

Liquid/liter solution in distilled water

Complete medium
    Glucose   10 gm
    Yeast extract   3 gm
    Beef extract   1 gm
    Peptone   10 gm
    Malt extract   3 gm
    For agar plates, add 25–30 gm of agar.

Minimal medium
    Glucose   10 gm
    Concentrated salts, 50 ml
    For agar plates add 25–30 gm of agar

Concentrated salts/liter final solution
    Sodium nitrate ($NaNO_3$) 20 gm
    Potassium phosphate (KH_2PO_4) 10 gm

        Potassium chloride (KCl) 5 gm
        Magnesium sulphate ($MgSO_4$) 5 gm
        Calcium chloride ($CaCl_2$) 1 gm
        Ferrous sulfate ($FeSO_4$) trace (1–2 crystals)
        Thiamine (aneurin hydrochloride) 1.0 mg
        Normally add 50 ml to a 250-ml flask or 20 ml to a 125-ml flask

Liquid YDP medium/liter solution
        Glucose, 100 gm
        Peptone, 20 gm
        Yeast extract, 10 gm

These sporidia are highly aerobic and are best maintained as shaken cultures at 15°–25°C. Growth ceases above 30°C, and the cultures rapidly lose viability at the elevated temperature. The doubling times of most strains are 4–8 hours at 18°–22°C in complete medium and 12–20 hours in minimal medium at 18°–22°C. The stationary stage is reached at a very high density of sporidia (about 10^9/ml) in complete medium. Shaken stationary-phase cultures remain viable for up to about 10 days.

2. Chemostat Cultures

Chemostat cultures of wild, auxotrophic, and diploid strains of sporidia have been maintained. We employ the Bio-flow chemostat from New Brunswick Scientific Company, New Brunswick, New Jersey. Cultures grown on complete medium are maintained at a dilution rate about 0.20 hr^{-1} in a 350-ml working volume at 22°C. The cell density is maintained at about 1 to 5 × 10^8 ml^{-1}, and at that density problems with foaming or clumping are not encountered as they are with many other fungi. For routine operation the apparatus is dismantled, sterilized, and inoculated on Friday, so that by the following Monday, cell density equilibrium is reached and the chemostat cultures are available as working material throughout the week. Bacterial contamination of the rich complete medium is encountered unless good care is used in setting up and inoculating the chemostat. Chemostat cultures are preferred for synchrony studies using zonal density-gradient centrifugation. They have great density and size homogeneity that is achieved by maintaining the cultures in a steady state for several generations. Using chemostat cultures, it is possible to achieve an extremely high level of reproducibility in the zonal separation of sporidia.

3. Maintenance and Preservation of Stocks

Sporidia are maintained in a dry, frozen state. Screw-cap vials containing a thick layer of dry sterile silica gel (30 to 120-mesh) are prepared, and a mound of actively growing sporidia is removed from a petri dish using a sterile transfer loop. The mound of sporidia is placed on the glass on the inner surface of the vial well above the layer of silica gel. The vial is capped and allowed to dry for one day at room temperature; then the vials are transferred to a freezer. Stocks prepared in this manner remain viable for several years. Colonies are reactivated by transferring a small clump of dry sporidia to complete medium and incubating it for several days at 22°C (Day and Jones, 1968).

Brandspores are maintained by drying flower buds from infected plants in small vials containing silica gel. These brandspores are maintained in the frozen state and are viable for several years. The brandspores are reactivated by crushing the anthers from an infected bud in sterile distilled water and spreading the black spores onto a complete medium plate (Day and Jones, 1969). Brandspores and sporidia can be stored indefinitely in liquid nitrogen without influencing the mutation rate or the frequency of intragenic recombination (Day et al., 1972).

III. Cell Biology

A. Microscopy

1. LIGHT MICROSCOPY

a. Phase-Contrast Microscopy of Living Cells. The culture chamber suggested by Cole and Kendrick (1968) can be used for the time-lapse photography of living cells. Log-phase cells are inoculated by means of a fine glass fiber into a small droplet of complete medium (containing 18 %, w/v, gelatin instead of agar in the culture chamber). This method has been useful in studying the duration of mitotic phases (Poon, 1974) and the progress of sexual morphogenesis (Poon et al., 1974).

b. Giemsa and Acetoorcein Staining. The staining technique recommended by Robinow (1970) can be used for staining the chromatin. Diploid or triploid cells attached to cover slips with fresh albumin are fixed for 30 minutes in either half-strength Schaudinn or Helly's fluid, rinsed once in 70 % ethanol, then extracted for 60 minutes in 1 % (w/v) NaCl at 60°C and hydrolyzed for 5–6 minutes with 1 N HCl at 60°C.

For Giemsa staining, the cells on cover slips are placed in buffered Giemsa (0.5 ml in 10 ml of phosphate buffer, pH 6.9) for 30 minutes then destained for a few seconds in acidified distilled water (a loopful of glacial acetic acid in 10 ml of water). Stained cells are mounted in diluted Giemsa (2 drops of Giemsa in 10 ml of buffer), squashed, and double sealed with glyceel (obtainable from Searle Scientific Services, High Wycombe, Bucks, England) and nail varnish. Specimens are stored for several days at room temperature to bleach the stain until good differentiation is obtained.

For acetoorcein staining, hydrolyzed cells were rinsed with water then 60 % glacial acetic acid. The cells are mounted in a drop of the stain on a slide and squashed; the cover slip was then sealed with glyceel or nail varnish (Poon, 1974). Warming of the cells in acetoorcein intensifies the stain but causes artifacts in the form of contracted chromatic granules (Poon and Day, 1975b).

c. Hot Acetoorcein Staining. A technique similar to that of Day and Jones (1972) is used. Cells attached to cover slips are fixed 15–30 minutes in Carnoy's fluid (1 part glacial acetic acid in 3 parts ethanol) in a Columbia staining jar, then rinsed once in 70 % ethanol and once in distilled water. After hydrolysis for 3–8 minutes in 1 N HCl at 60°C, the cells are rinsed once in distilled water, then again in 45 % glacial acetic acid. To stain, cells are mounted in 0.5 % (w/v) orcein in 45 % glacial acetic acid. The preparation is heated over a hot plate (set temperature at 90°C) for 60 seconds; then the cells are flattened. Traces of stain are removed by infiltrating drops of 45 % glacial acetic acid from one side of the cover slip, and the fluid is removed with absorbing paper from the other side until the preparation became clear. The preparations are then sealed (Poon, 1974).

d. Staining for Microtubules, Spindle Pole Bodies, Spindle, and Nucleoli. Helly-fixed cells are hydrated through a descending ethanol series, and are finally rinsed with 1 % glacial acetic acid. They are stained 15 minutes or longer in a 1 : 1000 dilution of acid fuchsin in 1 % glacial acetic acid, and the cover slip is sealed without squashing (Poon, 1974).

e. Acridine Orange Fluorescence. Acridine orange staining is the most rapid technique for detecting mitotic stages, as it can be used with living cells (Poon and Day, 1974a). Acridine orange [3,6-bis(dimethylaminoacridine)] 75–250 mg/liter, is incorporated into complete medium or into low molarity (i.e., 0.05 μm) phosphate or Tris buffer, pH 6.3. The preparation is examined using a good microscope with fluorescence equipment. In the living sporidia chromatin stains green, and the nucleolus, spindle pole body, and cytoplasm stain orange (Poon and Day, 1974). The mitotic index can be determined very rapidly and accurately from such preparations.

All the techniques listed above have been useful in determining the behavior of the nucleolus, chromatin granules, nucleoplasm, and spindle pole body during division. Estimates from time-lapse studies indicate that sporidial nuclear division requires about 45 minutes and that "prophase takes about 12 min, metaphase 5 min and anaphase–telophase about 30 min" (Poon, 1974).

2. ELECTRON MICROSCOPY

a. General Procedure. The technique used in our laboratory is as follows: Cells are fixed with 2–4 % glutaraldehyde for 1.5 hours at 4° or 27°C, followed by a postfixation treatment with 1 % osmic acid for 2 hours. The cells are then prestained with 0.5 % uranyl acetate for 4–18 hours, dehydrated with an ascending acetone or ethanol series, and embedded in Epon, Araldite, or Spurr's resin (Ladd Co.). Sections are mounted on 150-mesh grids coated with Formvar/carbon and stained with lead citrate (Reynolds, 1963) for 10 minutes. Serial sections are prepared according to the technique of Moens (1970). They are cut with a diamond knife on a Sorvall Porter-Blum MT-11 microtome and mounted on single-slot grids coated with Formvar and carbon (Poon, 1974; Poon and Day, 1975a,b). The above technique is not satisfactory for brandspore preparations.

b. Shadow-Casting or Negative-Staining. Cells from the culture to be tested are harvested, washed at least four times with particle-free distilled water, and suspended at low density in this water (about 10^6 cells/ml). Chemical treatments or a heat treatment can be applied to the cells at this stage. After the treatment is finished, the cells are washed three more times with distilled water. Formvar-coated 150-mesh grids are rinsed in dilute Kodak Photoflo 200 (1 drop in 10 ml of distilled water), then washed once with distilled water. A droplet of cell suspension (about 10^6 cells/ml) is then placed on the grid and allowed to stand for about 30 seconds. The bulk of the droplet is drawn off with a sliver of filter paper, and the cells are allowed to air dry on the grid.

To test for stain affinity, a droplet of 0.5 % or 1.0 % aqueous filtered phosphotungstic acid (PTA) (pH 7.4) is added to the air-dried cells on the grid (Bradley, 1962).

For shadow-casting the grids are coated with tungsten oxide at an angle of 18°–22°C in a vacuum chamber.

The cells are examined with a Philips E.M. 200 at 60kV, and photographs are taken on 3 × 4 plate film. Using these methods, many long (up to 10 μm) fine surface hairs or fimbriae have been demonstrated to be present on the exterior walls of the sporidia of *U. violacea* (Poon and Day, 1975). The role of these fimbriae in conjugation has been discussed by Day and Poon (1975). A cell with fimbriae is shown in Fig. 1.

B. Cell Fractionation and Analysis of Macromolecules

1. CELL BREAKAGE AND FRACTIONATION

The tough outer coat of the sporidium provided a technical obstacle to the preparation of good cell fractions and macromolecules in bulk. It has not yet been possible to prepare protoplasts of these sporidia using the range of lytic preparations now available. Satisfactory though rather small-scale preparations can be obtained by breaking the sporidia using a Mickle disintegrator. A variety of media give satisfactory cell breakage, but it is the ratio of glass beads (No. 12 Pyrex) to cells and the duration of vibration that are critical. We have found that 10^{10} cells (about 0.75 ml packed volume)

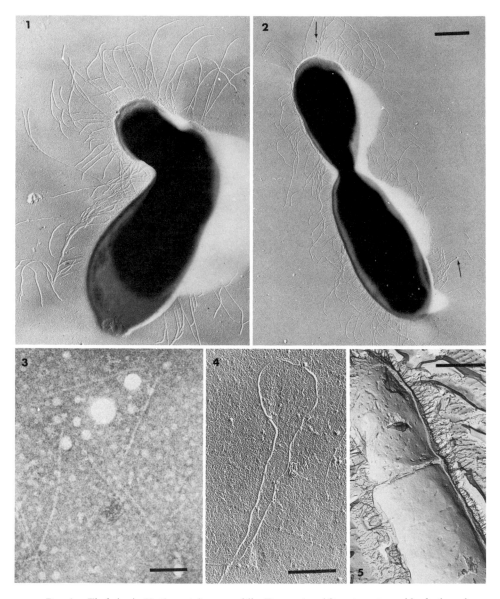

FIG. 1.　Fimbriae in *Ustilago violacea* sporidia. Frames 1 and 2 are tungsten oxide-shadowed sporidia showing numerous fimbriae, some of which are knobbed. Frame 3 shows a phosphotungstic acid preparation of detached fimbriae. Bar represents 1000 Å. Frame 4 shows detached fimbriae. Frame 5 is a freeze-etch micrograph of part of the promycelium produced after germination of the sexual brandspore; note the fimbriae around the edges of the cell. From Poon and Day (1975).

suspended in 4.5 ml final volume added to 7.5 gm of beads gives essentially complete cell breakage after 20 minutes of vibration. Decreasing cell number has little influence on breakage, whereas increasing cell number prevents breakage, and variation in volume or bead mass tends to result in unsatisfactory breakage. Large-scale preparation of RNA, protein, or sheared DNA can be achieved by passing cells through a French press at 15,000 psi.

After cell breakage using the Mickle procedures, satisfactory preparation of mitochondria, ribosomes, and polysomes can be achieved using standard isolation procedures. The nuclei remain intact after Mickle breakage but have not yet been isolated in pure state. Mitochondria prepared by Mickle cell breakage are best further purified by banding them in 15–50% linear sucrose gradients prior to isolating their nucleic acids.

2. PREPARATION OF DNA

Bulk preparation of DNA is best achieved by cell breakage in the French press. The extraction buffer contains 0.06 M EDTA, 0.20 M Tris, pH 8.9, and 1% sodium dodecyl sulfate (SDS). Add 3 ml of buffer for each milliliter of packed cells. Following cell breakage add a half volume of chloroform–5% octanol and a half volume of 90% phenol (90 gm of phenol–100 mg of 8-hydroxyquinoline per 100 ml) and mix well at room temperature. Centrifuge for 10 minutes at 15,000 g. Remove supernatant and repeat extraction. Add 2 to 3 times volume of ethanol ($-20°C$) and centrifuge for 10 minutes at 15,000 g. The pellet contains RNA polysaccharide and DNA. This pellet is taken up in 10 ml of 0.12 M phosphate buffer, pH 6.8, containing 0.03 M EDTA and passed through a fine syringe needle to shear it. The tube contents are added to a 3–4 cm × 1 cm hydroxyapatite column run at 65°C, washed through, and 0.12 M phosphate buffer is washed through until the optical density at 260 nm drops below 0.1. The DNA is eluted from hydroxyapatite with 0.48 M phosphate buffer and is free of RNA and polysaccharide (Britten, et al., 1970).

For preparation of highly labeled DNA, small numbers of labeled sporidia can be broken using the Mickle in a medium containing 8 M urea, 0.24 M phosphate buffer, 1% SDS, and 0.03 M EDTA (Meinke et al., 1974). After breakage, as described above, add enough extraction medium to make 20 ml and centrifuge at 500 g for 5 minutes to remove glass beads and unbroken cells. Shear the preparation by passing it through a fine syringe needle. Purify the preparation on a hydroxyapatite column as described above.

DNA prepared in the above manner bands well in a cesium chloride gradient with a major band corresponding to a 58% GC content and a minor band (about 10% of the total DNA) of 36% GC content. DNA obtained from sporidial mitochondria purified on sucrose gradients has a GC content of 36%. DNA prepared as above is useful in hybridization studies.

3. Preparation of RNA

High-molecular-weight RNA preparations are readily achieved at present by methods that require extensive preliminary trial. About 10^{10} cells are suspended in 4.5 ml of extraction buffer made up of 0.25 M sucrose, 0.05 M Tris pH 9.0, 0.02 M MgCl$_2$, and 0.015 M KCl (Lin and Key, 1967). The cell suspension is deposited on the glass beads of the Mickle disintegrator as described above, 50 μl of pyrocarbonate are added, and the cultures are vibrated for at least 20 minutes while the Mickle apparatus is in a freezer room ($-20°$). To the slurry of broken cells is added 7 ml of the extraction buffer, 1 ml M LiCl and 1 ml 10% sodium laurel sulfate. Phenol, 2.5 ml (90 gm phenol–100 mg 8-hydroxyquinoline) and 2.5 ml of chloroform (pentanol, 4%) are rapidly added, and the contents of the tubes are mixed well at 3°C. The mixture is centrifuged for 5 minutes at 500 g at 3°C. The supernatant is again extracted with 2.5 ml of phenol and 2.5 ml of chloroform–pentanol and centrifuged for 10 minutes at 15,000 g at 3°C. A 3 × volume of ethanol ($-20°$C) is added to the supernatant, and the mixture is centrifuged for 10 minutes at 15,000 g at 3°C. The ethanol is decanted, leaving a small translucent RNA pellet. This RNA yields cytoplasmic ribosomal RNA bands of 0.76 and 1.40 × 10^6 daltons and mitochondrial RNA of 0.60 and 1.26 × 10^6 daltons after electrophoresis in polyacrylamide gels using the method of Loening (1967).

4. Preparation of Protein for Gel Electrophoresis

Analysis of protein by gel electrophoresis is proving to be a useful contribution to the understanding of morphogenetic processes. We have extracted proteins from various cell fractions and whole broken cells using relatively simple procedures. Soluble proteins are precipitated from solution using cold 5% trichloroacetic acid and centrifugation for 10 minutes at 15,000 g. Proteins are dissolved in sufficient solubilization buffer to give 0.4–5.0 mg/ml final concentration. That buffer contains 0.08 M Tris pH 7.4, 0.04 M sodium acetate, 4 mM EDTA, 50 mM mercaptocthanol, 20% glycerol, 1% SDS and 10 μg of pyronine dye markers per milliliter. Dry urea is frequently added to make a 9 M final concentration. These preparations are heated to 100°C for 10 minutes shortly before delivering them onto the gel.

Proteins prepared in the above manner have been resolved on both Fairbanks et al. (1971) disk and flat-bed systems (Laemmli, 1970) of acrylamide electrophoresis.

C. The Morphogenesis of Mating

Synchronous mating is easy to initiate in this fungus, and it is a technique that is equally suitable for microscopic samples as for gram quantities of cells. Equal numbers of cells of opposite mating type are mixed and plated on water agar at 20°–22°C (Day and Jones, 1968); small samples are scraped off the surface of the mating mass and examined using a phase-contrast microscope.

1. Morphology

Approximately 1–1.5 hours after wild-type cells are mixed and plated on mating medium (2 % water agar), they form loosely bound couplets in which the two cells are connected and move together even though they may be separated by as much as the width of a cell (about 3 μm). No strands connecting them have been seen with the light microscope, but electron microscopic preparations of shadow-cast, thin-sectioned and freeze-etched cells show that the glycocalyx is "hairy" (Poon et al., 1974; Day and Poon, 1975). At about 2 hours after the initiation of mating, the cells appear to be aggregated into clumps, which later break up into pairs of closely attached cells. Electron micrographs of this stage show that the cell walls have fused, but that the plasma membranes are still intact. The copulatory organelle is assembled within the next 45–60 minutes. First small bumps appear on each cell at the site of contact, then each bump elongates into a peg, and finally the fused cell walls and plasma membranes at the point of contact of each peg break to form a connecting tube between the cells. Further elongation of this tube takes place over the next 4–8 hour until it is approximately 5 μm long, although sometimes it may reach a length of 15–20 μm. These stages are illustrated in Fig. 2.

Auxotrophic strains designated 1.C2 (histidine requiring, his_1 and a_1 mating type) and 2.716 (lysine requiring, lys_1 and a_2 mating type) develop conjugation tubes at a similar rate to wild-type cells but take about 1.5–2 hours longer to initiate mating. Thus intimately paired couplets are not seen until, at 3.5 hours mating is initiated between these auxotrophic strains.

In about 2–5 % of copulations involving haploid cells, multiple conjugations are observed. In crosses between a diploid, triploid, or tetraploid strain with a haploid strain, multiple conjugations account for more than half of the conjugations. A single triploid cell may copulate with as many as six haploid partners simultaneously (Day, 1972). Figure 3 illustrates the kinetics of copulation.

2. Environmental Factors

Critical factors in mating include the absence of organic nutrients, salts, temperature, oxygen, and supporting medium. Rich media inhibit mating, probably because divalent cations are inhibitory to copulation. The temperature optimum for copulation is 15°C, higher temperatures prevent conjugation and there is a rapid decrease in mating ability between 25° and 27°. Suboptimal temperatures slow assembly of the copulatory organ-

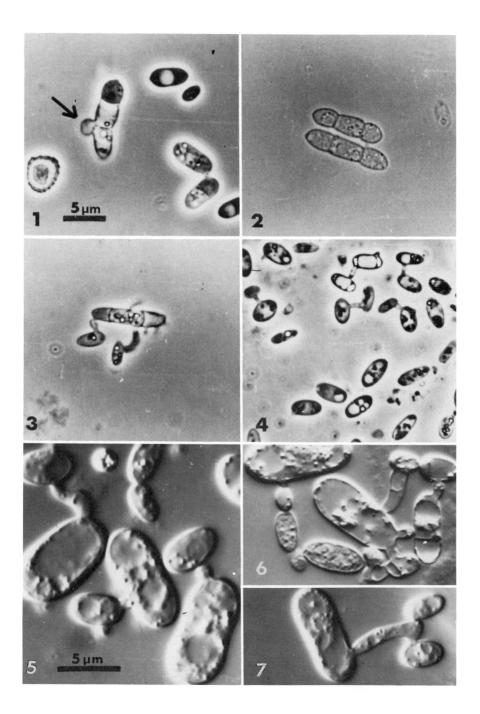

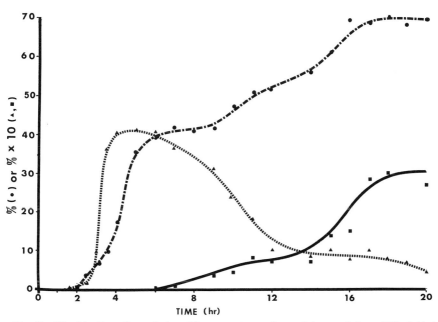

FIG. 3. The kinetics of copulation. The percentage of complete copulations (●) plotted against time from plating. The curves for rate of appearance of conjugation pegs (incomplete conjugations) (▲) and multiple conjugations (■) are plotted as 10 × (percentage occurrence). These kinetics were observed using an asynchronous population, but when a G_1 population of a_1 mating-type is mated with a synchronous or asynchronous population of a_2 mating type, the maximal percentage of mating is reached within 8 hours after plating. The difference in mating kinetics between synchronous and asynchronous a_1 populations is caused by the slow passage of G_2- and S-phase cells into the G_1-phase of mating competence. From Poon et al. (1974).

elle without greatly influencing the percent conjugation (Day and Day, 1974). Induction of the copulatory organelle is highly dependent upon a good supply of oxygen. The nature of the supporting medium is critical, and the only medium presently available that gives consistently high levels of mating is 2 % agar. Filters of glass, paper, glass fiber, and cellulose nitrate have proved to be unsatisfactory support media. For tracer studies, it is best to employ very thin layers of agar to conserve labeled compounds. Highly reproducible tracer uptake and incorporation studies can be conducted by adding isotope to concentrated cell suspensions, which are then deposited on a 0.5-ml layer of water agar in 16 × 150 mm test tube or by

FIG. 2. Phase-contrast photographs of conjugations in *Ustilago violacea*. Frame 1. Conjugation between adjacent cells of a promycelium. Frame 2. Conjugation between cells of different promycelia. Frame 3. Conjugation between a sporidium and a promycelial cell. Frame 4. Conjugation between sporidia. Frames 5–7. Nomarski interference contrast micrographs of conjugation between diploid cells $(a_1 a_1)$ and haploid cells (a_2) showing normal and multiple conjugations. From Poon et al. (1974).

adding isotope to the cell suspension on the water layer (Cummins and Day, 1974a). Figure 4 illustrates the sequence of gene transcription and translation that takes place during synchronous mating as it is deduced from inhibitor treatments, ultraviolet treatments, and tracer incorporation studies (Day and Cummins, 1974; Cummins and Day, 1974b).

IV.   Genetic Methods

A.   Preparation of Mutants

1.   MUTAGENESIS

Two mutagens are routinely employed in this laboratory, these are ultraviolet light (UV) and N-methyl-N'-nitro-N-nitrosoguanidine (MNNG). For UV mutagenesis a sterile petri dish is filled with a 10 ml of water suspension of sporidia, the contents of the plate are agitated during irradiation. Irradiation for 45–60 seconds with a lamp yielding 26 ergs/mm^2 per second

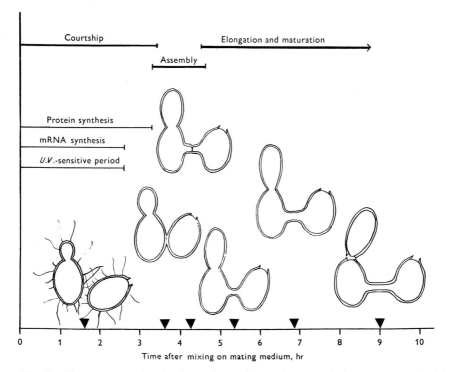

FIG. 4.   The program of molecular and morphogenetic events during copulation in the auxotrophic strains, 1.C_2 and 2.716. The series of temporal events that unfolds during conjugation is schematically represented above (Cummins and Day, 1974b).

at 254 nm normally gives 90–99 % kill and the greatest yield of mutants (Day and Jones, 1968). After irradiation, 200 viable sporidia are spread on each of a series of plates of complete medium and the plates are incubated for 1–2 weeks. The protocol for MNNG mutagenesis is first to centrifuge 20 ml of a young log-phase culture of *Ustilago* in a sterile capped centrifuge tube, then add 10 ml of sterile Tris buffer, 0.05 M, pH 7.5, along with 2 ml of a 2 mg/ml MNNG stock solution. Incubate 10 or 20 minutes at room temperature, then add 1 ml of 4 M acetate buffer, pH 5.5, to destroy MNNG. Dilute as quickly as possible in water. The 10-minute incubation yields 50 % to 80 % killing, and the 20-minute incubation gives 80 % to 95 % killing. The main precaution is that the sporidial population should be actively growing (30–50 % budded) prior to MNNG mutagenesis.

2. SELECTION OF MUTANTS

Auxotrophic mutants are detected by replica plating the colonies from complete medium to minimal medium. The biochemical requirements of the mutants are identified by auxanography using strictly conventional techniques (Fincham and Day, 1971). About 0.5 % of the colonies surviving irradiation are auxotrophic, and the commonest mutants are those requiring arginine, lysine, methionine, or histidine (Day and Jones, 1968). UV-sensitive mutants can be selected by replica plating to complete medium and using a 15-second dose of UV light to detect sensitive colonies. About 0.3 % of surviving colonies are UV sensitive (Day and Day, 1970). Temperature-sensitive mutants can be detected following MNNG mutagenesis and growth of the colonies on complete medium at 15°C, followed by replica plating and growth of a sensitive temperature, 25°C. About 0.5 % of the colonies are temperature sensitive following MNNG mutagenesis.

B. Sexual Analysis

1. HOST PLANTS

Crosses between haploid stock cultures are obtained by sampling the diploid brandspores produced in the host plant *Silene alba*. Seed labeled *S. alba* obtained from several botanic gardens was found to be hybrid (*S. alba × S. dioica*). This hybrid is resistant to the physiologic race of *U. violacea* obtained from *S. alba*. Thus seed should be obtained from areas where *S. dioica* is rare, to ensure successful inoculations (Day and Jones, 1968).

2. THE INOCULATION OF THE PLANT

In *U. violacea* meiosis takes place in the diploid resting spores, or brandspores, which are formed only in the anthers of a host plant. As the brandspores are produced in the anthers of the plant, it is necessary to wait for the

The analysis of papillae from many diploids shows that a wide range of genotypes, both parental and recombinant, are obtained from each diploid. The types and range of genotype are characteristic of haploidization, rather than mitotic crossing-over or nondisjunction, as these latter types of segregation do not usually produce phenotypes expressing recessive characters from both parents simultaneously (see Pontecorvo and Käfer, 1958).

Analysis of the haploid genotypes provides the most efficient means of assigning genes to linkage groups, as haploidization proceeds independently of crossing-over and rarely coincides with it (Pontecorvo, 1956; Pontecorvo and Käfer, 1958). Genes on the same chromosome remain linked, whereas genes on different chromosomes recombine randomly. Two chromosomes tend to remain disomic in most cells of each papillum of *U. violacea* (Day and Jones, 1971).

3.   Mitotic Crossing-Over

Diploid segregants are obtained when a diploid is irradiated with UV using a technique identical to that used to obtain mutants in haploid strains as described ealier (Day and Jones, 1968, 1969).

D.   Production of Polyploid Strains and Complementation Tests

1.   Production of Diploids

The technique used for the synthesis of diploids of *U. violacea* is an adaption of the balanced heterokaryon technique (Roper, 1952). The sporidial pairs after conjugation (plasmogamy) are dikaryons. These dikaryons cannot grow vigorously on any artificial medium yet tested, and normally revert to the haploid condition of budding off haploid sporidia. However, if sporidia with different auxotrophic genes are allowed to mate and are then plated on minimal medium, the reversion to haploid budding is prevented and only the few diploid cells formed after rare fusion of nuclei grow vigorously (Day and Jones, 1969, 1969). Thus prototrophic diploid cells are selected from the haploid parental lines which are auxotrophic, and from the transient dikaryon, which is obligately parasitic.

Mating of compatible sporidia carrying auxotrophic genes is carried out as described in a previous section. In some cases where strains carrying several auxotrophic mutants are used, it is necessary to supplement the water agar with one or more of the required nutrients to allow sufficient growth for fusion. After incubation overnight, about 5×10^5 cells of the mixture of mated and single sporidia are scraped off the water agai with a sterilc wire loop, and spread thinly over a series of plates of minimal medium.

After incubation for about 2 weeks, many prototrophic colonies grow on the plate. These colonies are often of two sorts; small, firm, opaque colonies of typical haploid appearance; and much larger, slimy, translucent colonies. The small colonies had typical haploid cells, while the larger colonies had large, dumbbell-shaped sporidia full of granular, refractive cytoplasm. The large-celled isolates are formed only from pairings between complementary, compatible sporidia, but the small colonies occurred even when pure cultures of an auxotroph are spread on minimal medium, and are probably proto-trophic revertants.

The influence of environmental factors on karyogamy has been extensively studied by Day and Day (1974). About one in 10^4 pairs of cells produced dip-loids normally and that the temperature optimum for nuclear fusion (25°C) is greater than that for cell fusion (15°C). The frequency of diploid formation is about 10 times higher on complete medium than on minimal medium (but selection of diploids is difficult). Stationary stage cultures yield more diploids than log-phase cultures. Finally, it was noted that ultraviolet light (254 nm) induces a high rate of nuclear fusion among the surviving conjugants up to 100 % of these survivors being diploid after very high doses (Day and Day, 1974).

2. COMPLEMENTATION TESTS

Complementation tests provide a method for determining whether or not mutants are allelic. When the chromosome sets of two auxotrophic mutants occur together in the same cell, either separately as in a heterokaryon, or together as in a diploid, the prototrophic phenotype results if each mutant can supply the deficiencies of the other, but an auxotrophic phenotype is usually obtained if both "parents" are mutant at the same locus. Thus the failure to grow on minimal medium of a diploid or heterokaryon synthesized from two auxotrophic parents suggests strongly that the two mutations are allelic. However, growth does not necessarily indicate that the mutants are nonallelic, since intragenic complementation may have occurred (Fincham and Day 1971). Failure to produce diploid colonies in a well mated mixture of compatible sporidia bearing similar mutational lesions is evidence for allelism (Day and Jones, 1968).

3. PRODUCTION OF POLYPLOIDS

Polyploid strains are synthesized using the technique described previously for producing diploids. Strains homozygous for mating type are isolated by irradiating diploids to induce mitotic crossing over as described previously. About 2 % of surviving colonies are homozygous for mating type when tested individually for the mating reaction. Day (1972) has isolated triploid and tetraploid strains of *U. violacea* in this manner. It was noted that the cell length increased in direct proportion to the ploidy level in stationary stage cultures of the polyploid and that triploid and tetraploid strains tend to be unstable and thus must be reisolated frequently.

V.   Cell Cycle

Growth of *U. violacea* sporidia, as mentioned earlier, is similar to that of a typical yeast cell. Log-phase cultures contain 45–60% budded cells in complete medium. Variation in the duration of the cell cycle influences the percentage of budding. In a range of generation times from 5 to 8 hours, the main variation in the cell cycle is in the G_1 period, which varies from 1.5 to 4.5 hours. The S period commences at about the time of bud initiation and has a duration of 2 hours. The total duration of G_2 and mitosis which are difficult to separate, is close to 1 hour (Day and Cummins, 1975). Poon (1974) indicates that mitosis is quite slow (about 45 minutes) and that "prophase" takes about 12 minutes, metaphase 5 minutes, and anaphase-telophase about 28 minutes. The S and G_2 phase may therefore last as little as 15 minutes. *U. violacea* sporidia are arrested in the G_1 phase of the cell cycle, unlike another species of *Ustilago*, *U. maydis*, whose sporidia are arrested in G_2.

The DNA content of stationary phase sporidia is 3×10^{10} daltons while asynchronous log-phase cultures (45% budded) have 3.6×10^{10} daltons (Day and Cummins, 1975). These values are consistent with the cell cycle durations determined using synchronous cultures as shown by the mathematical procedures of James (1974).

A.   Methods of Synchronization

1.   STARVATION ENRICHMENT

The method originally described by Williamson and Scopes (1962), who used yeast, was applied to *U. violacea* by Day and Jones (1972). That technique was superseded by the more convenient procedures described below.

2.   STATIONARY STAGE OUTGROWTH

An early stationary culture (one day into stationary phase in complete medium) contains 90% unbudded cells. The cultures grow out synchronously when inoculated in fresh complete medium at 10^6 to 10^7 cells/ml. The cell cycle is typically about 5 hours and 70–80% budding is maximally achieved (seldom more) in the first cycle of outgrowth. Diploid cells are usually more synchronous in their outgrowth than haploid cells (Day and Cummins, 1975).

3.   SUCROSE GRADIENT TOP FRACTIONS

This method is modified from one described by Mitchison (1970). A 15–50% linear sucrose gradient in water is used to separate off a fraction of small unbudded cells (the "top" fraction). About 10^{10} log-phase cells are resuspended in 2 ml of water after their initial centrifugation from growth medium. This thick slurry of cells is gently layered over the surface of a 45-ml gradient in a 50-ml plastic centrifuge tube. The preparation is centrifuged in a swinging bucket rotor for 10 minutes at 300 g at room temperature. The top fraction (5–10 ml) from just above the thick band of cells is withdrawn and used to inoculate a 50-ml flask of fresh complete medium. The cells in this fraction are 85–95 % unbudded and grow out synchronously to a maximum of 60–80% budded cells. The cycle times vary from 6 to 8 hours. 10 to 30 % linear sucrose gradients are also satisfactory for preparing top fractions (Cummins and Day, 1973; Day and Cummins, 1975).

4. Sucrose Gradient Centrifugation in a Reorienting Zonal Centrifuge

The Sorvall SZ-14 zonal rotor has provided the means of obtaining both good synchrony and high yield of synchronous cells. Both linear and convex aqueous sucrose gradients are useful for preparing synchronous cultures, but the convex gradient provides the most uniform and consistent method of obtaining synchronous cell populations. Before describing the centrifugation technique we note that commercial grade sucrose is satisfactory and economical for preparing gradients and its use does not influence the synchrony or later outgrowth of living cells. The linear gradient is prepared by dynamically loading the rotor at 2500 rpm using the Sorvall gradient maker. A 15–50 % gradient is prepared and the centrifuge is accelerated to 3500 rpm, then 50 ml of a log-phase culture (preferably a chemostat culture) is loaded. The centrifuge is run for 7.5 minutes, then allowed to coast down to 1000 rpm, then slowly brought to a stop by adjusting the rate controller to a 25–30 minute deceleration. The rotor contains 1.55 liters of solution and an appropriate number 20- to 4-ml fractions are collected. The budding varies from 5 % at the top to 95 % at the bottom of a gradient. The convex gradient is prepared statically using the Sorvall gradient maker. A 15 % to 50 % convex gradient is prepared by placing 300 ml of 15 % sucrose in the mix chamber of the gradient maker and 1200 ml of 50 % sucrose in the reservoir. When the rotor is filled the centrifuge is slowly accelerated to 500 rpm then rapidly to 3500 rpm. Loading deceleration and fractionation are as for the linear gradient, the budding varies from 4 % to 98 % from the top to the bottom of a gradient. Forms of gradient other than linear and convex such as the exponential have been constructed but did not yield more satisfactory synchrony. The fractions prepared in this manner enter mitosis synchronously and with little apparent delay due to treatment when inoculated into complete medium. Figure 5 shows the results of a typical convex gradient preparation of synchronous sporidia.

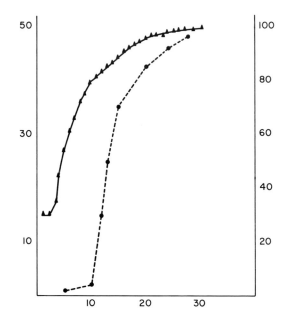

FIG. 5.   Synchronous sporidia prepared by sucrose gradient centrifugation in a reorienting zonal centrifuge using a 15%–50% convex gradient. Preparation of the gradient was as described in the text. Analysis of sucrose concentration was completed using a refractometer (▲———▲), and the percentage of budding (●---●) was by direct microscopic determination.

B.   Mating and Cell Cycle

Normally fractions from the sucrose gradient are diluted with water (to give about 10% sucrose) centrifuged and mated against an asynchronous or stationary stage population of the opposite mating type. Alternatively fractions from the synchronous population may be split, one portion being inoculated to complete medium while the other fraction is held in water at 3°C. The cold stored cells retain their ability to mate (provided that they are in the appropriate stage of the cell cycle) for an extended period without budding or replicating their DNA. At intervals during the outgrowth of synchronous cultures samples can be mated with cold stored samples of a population of the opposite mating type. Using this technique it was possible to confirm the stringent nature of the a_1 allele and the relaxed control of the allele a_2 (Day and Cummins, 1975).

VI. Mutant Cell Cycle Controls

Mutant cell cycle controls were first observed in the following manner. A fresh isolate of an a_1a_2 diploid originally mated with a_1 partners during its log phase of growth. After about 4 subcultures over a period of 6–8 weeks, very few cells (5%) could mate with the haploid a_1 partner. This neutral culture appeared to be still diploid by other criteria such as cell size, colony morphology, and solopathogenicity (i.e., capability as a pure culture to produce diploid sexual spores in a host plant) (Day and Jones, 1968). To discover whether this neutral strain was truly sterile, i.e., had lost function at one or both of its alleles, or whether it had an altered cell cycle control system, it was inoculated into a host plant to produce the sexual brandspores. The haploid segregants produced after a meiotic division in these brandspores were collected and analyzed first for mating type and second for cell cycle control properties.

It was found that all haploid sporidia derived from these neutral diploid strains were capable of mating and that both mating types were equally represented. Thus the neutral diploid was heterozygous for two fully functional mating-type alleles.

Cultures of each of two randomly chosen cultures from both the a_1 and a_2 haploid segregants of the neutral diploid were synchronized in a sucrose gradient in the zonal rotor and the different fractions were tested for mating ability with appropriate standard cultures of the opposite mating type. The possible outcomes and theoretical basis for the predictions are illustrated in Fig. 6.

Both a_1 cultures tested retained their stringent control and did not differ in any way from the standard wild-type and auxotrophic stocks described earlier (Cummins and Day, 1973). However, both a_2 cultures mated only in the G_1 phase of the cell cycle, and therefore were now associated with a stringent control, quite different from the relaxed control of the original a_2 cultures. Microscopic observation of these two strains and four other randomly chosen a_2 strains derived from the neutral diploid confirms this conclusion, as it was found and that they only mated in the unbudded G_1 condition.

Thus it is clear that during the growth of a freshly isolated $a_1 a_2$ diploid an event occurred which resulted in mutation of the a_2 control system from relaxed to stringent. As a further confirmation of these conclusions diploid cultures were resynthesized from the haploids derived from the sterile diploid, i.e., a_1 (stringent control) and a_2 (stringent control). As expected, these diploids were neutral (less than 5% mating with a_1 haploids) when freshly isolated.

About one-fourth of freshly isolated heterozygous diploids are found to be neutral and about the same proportion has been observed in diploids

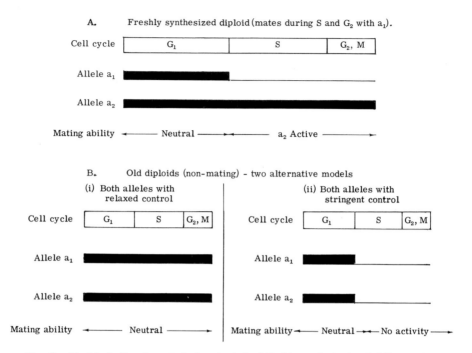

FIG. 6.   Model of cell cycle control of mating in both freshly synthesized and old heterozygous $(a_1 a_2)$ diploids. Model assumes that (1) the period of inducibility of each allele is the same as for the corresponding haploid; (2) the cell does not mate when both alleles are active simultaneously. From Day and Cummins (1975).

preserved from 2 to 5 years. Presumably altered cell cycle controls arise through mitotic crossing-over or gene conversion during the initial divisions of a newly formed diploid (Day and Cummins, 1975). Neutral diploids serve as a rich source of mutants bearing altered cell cycle regulation, and such mutants can be recovered using the procedures outlined above. To recapitulate the protocol for isolating mutants of the cell cycle control system: first diploid clores are screened to recover nonmaters (neutrals). Next, haploids are recovered from the neutral diploids. These haploids are putative mutants of the cell cycle control system. They are synchronized, and their ability to mate, whether during G_1 (a stringent control) or throughout the cell cycle (a relaxed control), is determined.

Acknowledgments

Much of the work described in this article has been supported by the Canadian National Research Council. We express sincere appreciation to the many "summer research students" who have contributed to the development of methodology in our laboratory and to Mrs. Vera Marnot for her continuing assistance.

REFERENCES

Baker, H. (1947). *Ann. Bot. (London)* [N.S.] **11**, 333.

Bradley, D. (1962). *J. Gen. Microbiol.* **29**, 503.

Britten, R., Pavich, M., and Smith, J. (1970). *Carnegie Inst. Washington, Yearb.* **68**, 400.

Cole, G., and Kendrick, W. (1968). *Mycologia* **60**, 340.

Cummins, J., and Day, A. W. (1973). *Nature (London)* **245**, 259.

Cummins, J., and Day, A. W. (1974a). *In* "Cell Cycle Controls" (G. M. Padilla, I. L. Cameron, and A. M. Zimmerman, eds.), p. 181. Academic Press, New York.

Cummins, J., and Day, A. W. (1974b). *J. Cell Sci.* **16**, 49.

Day, A. W. (1968). Ph.D. Thesis, University of Reading, England.

Day, A. W. (1972a). *Nature (London), New Biol.* **237**, 282.

Day, A. W. (1972b). *Can. J. Genet. Cytol.* **14**, 925.

Day, A. W., and Cummins, J. (1973). *Nature (London)* **245**, 260.

Day, A. W. and Cummins, J. (1974). *J. Cell Sci.* **14**, 451.

Day, A. W., and Cummins, J. (1975). *Genet. Res.* **25**, 333.

Day, A. W., and Day, L. (1970). *Can. J. Genet. Cytol.* **12**, 891.

Day, A. W., and Day, L. (1974). *J. Cell Sci.* **15**, 619.

Day, A. W., and Jones, J. K. (1968). *Genet. Res.* **11**, 63.

Day, A. W., and Jones, J. K. (1969). *Genet Res.* **14**, 195.

Day, A. W., and Jones, J. K. (1971). *Genet Res.* **18**, 299.

Day, A. W., and Jones, J. K. (1972). *Can. J. Microbiol.* **18**, 663.

Day, A. W., and Poon, N. H. (1975). *Can. J. Microbiol.* **21**, 547.

Day, A. W., Wellman, A., and Martin, J. (1972). *Can. J. Microbiol.* **18**, 639.

Fairbanks, G., Steck, T., and Wallach, D. (1971). *Biochemistry* **10**, 2602.

Finchman, J., and Day, P. (1971). "Fungal Genetics," 3rd ed. Blackwell, Oxford.

Fischer, G., and Holtón, C. (1957). "Biology and Control of the Smut Fungi." Ronald Press, New York.

Holiday, R. (1961). *Genet. Res.* **2**, 204.

James, T. (1974). *In* "Cell Cycle Controls" (G. M. Padilla, I. L. Cameron, and A. M. Zimmerman, eds.), p. 31. Academic Press, New York.

Laemmuli, U. (1970). *Nature (London)* **227**, 680.

Lin, C., and Key, J. (1967). *J. Mol. Biol.* **26**, 237.

Loening, U. (1967). *Biochem. J.* **102**, 251.

Meinke, W., Goldstein, D., and Hall, M. (1974). *Anal. Biochem.* **58**, 52.

Mitchison, J. (1970). *In* "Methods in Cell Physiology" (D. Prescott, ed.), Vol. 4, p. 131. Academic Press, New York.

Moens, P. (1970). *Proc. Can. Fed. Biol. Sci.* **13**, 160.

Pontecorvo, G. (1956). *Annu. Rev. Microbiol.* **10**, 393.

Pontecorvo, G., and Käfer, E. (1958). *Adv. Genet.* **9**, 71.

Poon, N. H. (1974). Ph.D. Thesis, University of Western Ontario, Canada.

Poon, N. H., and Day, A. W. (1974). *Nature (London)* **250**, 648.

Poon, N. H., and Day, A. W. (1975). *Can. J. Microbiol.* **21**, 537.

Poon, N. H., Martin, J., and Day, A. W. (1974). *Can. J. Microbiol.* **20**, 187.

Poon, N. H., and Day, A. W. (1976a). *Can. J. Microbiol.* **22**, 495.

Poon, N. H., and Day, A. W. (1976b). *Can. J. Microbiol.* **22**, 507.

Roper, J. (1952). *Bot. Rev.* **18**, 447.

Reynolds, E. (1963). *J. Cell Biol.* **17**, 208.

Robinow, C. (1970). *In* "Methods in Cell Physiology" (D. M. Prescott, ed.), Vol. 4, p. 165. Academic Press, New York.

Williamson, D., and Scopes, A. (1962). *Nature (London)* **193**, 256.

Subject Index

A

Acathamoeba
  uptake of latex beads by, 318
Acanthamoeba castellanii, 303
  alternative procedure for isolating
    plasma membranes from, 312
  culture methods, 304
  growth curve for, 306
  mitochondria of, 184
  separated mitochondria of, 195
Acetoorecin staining
  hot, 450
Albumin (BSA) gradient, 59
Alcohol dehydrogenase, 122
Algae
  synchronized cultures of, 201–219
Algae cells
  methods for breaking, 202
Amaranthus, 187
Amido black
  staining ability of, 155
Anacystis, 202
Antioxidants, 180
Antirrhinum, 186, 193
Antirrhinum majus, 197
Aqueous two-phase partition, 268
Auto-Densi Flow Collector, 35
Azaguanine resistance, 367
Azaguanine-resistant rat line, 327
Azaguanine-sensitive line, 367

B

Bacto-peptone
  growth-promoting activity of, 425
Beta, 186, 193
  isolated mitochondria of, 195
Brandspores
  harvesting of, 460
Bromodeoxyuridine-resistant mouse line,
    327
Bumilleriopsis filiformis, 202
Bungarotoxin, 292

C

Carbonic anhydrase, 122
Cell cycle
  mutant controls, 467
Cell cycle analysis
  of smut fungus, 445
Cell density, 75
Cell disruption, 205, 247
  techniques for, 251
Cell fractionation
  in silica sols, 181
Cell fusion, 106
  effects of polyethylene glycol concentra-
    tion on, 331
  with polyethylene glycol, 328
Cell growth
  in serum-free medium, 429
Cell hybrid(s)
  selection of, 327
Cell hybridization
  induction of by polyethylene glycol,
    325–338
Cell monolayers
  measurement of the growth of, 407–419
Cell nuclei
  enumeration of, 409
Cell number
  determination of, 98, 411
Cell separation methods, 50
Cell separations, 1–14
Cell suspensions
  method of preparation, 19, 57
  treatment with polyethylene glycol, 336
Cell synchronization, 112
Cell volume distribution, 13
Cellogel electrophoresis, 151, 157
Cellogel electrophoretic analysis
  apparatus for, 154
Cellogel strips, 154
Cells
  made permeable with toluene, 371–380
Cellulase, 189

CONTENTS OF PREVIOUS VOLUMES

Volume I

Volume V

Volume VI

Volume VIII

Volume IX

Volume XIII

Volume XIV

A B C D E F G H I J

7 8 9 0 1 2 3 4 5